Digital Circuits
and Devices

PRENTICE-HALL NETWORKS SERIES

ROBERT W. NEWCOMB, *editor*

Digital Circuits and Devices

Teuvo Kohonen

Chairman, Department of Technical Physics
Helsinki University of Technology

PRENTICE-HALL, INC., *Englewood Cliffs, N.J.*

10 9 8 7 6 5 4 3 2 1

ISBN: 0-13-214122-1

Library of Congress Catalog Card Number: 79-38045

Printed in the United States of America.

PRENTICE-HALL INTERNATIONAL, INC. *London*
PRENTICE-HALL OF AUSTRALIA PTY. LTD., *Sydney*
PRENTICE-HALL OF CANADA, LTD., *Toronto*
PRENTICE-HALL OF INDIA PRIVATE LIMITED, *New Delhi*
PRENTICE-HALL OF JAPAN, INC., *Tokyo*

Contents

Part Two

DIGITAL ELECTRONICS

Preface

This book has been written for those who want to know how logic circuits can be put together to form working digital computing systems. The fundamental problem is how to describe the lengthy, mutually interlocking digital and logic operations without overwhelming the reader with words. It is striking that rather complex concepts and operations can be explained and understood in terms of only a few and much simpler ones, if the purpose and main functional structure of the latter have been defined, even without showing how their details are implemented. In circuit and system theory, block schemes are frequently used for this purpose. On the other hand, special languages have been developed for the definition of automatic computing operations. Both of these methods are useful in the explanation of internal operations of information-processing devices. This is called the system-theoretical approach.

The main purpose of this book is to train designers of digital circuits and devices; readers more interested in the mathematical theory of computing machines are advised to follow the collateral texts mentioned in the Bibliography. Although some parts of this book may appear too formal on first reading, it is probable that such topics will sooner or later be encountered by the practicing engineer in the computer field.

It has been said that the worst enemy of good things are the best things, and this is very true in electronic engineering. A book could never be completed if it had to contain all of the important and up-to-date material in the field. Such completeness is not necessary, because the electronic industry is continuously delivering new data as well as applications reports, and the reader is urged to follow up this kind of information. One purpose of this book is to tie up such sources of information in a unified presentation to give at least a birds-eye view of the field. The author believes that if a reader becomes acquainted with this book he ought to be able to manage most standard design jobs associated with digital devices.

This text has been specially designed for an independent course in digital engineering. Some previous knowledge in semiconductor components and basic electronic circuits will suffice as a prerequisite. It is hoped that the book or parts of it could be used as a text for college courses in digital circuits or digital systems. Accordingly this book could be used for collateral reading in related fields in electrical engineering, too.

Especially nowadays the semiconductor circuit technology is in a state of rapid development, and the component industry continuously announces new circuit packages for users: although the simplest discrete components such as resistors, capacitors, diodes, and transistors are still needed, the major part of contemporary electronic devices consists of *integrated circuits* which are circuit assemblies manufactured, tested, and vended as indivisible units. It is, in principle, a straightforward task to interconnect them. The new technologies are often identified according to degree of integration: *medium scale integration* (MSI) is a fabrication technique in which active (and also passive) components from a few to several hundreds are automatically produced on the same semiconductor substrate and interconnected in such a way that they form one or a few elementary operational units in each package, ready to be interconnected into larger assemblies. *Large scale integration* (LSI) is the name for fabrication techniques in which thousands of components or elementary circuits are manufactured on the same substrate and interconnected in such a way that whole functional units (registers, computing circuits etc.) may be built in integrated packages. Owing to developments in the automated production, the optimal number of circuits per package is continuously increasing, and the designer must face the problems of electronic design from a more general point of view than before. The elementary circuit design now is and remains largely a problem of the manufacturers of these devices, and the application of the integrated circuits becomes a system design problem to the user.

This book can be considered in two parts of which the first one, Chaps. 1 to 5, forms a whole and is devoted to logic design. The latter part is more engineering-oriented. There is a definite purpose in placing all logic circuits after extensive discussion of logic design: it is to motivate the choices encountered in commercial lines of circuits, and to show that logic design always precedes circuit design.

The introductory chapter has been written for those who have no previous experience in the field, to give an idea how digital computing is carried out in reality. The most important point is that in most digital computers, operations are performed in successive steps. Information is stored in several storage locations from which it is picked up one operand at a time and similarly loaded back when numerical transformations have been carried out.

Chapter 1 is an introduction to the numerical representation of information in machines. Digital representation relies on symbols and thus the coding of numbers in computers is the central theme of this chapter. The representation of codes by machine variables completes this chapter.

In applied electronics, the terms *logic circuits* and *digital circuits* appear intermittently. From the point of view of circuit technology, both use the same basic electronic solutions, i.e., switching circuits. The difference comes in the function for which these circuits are planned: digital circuits are primarily meant for the manipulation of numbers, i.e., mathematical variables, whereas logic circuits are used to simulate logic decisions. All digital circuits make use of logic circuits, but logic circuits are used for other purposes, too; for example, in the control of interlocking operations in complex machines. All of these operations can be described in terms of propositional calculus which is related to the special algebra of bivalent variables, called Boolean algebra. Its application to basic logic circuits, called combinational circuits, is discussed in Chap. 2. Combinational circuits are logic circuits in which output signals are single-valued functions of input signals.

Chapter 3 is a logical continuation of Chap. 2. It deals with logic circuits in which there may be several possible output signals for a given input signal combination. The occurrence of particular output signals depends on the way input signals have been changed in the past. In other words, the circuit has a kind of memory which may hold and detect input signal sequences. This is a feature which is desirable in computing since digital variables are operated on in succession and the result of a previous operation is used as the operand for the next one. In this chapter the general theory and the most important kinds of sequential circuits are discussed.

Chapter 4 forms one of the most important parts of this book, and its purpose is to introduce operational units by which digitally represented numbers are transformed to yield new numbers according to arithmetic rules. Some computing circuits, e.g., counters and adders, are frequently used in measuring and control equipment, and they also comprise the central part in digital computers. Some circuit examples shown in this chapter are to be understood as simplifications and not necessarily representatives of optimal commercial solutions. Again, a compromise had to be made between the pedagogic and practical value of the examples discussed. At the end of this chapter, some special computing circuits encountered in instrumentation are discussed. The main idea in this section, called incremental computing, can be utilized with physical systems where the speed of computing is not of primary importance.

The first part of the book ends up with Chap. 5 where it is explained how working systems are made of operational units. The operation of digital circuits must be controlled manually as well as without human intervention. This means that if there are several alternative automatic operation sequences of which only one has to be selected, the system must be able to test the branching condition and to make a decision which way it will go by itself. Such sequences were already discussed in Chap. 3. Here a proper meaning is given to such sequences in terms of real computing operations. This chapter starts with the introduction of simple control circuits and ends up with program control which is so important in all general purpose digital computers.

The first part of the book does not rely upon particular electronic implementations but leaves many alternatives for realization. For the actual circuit engineering, it is necessary to obtain an idea how nonlinear electronic circuits operate and such ideas are presented in Chap. 6. The switching of digital signals, i.e., of currents as well as of voltages by semiconductor components, is explained. This chapter is rather independent and may be skipped by those already familiar with semiconductor circuits.

In the commercial digital circuits several families are encountered, each consisting of circuits which are standardized and mutually compatible. It is the purpose of Chap. 7 to list and describe the electrical properties of the different families. There are two basic active components, namely the bipolar transistor and the field-effect transistor, which compete in relative importance. As of this writing, it is too early to predict which one of them will survive, or possibly both will. Some contemporary circuit examples are given for both of these technologies.

A primary time reference for computing and control operations must be provided by dynamic switching circuits. The logic circuit modules provide a means to implement timing circuits more easily than before, and Chap. 8 begins with systematic examples of timing circuits implemented by standard logic circuits. In order to give a more detailed explanation of these, a few more detailed classical timing circuit examples are analyzed. Finally, precision timing with the aid of operational amplifiers is introduced.

Contemporary computers are designed around memories, and the basic types of memories are discussed in Chap. 9. This chapter is more encyclopedic than the rest of the book.

The last chapter, Chap. 10, is devoted to the special problems associated with the interconnection of digital circuits and the transmission of digital signals. For pulsed signals, the distributed nature of line parameters must be taken into account in order to avoid unwanted effects. These considerations are more important the longer the transmission lines become. This discussion ends up with transmission methods used over long distances.

If this book is to be used as a text in digital circuits or digital systems, it might be advisable to skip Chaps. 6, 8, and 10. On the other hand, for courses in pulse circuits or equivalent, these chapters provide more relevant material.

I would like to thank my colleagues at the Helsinki University of Technology as well as at the University of Washington where I had a pleasant stay during 1968–69, for encouraging me to initiate this work. During the writing I had valuable help from the staff members of the university, especially from Mr. Seppo Haltsonen, Mr. Matti Kilpi, Mr. Heikki Laine, Mr. Yrjö Neuvo, and Mr. Olof Staffans. The tireless efforts of Mrs. Rauha Tapanainen in the typing of this manuscript and preparing of illustrations are inestimable. Last, but not least, I would like to thank my wife and family for their patience and support during the lengthy writing periods.

TEUVO KOHONEN

Introduction to Digital Computing

I.1. General

In this book we shall be dealing with electronic devices that are intended for the processing of information. A *computer* is an operational unit which accepts numbers in some automatically recognizable form and produces new numbers after performing a series of calculations on the source information. More generally, a computer is a problem-solving device.

The two basic classes of computers are called *analog* and *digital*. In this book we shall consider only digital computers.

Most computers these days are *general-purpose computers*; they have programming provisions for a very wide class of mathematical problems. The fact that many general-purpose computers are used for specialized purposes (e.g., for the control of industrial processes), is a result of mass production which makes these computers sufficiently inexpensive. There are many types of general-purpose computers, ranging from portable desk computers to huge centralized computing facilities. At first glance it may seem that they have nothing in common, but actually large computing systems are often composed of multiple similar units, and, although their combined operation involves rather complex control conditions, the *computer hardware* itself has been implemented by much more basic techniques. The way in which these systems are operated is defined by *programs*, the *software*, which resides as transferable information within the computer's memories. We shall not consider the programming of computers in this book.

With the advent of integrated electronic-circuit assemblies, it has become obvious that even in small specialized calculators and instruments, the processing of information by digital methods is usually more efficient than by the previously used analog methods. For example, when trigonometric

1

functions must be generated and spectrum analysis performed, digital techniques provide means for the design of effective *special-purpose computers*.

Digital techniques offer features that are advantageous in the processing of measurements or other data:

1. Accuracy of representation does not rely on the ultimate stability of elementary circuits, because longer numbers only require more space in the computing circuits.
2. Variables can be stored in the memories for indefinitely long periods.
3. The speed of calculation is very high; in smaller tasks the speed is many orders of magnitude higher than is necessary.
4. Complex special functions can be computed using reasonably short algorithms developed for these purposes.

When speaking of *digital circuits*, we often mean electronic circuits which operate in the *switching mode* and in which the input–output transfer relations of signals obey simple rules of logic or arithmetic. Some elementary digital circuits are called *logic circuits*, because they can be used to implement logic decisions. Such circuits are needed in the interlocking control of industrial plants and for control of elevators, automata, etc. In the design of these devices, standard methods, as discussed in Chapters 2 and 3, can be used. But the scope of this book is wider; we shall lay the foundations for the design of more complex computing circuits, which constitute the hardware of modern digital calculators and computers.

I.2.   Anatomy of a Digital Computer

The digital computer is a *sequentially operating* device. It has a bank for all the information that it deals with. Such a bank is called a *memory* or a *memory system*, and numbers or other information is usually inserted in and taken out of it one variable at a time. The loading of this bank is done in a separate operation, or intermittently during computations; results of accomplished operations are, similarly, taken out of the bank for displaying when they are available, or after all computations have been completed.

When a computer is working, arithmetical or other transformations are carried out on only a few variables in a single operation. Most operations are of the following types:

1. Call in a number from memory to arithmetic circuits.
2. Carry out an arithmetic operation (e.g., addition, subtraction,

multiplication, or division) on two numbers temporarily existing in arithmetic circuits.

3. Load a number (e.g., a result of an arithmetic operation) into memory.

4. Write out a part of the contents of the memory.

During operation on some variables, the rest of the stored information is left undisturbed. In this way the computer often performs very long operations sequences (e.g., evaluation of algebraic expressions or integrals). Operations must be programmed as separate calculational steps, as would be done in numerical mathematics when preparing problems for desk calculators. It is a fundamental property of automatic digital computers that the directions for the operation of arithmetic circuits and memory are given automatically without human intervention, according to a *program code* stored in the computer's internal circuits, or memory. A computer can also modify its behavior by selecting one of several alternative programs according to what results it has produced in the last operation performed. This feature allows it, for example, to repeat similar program loops a given number of times. An electronic computer works extremely fast, frequently performing one elementary operation in less than 1 μsec.

    A rather general organization of general-purpose and even special computers, simplified for illustrative purposes, is given in Fig. I-1. A computer consists of five major parts:

1. Input devices
2. Memory
3. Arithmetic circuits
4. Output devices
5. Control circuits

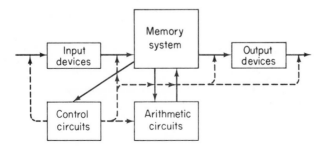

Fig. I-1. Major parts of a computer.

It is usually difficult to point out *control circuits* in a computer, as they are distributed all over the system and embedded with other circuits. We shall leave their description to Chapter 5. In the functional respect, however, we may think in terms of a separate block for control circuits which produces *control signals*. The control signals are ordered properly in time at different places so that the operational circuits are permitted to operate in the desired sequence. In Fig. I-1, solid arrows between blocks indicate that numbers or other variables can be transferred, one at a time, in the direction shown by the arrow. The diagram does not, however, indicate the time at which these transfers occur; that is decided by the control circuits. Dashed arrows emanating from the control circuit represent signals that activate transfer paths and perform some extra operations in the arithmetic circuits.

The solid arrow pointing from the memory system to the control circuits tells us that the program code is usually stored in the memory system, from which operational instructions are transferred to the control circuits. The computer is thus self-controlling. This principle, discussed in more detail in Chapter 5, is called the *principle of stored program*.

Input and Output Devices. Every computer needs an interface through which it communicates with the outside world. The interface is provided by the input and output devices. In the very simplest special-purpose computers, which may be built-in units of other larger electronic systems, the input information is fed in by switches, push buttons, or in the form of electrical signals; the output information is displayed in the form of electrical signals or by indicator lamps. In small general-purpose computers, communication with the machine is normally carried out through a teletypewriter terminal having a keyboard, a typing mechanism, and provision for reading and punching paper tape. Initial directions to the computer are given by the front panel controls.

Most large-scale computers have a number of input–output devices, of which the following are examples:

INPUT DEVICES

1. Keyboards and consoles
2. Punched card readers
3. Punched tape readers
4. Optical or magnetic character readers
5. Light pens

OUTPUT DEVICES

1. Indicator lamps
2. Typewriters

3. Line printers
4. Card punches
5. Tape punches
6. Cathode-ray tubes
7. Plotters

Moreover, there are other graphic input–output devices. For control of industrial processes, computers are directly connected to signal transducers and also give control signals for process actuators. Very often the input and output terminals are located remotely from the central processors, and public telephone lines can be used for the transmission of input and output signals. A convenient device through which a common user can be in contact with large computers is called *acoustical coupler*, or *dataphone*. This device closely resembles a telephone. A user dials the number of a computer and reserves the line. One of the simplest remote terminals is the *touch-tone* unit, which consists of a keyboard and a speaker: the user signals in numbers and instructions using push buttons, and the computer responds by giving the results in a voice resembling that of a human being.

Program codes and data must be prepared manually in a form that is automatically recognizable by the computer. It is usual to prepare programs and data as punched cards or as a tape at a separate station and then feed them into the machine.

Computers may also be coupled together and information exchanged through direct electrical interconnections. This is the fastest mode of communication; the speed is usually limited only by the quality of lines if computers are located far apart. Computers in the same room may be directly interconnected by cables.

Memories. The most important of computer memories is the *main memory*, frequently also called *mainframe* or *working memory*. In most contemporary computers the main memory consists of a ferrite-core stack in which one annular core is reserved for each binary symbol (*bit*). The cores are magnetized into saturation, and the direction of the saturation flux is used to represent the binary symbols 0 and 1. The cores are organized into sets called *memory locations*; each location can store one binary number or other type of information. The contents of one location are also called a *word*; word lengths usually vary from 8 to 60 bits. Each location has a numerical *address* (i.e., a code by which the location can be identified). The size of a memory is frequently expressed by telling the number of addressed locations, in which case word length must also be specified. Small computers may have core memories that consist of as many as 4096 words, or 4K ($1K = 2^{10} = 1024$), and very large computers may store 1 million words in their main memory.

The ferrite cores are provided with control wires threaded through them by which information is written and read (see Chapter 9 for more details).

There are some indications that ferrite-core memories will be augmented or partially replaced by faster *active memories* (i.e., electronic circuits organized in the same way as main memories). In many cases, a small active memory is used as a *scratch-pad memory* for the temporary storage of intermediate results.

The *access time* (i.e., the time needed to pick up a particular word from the memory) is less than 1 microsecond for ferrite-core memories, but active memories may be 10 to 100 times faster. The memory capacity provided by main memories is rather expensive per bit; for a large number of data or programs, extension of main memory is a costly solution. An optimal solution is to have a moderately sized main memory and much slower but cheaper *backing* or *auxiliary memories*, which can store masses of information. Most backing-storage systems are based on the use of magnetized surfaces onto which information can be written and from which it can be read in the same way as in tape recorders—in the form of special codes. The conventional backing memories are *magnetic drum*, *magnetic disc*, and *magnetic tape units*. Millions of words can be stored in a moderate-sized backing storage system. The data are usually stored in blocks of contiguous words, separated by larger gaps. Information is transferred between the backing storage and the main memory by the blocks, because during reading and writing the memory is scanned by mechanical movement and cannot be stopped arbitrarily. Although the number of words in one block is not fixed, backing storage systems are not used unless the blocks are reasonably large—hundreds to thousands of words.

In large computer centers, backing memories, especially magnetic tape stations, often occupy most of the space. This is because the many customers using the system are serviced in turn, and all their programs and data must be available in files stored in the backing memories.

Arithmetic Circuits. We use a general name, *arithmetic circuits*, for that part of the computer in which operands are transformed. In principle, the arithmetic circuits constitute the calculating machine itself, the most usual arithmetic operations being addition, subtraction, multiplication, and division. Additional useful operations are the formation of the negative of a number or complementing of it, and the shifting of digits into other positions. Mathematical functions are usually calculated by programmed algorithms. It is obvious that an arithmetic unit must contain places for temporary storage of operands. These places are called *registers*, and they are usually one word in length. The number of registers in an arithmetic unit varies. A three-register system is discussed in Chapters 4 and 5.

Figures I-2 and I-3 present the operation of an arithmetic unit. With the aid of control signals defining a desired operation, the processing circuits

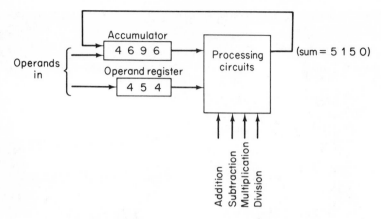

Fig. I-2. Simplified illustration of an arithmetic unit.

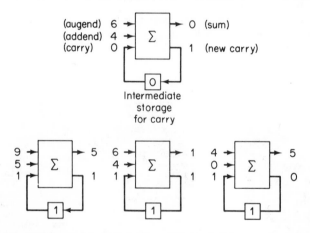

Fig. I-3. Successive steps in the addition of digits, starting with the least significant digits.

are set into a state in which they perform, say, the addition of two numbers. Let these numbers be stored in an *accumulator* and an auxiliary *operand register*. For illustrative purposes, let us assume that the numbers can be represented in decimal form with digits in respective sections of registers. What we want is for the arithmetic circuits to form output signals that represent the sum of the operands. The output signals are then used to replace the previous contents of the accumulator by new contents, the sum. Usually the operation is carried out *by digits*, starting with the least significant one; a *carry*, formed of the addition of two digits, must be stored and taken into account in the formation of the next sum digits; and so on. (Notice that the carry may be a 0 or a 1 only.) What is needed is a sort of *sum table* from which the sum digit and a new carry, corresponding to two entry digits and

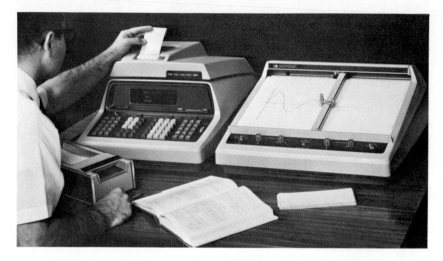

Fig. I-4. Digital desktop calculator system, showing marked card reader (lower left), calculator, printer (top), and X-Y plotter. (*Courtesy of Hewlett-Packard Co.*)

Fig. I-5. Birds-eye view of a large computer system (IBM System 360, Model 85) showing central processor (cabinets in front), operating console (middle), card readers (right), typewriter console (middle, behind), line printers (left), small magnetic disk units (rear left), and large magnetic disk units (rear right). Large portions of the central processor, as well as the 12 magnetic tape stations of this installation, are not visible. (*Courtesy of International Business Machines Corp.*)

a carry, can be picked up. Such tables can be stored in special small memories, but they can also be implemented by *logic circuits*.

I.3. Examples of Digital Systems

Not all digital-circuit designers of the future will be engaged in the design of computer modules. On the contrary, we would guess that, in practice, most problems are centered around the interfacing of computers to other systems. Also, separate digital circuits are needed in laboratories or test stations, where engineers are often asked to design electronic parts for experimental setups. In the era of digital integrated circuits, rather exacting requirements can be met at reasonable cost.

In the following series of photographs, Figs. I-4 to I-8, we have shown a few examples of computers and digital systems.

Fig. I-6. Digital data-acquisition system. Shown are a teletypewriter (left), magnetic tape memory (top left), data logger including a digital voltmeter (three plug-in modules in the middle of the left cabinet), and paper tape punch (bottom of the left cabinet). A general-purpose computer is housed in the right cabinet, and a paper tape reader plugs in below it. (*Courtesy of Hewlett-Packard Co.*)

Fig. I-7. Digital Fourier analyzer, including a general-purpose computer system (left cabinet) and input–output interface devices. The units in the right cabinet are, from the top, an *X-Y* plotter, cathode-ray-tube display unit, analog-to-digital converter (middle left), and operating console (middle right). (*Courtesy of Hewlett-Packard Co.*)

Fig. I-8. The basic building block of the Saturn V and Apollo backup computers is the unit logic device held at the left. There are 8918 of these devices in each computer and data adapter in the guidance system. The devices are mounted by infrared solder reflow on the interconnection cards shown here, which, in turn, are interconnected back to back to form page assemblies. Each computer has 150 page assemblies. (*Courtesy of International Business Machines Corp.*)

part one

LOGIC DESIGN

chapter 1

Numbers, Codes, and Machine Variables

1.1. Numerical Representations

Analog and Digital Representations. The two basic ways of representing quantities are called *analog* and *digital* (or *numerical*). In *analog representation*, a quantity is proportional to another quantity, which is used in place of the first one. Examples of analog representations are graphs and diagrams in which the length of a line, or the distance of a point from an axis, are analog quantities relating to the object represented. *Models* are another type of analog representation, in which a substitute object, usually made in a reduced scale and subjected to test conditions in a laboratory, is supposed to behave according to similar mathematical laws as the original object in its actual environment. Models are used extensively in hydro- and aerodynamics and in structural engineering and elasticity studies. *Dynamical models* are dynamical systems that are described by similar differential equations in time as the object system. It is customary to use electrical models for dynamical systems and to organize them into the form of an easily operable device, called an *analog computer*. The variables are represented as electrical voltages or currents which can be easily measured and recorded. Analog computing means simply that the system variables are measured or plotted versus time; the initial conditions may be varied to study their effects on the characteristic behavior of the system.

In *digital representation* (the Latin *digitus* = finger, toe) the numbers are not represented by proportional quantities but *symbolically*, with the aid of *digits*. The digital representation is usually based on a particular *number system*. In a general number system, let R units form a measure called the *base* of this system; let R such measures or R^2 units form a larger measure, the next larger measure being R^3, etc. Also, fractions can be described using

13

descending measures—R^{-1}, R^{-2}, etc. A number is evaluated digitally by comparing it with a set of multiples of different measures. This is equivalent to weighing a body with a set of standard weights and counting the numbers of different weights used. Let a number X be expressed as

$$X = \sum_{i=-m}^{n} d_i R^i \qquad [1\text{-}1]$$

where the d_i's are natural numbers from a set $\{0, 1, 2, \ldots, R - 1\}$ (digits). The base R of this system is also called the *radix*. People have generally accepted $R = 10$ (decimal number system) as the basis of numerical mathematics, although 5, 12, and 20 are bases that have sometimes been used. In this age of computers, $R = 2$ (binary number system) has become commonplace, particularly in automatic data processing. Only two digit symbols, 0 and 1, called *bits* (*bi*nary dig*its*) are needed in the binary number system.

In general, if the base is known from context or is otherwise obvious, it is customary to indicate the digits by *positional notation*, introduced in India about 400 A.D. In this notation the digit symbols (*numerals*) associated with the different powers of R are juxtaposed, with the powers descending from the left to the right. Thus we denote

$$X = d_n \cdots d_2 d_1 d_0 . d_{-1} d_{-2} \cdots d_{-m} \qquad [1\text{-}2]$$

The point between d_0 and d_{-1} in Eq. [1-2], the *fractional point* or *radix point*, is needed in the compact symbolic notation to show the place of R^0.

When different number systems are mixed, a particular system can always be specified with a subscript (always expressed as a decimal number):

$$X = X_R \qquad [1\text{-}3]$$

For example, *one thousand* may be written 1000, 1000_{10}, or 1111101000_2.

Number Systems of Digital Technology. The conventional number systems of digital technology are shown in Table 1-1. The most common systems are the binary, octal, and decimal.

Table 1-1. Number Systems

Radix	Number System	Digit Symbols
2	binary	0, 1
3	ternary	$\bar{1}, 0, 1$ ($\bar{1} = -1$)
8	octal	0, 1, 2, 3, 4, 5, 6, 7
10	decimal	0 through 9
16	hexadecimal	0 through 9, A, B, C, D, E, F

In every number system there are just enough symbols to express all digits. The ternary system is in a special position; it has three symbols, of which one ($\bar{1}$) is negative. This system is sometimes selected for special devices because of the simplicity of arithmetic operations of negative numbers. For the binary, octal, and decimal number systems we use arabic numerals; only 0 and 1 are used in the binary system, 0 through 7 in the octal system. In the hexadecimal system, arabic numerals are used for digits 0 through 9; for 10 through 15, the numerals must be augmented by the letters A, B, C, D, E, and F (in that order). The digit symbols of the hexadecimal system are called *alphanumeric* or *alphameric digits*.

Conversions Among Binary, Octal, and Hexadecimal Numbers. Binary numbers can be easily converted into the octal or hexadecimal system and expressed with octal or alphanumeric digits by grouping the bits into sets of three or four bits, starting at the radix point, and assigning new symbols to the groups. Octal notations for binary numbers are extensively used in computers.

Example 1-1. Let us express the following binary number with octal and alphanumeric digits:

$$110101110101$$

Grouping into three-bit sets we get

$$110 \ 101 \ 110 \ 101 = 6565_8$$

Grouping into four-bit sets we obtain

$$1101 \ 0111 \ 0101 = D75_{16}$$

Example 1-2. Let us express 705_8 and $A6F_{16}$ as binary numbers:

$$705_8 = 111 \ 000 \ 101 \quad = 111000101_2$$

$$A6F_{16} = 1010 \ 0110 \ 1111 = 101001101111_2$$

Example 1-3. The following endless binary fraction is easily converted into octal:

$$(0.011011011 \ldots)_2 = (0.333 \ldots)_8$$

To convert binary numbers into decimal, it is advisable to use a table for powers of 2; such a table is shown as Table 1-2. The powers of 2 of a binary number indicated by bit value 1 are simply summed up.

Table 1-2. Powers of Two

2^n	n	2^{-n}
1	0	1.0
2	1	0.5
4	2	0.25
8	3	0.125
16	4	0.062 5
32	5	0.031 25
64	6	0.015 625
128	7	0.007 812 5
256	8	0.003 906 25
512	9	0.001 953 125
1 024	10	0.000 976 562 5
2 048	11	0.000 488 281 25
4 096	12	0.000 244 140 625
8 192	13	0.000 122 070 312 5
16 384	14	0.000 061 035 156 25
32 768	15	0.000 030 517 578 125
65 536	16	0.000 015 258 789 062 5
131 072	17	0.000 007 629 394 531 25
262 144	18	0.000 003 814 697 265 625
524 288	19	0.000 001 907 348 632 812 5
1 048 576	20	0.000 000 953 674 316 406 25
2 097 152	21	0.000 000 476 837 158 203 125
4 194 304	22	0.000 000 238 418 579 101 562 5
8 388 608	23	0.000 000 119 209 289 550 781 25
16 777 216	24	0.000 000 059 604 644 775 390 625
33 554 432	25	0.000 000 029 802 322 387 695 312 5
67 108 864	26	0.000 000 014 901 161 193 847 656 25
134 217 728	27	0.000 000 007 450 580 596 923 828 125
268 435 456	28	0.000 000 003 725 290 298 461 914 062 5
536 870 912	29	0.000 000 001 862 645 149 230 957 031 25
1 073 741 824	30	0.000 000 000 931 322 574 615 478 515 625
2 147 483 648	31	0.000 000 000 465 661 287 307 739 257 812 5
4 294 967 296	32	0.000 000 000 232 830 643 653 869 628 906 25
8 589 934 592	33	0.000 000 000 116 415 321 826 934 814 453 125
17 179 869 184	34	0.000 000 000 058 207 660 913 467 407 226 562 5
34 359 738 368	35	0.000 000 000 029 103 830 456 733 703 613 281 25
68 719 476 736	36	0.000 000 000 014 551 915 228 366 851 806 640 625

1.2. Conversion of Numbers Between Arbitrary Bases

The following algorithms can be used for the conversion of numbers expressed in base P to representations in base R, provided that the arithmetic operations in the system with base P are known. In practice, the most frequent case is $P = 10$ and $R = 2$ (*decimal-to-binary conversion*).

Different algorithms are used for integers and for fractions. For a general number, separate conversion operations are carried out for the integral and for the fractional parts, respectively.

Conversion Algorithm for Integers. Let X_P be an integer in a base P. Dividing by the radix R we get

$$\frac{X_P}{R} = A + \frac{d_0}{R} \qquad [1\text{-}4]$$

where A is the quotient and d_0 the remainder ($\leq R - 1$). Successive divisions are continued until the last quotient is 0:

$$\frac{A}{R} = B + \frac{d_1}{R}$$

$$\frac{B}{R} = C + \frac{d_2}{R}$$

$$\cdot$$
$$\cdot$$
$$\cdot$$

$$\frac{Z}{R} = 0 + \frac{d_n}{R}$$

The expression sought is

$$X_P = (d_n d_{n-1} \cdots d_1 d_0)_R \qquad [1\text{-}5]$$

Proof. By direct substitutions starting with Eq. [1-4] we obtain

$$X_P = d_0 + RA = d_0 + Rd_1 + R^2B = \cdots$$
$$= d_0 + Rd_1 + R^2d_2 + \cdots + R^nd_n$$

where d_i's ($i = 0$ to n) are from the set $\{0, 1, \ldots, R - 1\}$.   Q.E.D.

Example 1-4. Convert 52_{10} to its binary equivalent.

$$\frac{52}{2} = 26 + \frac{0}{2} \qquad d_0 = 0$$

$$\frac{26}{2} = 13 + \frac{0}{2} \qquad d_1 = 0$$

$$\frac{13}{2} = 6 + \frac{1}{2} \qquad d_2 = 1$$

$$\frac{6}{2} = 3 + \frac{0}{2} \qquad d_3 = 0$$

$$\frac{3}{2} = 1 + \frac{1}{2} \qquad d_4 = 1$$

$$\frac{1}{2} = 0 + \frac{1}{2} \qquad d_5 = 1$$

$$52_{10} = 110100_2$$

Conversion Algorithm for Fractions. Let Y_P be a fraction expressed in a base P. Multiplying by the radix R we obtain

$$RY_P = d_{-1} + A' \qquad \text{[1-6]}$$

where d_{-1} is the carry digit propagated to the integer part $(d_{-1} \leq R - 1)$ and A' is the new fraction. Successive multiplications are continued until a desired number of digits have been calculated or until the number is automatically truncated by the fact that the next faction is zero:

$$RA' = d_{-2} + B'$$

$$RB' = d_{-3} + C'$$

$$\cdot$$
$$\cdot$$
$$\cdot$$

$$RZ' = d_{-m} + \epsilon \qquad \text{(where } \epsilon < 1 \text{ and } R^{-m}\epsilon \text{ is the truncation error)}$$

The expression sought is

$$Y_P = (0.d_{-1}d_{-2} \ldots d_{-m})_R + R^{-m}\epsilon \qquad \text{[1-7]}$$

Proof. By direct substitutions starting with Eq. [1-6] we obtain

$$Y_P = R^{-1}d_{-1} + R^{-1}A' = R^{-1}d_{-1} + R^{-2}d_{-2} + R^{-2}B' = \cdots$$
$$= R^{-1}d_{-1} + R^{-2}d_{-2} + \cdots + R^{-m}d_{-m} + R^{-m}\epsilon$$

where the d_{-i}'s $(i = 1$ to $m)$ are from the set $\{0, 1, \ldots, R - 1\}$.   Q.E.D.

Example 1-5. Find the binary expression for 0.3_{10}.

$$2 \times 0.3 = 0 + 0.6 \qquad d_{-1} = 0$$

$$2 \times 0.6 = 1 + 0.2 \qquad d_{-2} = 1$$

$$2 \times 0.2 = 0 + 0.4 \qquad d_{-3} = 0$$

$$2 \times 0.4 = 0 + 0.8 \qquad d_{-4} = 0$$

$$2 \times 0.8 = 1 + 0.6 \qquad d_{-5} = 1$$

$$2 \times 0.6 = 1 + 0.2 \qquad d_{-6} = 1$$

$$\cdot$$
$$\cdot \qquad \text{(repeating)}$$
$$\cdot$$

$$0.3_{10} = (0.0100110011\ldots)_2$$

1.3.   Codes

Binary Coding of Numbers. To achieve the required reliability regarding stability and noise immunity in electronic digital computers, only bivalent circuit elements are of practical importance. A digital representation of numbers or other discrete variables which is based on bivalent symbols is called *binary representation.* Any combination of bivalent symbols is a binary representation; the *binary numbers* form a small, but very important, subset.

According to common usage, we shall, in the following, call bivalent or binary symbols *bits* (*bi*nary un*it* of information), with a numerical value 0 or 1, although they would have nothing to do with the binary number system. A particular bit combination assigned to an object is called the *code* of this object; a code is thus a kind of name that is automatically recognizable. The digital machine variables are always represented by number codes. In telecommunication, and in peripheral units of computers, codes are also assigned for letters, special characters, and instructions controlling the receiving unit.

Coding is used to represent digit symbols in other than binary number systems. For example, to represent the 10 decimal digits, at least 4 bits are needed, because $2^3 = 8$, but $2^4 = 16$. Of the 16 possible combinations of 4 bits, any 10 taken in any order can be assigned to the 10 arabic numerals $0, 1, \ldots, 9$. Obviously there are $16!/6! \simeq 2.9 \times 10^{10}$ ways to do this.

Weighted and Unweighted Binary Number Codes. In the juxtaposition of bits we can assign a fixed numerical value or *weight* for each bit position, and the numerical value of this number code is then the sum of the weights that are indicated by the 1 bits. If the bits are denoted by d_i and the corresponding weights by w_i and we restrict ourselves to integers, we have, for the expansion of the number X,

$$X = \sum_{i=0}^{n} d_i w_i \qquad d_i = 0, 1 \qquad [1\text{-}8]$$

In the binary number system $w_i = 2^i$. The weights are usually integers, and some of them may be negative. Codes in which a weight is assigned for each bit position are called *weighted codes.* All other codes are *unweighted* codes.

We have shown in Table 1-3 a few binary codes for decimal digits. Notice that there may be several alternative codes in the same system for the same digit. All of these codes are called *binary-coded decimal,* or BCD, codes.

The 8421 code is equivalent with the representation in the binary number system, except that the nonsignificant zeros in front of the number are always included. Of the 2421 codes, the one shown first has the property that the first bit is 0 for the digits 0 through 4, and 1 for the digits 5 through 9.

Table 1-3. Examples of BCD Codes

Decimal Digit	Weighted			Unweighted
	8421	2421	5043210 (Biquinary)	XS3 (Excess-3)
0	0000	0000	0100001	0011
1	0001	0001	0100010	0100
2	0010	0010 or 1000	0100100	0101
3	0011	0011 or 1001	0101000	0110
4	0100	0100 or 1010	0110000	0111
5	0101	1011 or 0101	1000001	1000
6	0110	1100 or 0110	1000010	1001
7	0111	1101 or 0111	1000100	1010
8	1000	1110	1001000	1011
9	1001	1111	1010000	1100

This particular 2421 code is usually denoted by 2*421. The *biquinary* code is formed of seven bits, of which the first two form a group for the weights 0 and 5 and the remaining five bits have the weights 0 through 4. There is always one 1 in each group. The *excess*-3, or XS3, code is obtained by first adding three to the digit and then taking the 8421 code of this number. There is no possibility of assigning weights to the bit positions of this code.

Unary Numbers. There is a simple digital representation which also belongs to the number systems. Suppose that an integer is represented by an equivalent number of units placed in a row. This is obviously a representation in a number system with a base $R = 1$. The representation where the number of units or quanta are explicitly shown is called a *unary number*. We meet

Table 1-4. Comparison of Binary and "Excess-1" Unary Numbers

Binary	Excess-1 Unary
0	1
1	11
10	111
11	1111
100	11111
101	111111
110	1111111
111	11111111
...	...

unary numbers in systems in which the number of discrete impulses is used to transmit digital information. Notice that if the number "zero" has to be represented, an "excess-1" unary system is needed, as shown in Table 1-4.

Reflected Number Systems and Reflected Codes. Flores [132] has suggested the *reflected number system,* in which any two successive numbers can differ only by one digit. In this system there is a digit for each power of the radix R as in the usual notation, but the digit symbols have been used in a special way; one time they are assigned to the numbers $0, 1, 2, \ldots,$ $R - 1$ in the normal order, but another time they are assigned in the reversed order. Reflected number systems, or *reflected codes,* as they are also called, are shown in Table 1-5.

Table 1-5

Normal Decimal	Reflected Decimal	Normal Binary	Reflected Binary
0	0	0	0
1	1	1	1
2	2	10	11
3	3	11	10
4	4	100	110
5	5	101	111
6	6	110	101
7	7	111	100
8	8	1000	1100
9	9	1001	1101
10	19	1010	1111
11	18	1011	1110
12	17	1100	1010
13	16	1101	1011
14	15	1110	1001
15	14	1111	1000
16	13	10000	11000

Since the idea in constructing reflected codes for successive numbers is to change only one digit at a time, we shall proceed according to the following rule:

In the ascending sequence of numbers starting from 0, the digit symbols for the least significant digit are first used in the normal order. After the Rth count the symbols for the last digit are used through the symbol list in the reverse order. After the (2R)th count the symbols for the last digit are again used in the normal order, etc. This rule is applied also to the other digit symbols when they are changed.

The binary code corresponding to the reflected binary number system is also called the *Gray code*. It is used in code disks and code plates for the digital indication of angular or linear position (see Fig. 1-1). Since the codes of two adjacent sections never differ by more than one bit, the indication error can never be larger than the width of one section, even if the bit detectors were slightly misaligned.

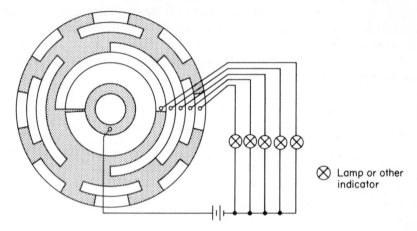

Fig. 1-1. Code disk. (Shaded sectors are conducting.)

Error Detecting and Correcting Codes. Digital systems are inherently more stable than analog systems, but an error in digital representation can be very serious, especially when it occurs in the most significant digits. Many attempts have been made to increase the reliability of computer hardware by adding redundancy (i.e., extra circuits).

The most usual way to increase the reliability of numerical representations is to use check marks which are manipulated at the same time as the numerical contents. The simplest check of a binary number is based on *parity bits*: to each word, a bit is added the value of which is such that it makes the number of 1's in the word either even or odd. If a single bit in the word is changed, the parity is changed and an error is detected. A drawback of this method is that it does not give any indication if an even number of bits are changed simultaneously. The probability for the occurrence of simultaneous multiple changes is usually negligible compared to the probability for single errors, unless multiple errors have the same origin.

If more extra bits can be used for each word, it is possible not only to detect multiple errors but also to correct them.

As an example of codes which are able to correct single errors, let us imagine the following thought experiment. For all occurring numbers we have assigned codes that differ from each other by exactly three bits. A word that differs from any of the legitimate codes by one or two bits is erroneous; if the occurrence of a single error would have been assumed, the code from which the erroneous code differs by one bit is the correct one.

In *Hamming codes*, which permit easy automatic error correction, there are several parity checks over specified bit groups. The added parity bits are always such that they make the parity of every check group *even*. A single erroneous bit in the code may violate the even parity in one or several check groups, as the case may be. It is now possible to show that if a certain minimum number of parity bits are used, it is possible to deduce the position of the erroneous bit from the particular type of parity violation. As an example we consider the Hamming codes of decimal digits. These are shown in Table 1-6, where the data bits are denoted by x_i and the parity bits by p_j. In these codes we have even parity over the following bit groups:

$$p_1, x_3, x_5, x_7 \qquad \text{(check 1)}$$

$$p_2, x_3, x_6, x_7 \qquad \text{(check 2)}$$

$$p_4, x_5, x_6, x_7 \qquad \text{(check 3)}$$

Table 1-6. Hamming Codes for Decimal Digits

	Position						
	7	6	5	4	3	2	1
Decimal Digit	x_7	x_6	x_5	p_4	x_3	p_2	p_1
0	0	0	0	0	0	0	0
1	1	0	0	1	0	1	1
2	0	1	0	1	0	1	0
3	1	1	0	0	0	0	1
4	0	0	1	1	0	0	1
5	1	0	1	0	0	1	0
6	0	1	1	0	0	1	1
7	1	1	1	1	0	0	0
8	0	0	0	0	1	1	1
9	1	0	0	1	1	0	0

Only one parity bit occurs in each check group. Also other combinations of the x_i's would have provided a unique detection of the errors. These bit combinations and the particular selection of the parity bit positions in the code have a remarkable property:

Let there be a single erroneous bit anywhere in the code. Denote a successful parity check by 0 and a failure (i.e., odd parity) by 1. The binary number resulting from the bits of check 3, *check* 2, *and check* 1 *in this order indicates the position of the erroneous bit.*

A proof for this surprising statement can be given in the following way:

1. Obviously, if the error occurs in a parity bit, one check fails and the two others succeed. The two successful parity tests which do not include this bit show that the error cannot be in the data bits x_i, since all data bits are always included in the two successful checks. If p_1 is erroneous, check 1 gives 1, and the two other tests give 0's, so the resulting binary number is 001, which directly indicates the position 1. Similarly, if p_2 is erroneous, the resulting binary number is 010, indicating the position 2; if the error is in p_4, the binary number is 100, indicating position 4.

2. If x_3 is erroneous, check 1 and check 2 fail but check 3 succeeds. Thus the resulting binary number is 011 ($= 3$). Similar proofs can be given if the error occurs in x_5, x_6, or x_7. Of course, if all checks succeed, the binary number is 000, indicating the correctness of the result.

Example 1-6. We have received a message 1101000, which is an erroneous Hamming code. The parity checks are check 1, failure; check 2, success; check 3, failure. The resulting binary number is 101, giving the erroneous position, which is 5 (the fifth from the right). The corrected code is 1111000, which is the legitimate code of 7.

If more than one erroneous bit is present in the message, this correcting procedure gives in general a false result, even if the corrected number would represent a legitimate code.

It can be shown that for the correction of single errors, the necessary number k of parity bits for j bits of data is obtained from the relation

$$2^k \geq j + k + 1 \qquad \text{[1-9]}$$

There also exist codes that correct multiple errors.

1.4. Storage and Transfer of Information in Machines

Machine Variables. Mathematical varibles are represented in computers by their physical counterparts, called *machine variables.*

In *analog computers*, machine variables are usually voltages (or actually charges) on computing capacitors which act as a kind of storage location. The mathematical variables are mapped on machine variables, let us say from -10 to $+10$ V, using a proper *scale factor*. The relative accuracy of this representation, however, is limited by the stability (constancy) of electrical components: the capacitors themselves and the associated circuitry. Technical difficulties become paramount if a relative accuracy better than, say, 0.01 percent, is required. For less accuracy, analog representation is advantageous.

In *digital computers*, the values of mathematical variables are stored as *discrete states of electronic circuits* or other components, which thus constitute the machine variables. Notice that a continuous variable has an infinite number of possible values, which cannot all be represented by a finite number of symbols. In numerical representation, variables are therefore always *quantized*.

In order to represent one digit in a base of R, we need an elementary system that has R distinguishable states; for the representation of N digits, N such elementary systems are needed. Almost invariably digital computers make use of elementary circuits or memory components (storage elements) with exactly two states, whereby the introduction of the binary number system is natural. If each storage element has two states, the number of different state combinations in an N-element system is 2^N. For example, a 36-element binary register is rather common in many computers, and there are $2^{36} = 68,719,476,736$ different states in this system. Hence we see a particular property of digital representations—the machine variables can be represented to an arbitrary accuracy by including a sufficient number of storage elements.

Storage Elements and Registers. The positional notation is used in electronic computers where physically adjacent elements represent bits. In Fig. 1-2 the storage elements, bit storages, are shown with a square and the bits are written inside the symbol. The states of these systems are used as machine variables, and a combination of elements is called a *register*. In a binary register [Fig. 1-2(a)] the adjacent storage elements are used to represent binary digits, but in a binary-coded-decimal (BCD) register [Fig. 1-2(b)], for example, a group of storage elements is needed to represent a decimal digit.

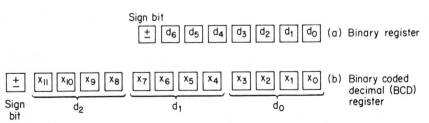

Fig. **1-2.** Registers and machine variables.

The sign is usually represented by the leftmost bit, which is 0 for positive numbers and 1 for negative numbers. There are also other representations for negative numbers, called *one's* and *two's complements*, which will be discussed in Chapter 4.

There is no way of showing the position of the radix point; its location must be known from the context. It is left to the programmer to prepare all numbers in such a way that the radix point is correctly taken into account.

Digital computing systems are essentially built around registers. Not only are registers used to store or temporarily hold ("buffer") digital information, but many registers are able to transform the information.

An important type of operational register is the *counter*. When the control circuits of the counter receive a counting command, the numerical contents of the register are incremented by one unit (forward counting, or addition of one), or decremented by one (backward counting, subtraction of one). There are several counter implementations for different number systems and for various purposes, as discussed in Chapter 4.

A basic operation in computing systems is the *transfer of information* from one register to another. Very often electronic copy gates are provided between all corresponding bit positions so that the contents can be transferred *in parallel*. Not as often, at least in larger devices, the transfer of information is accomplished *in series* by what is called *shifting*. A *shift register* is a register that is provided with direct copy gates between adjacent storage cells. When a shift command is simultaneously given to all copy gates, each bit jumps into the next position on the right or left, as the case may be, replacing the old information. Some registers can shift in both directions; more often the registers are exclusively *right* or *left shifting*. In a *right-shift register*, the contents of each bit position are simultaneously copied into the next position on the right, whereby the rightmost bit is lost. This is called *overflow*. A new bit must be copied from somewhere to the leftmost position. An *end-around right shift* means that during the shifting the rightmost bit is cyclically copied into the leftmost position and the register thus forms a ring. We define the *left shift* and the *end-around left shift* analogously.

Let us illustrate the operation of the shift registers and end-around shift registers in Fig. 1-3 by showing the contents before and after the shift command. The new information bit, located somewhere, is shown by a dashed square.

The operation of a shift register in the arithmetic unit of a computer may differ slightly from what has been described so far. First, *logical shifting* means that the complete register is operated. But in arithmetic operations, the leftmost bit is usually reserved for the indication of the sign (0 for $+$ and 1 for $-$), and it should sometimes be operated separately. In *arithmetic shifting*, only the numerical part (the rest of the register) is shifted in some

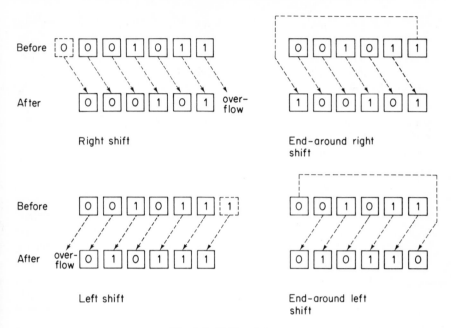

Fig. 1-3. Shift registers.

of the ways indicated, so in arithmetic left shifting, it is the next-to-leftmost bit that overflows. Also, in arithmetic-shifting operations the new bits filling the vacant ends may be ones or zeros, depending on the representation of negative numbers (See Chapter 4).

Shifting operations are needed in many computing operations; for example, if the register is used to represent a binary number with the least significant bit on the right, an arithmetic right-shifting operation is equivalent to dividing the previous contents by 2. If the rightmost bit was 1, it will cause a truncation error by the overflow since the remainder cannot be shown. If at least one of the leftmost bits of the numerical part was zero before, an arithmetic shifting to the left will be equivalent to multiplying the previous contents by 2.

Memories. Almost all but the simplest computing devices make use of various kinds of memories, which consist of a large number of storage elements. Each element stores one bit. Most memory systems are organized as a collection of fixed-length *memory locations,* each one of which is equivalent to an elementary register. The contents of one location is called a *word,* which is simply a set of bits.

Registers are, in fact, memories with one memory location. They are usually built of active electronic components and provided with auxiliary circuits for the manipulation of the contents, whereas in *mass memories,* often made of magnetic materials, information can only be *written* into and *read* from the locations. The most important feature of all memories is, however, that there is a simple *access* to each memory location. Every location has a unique *address* which can be specified by control signals. After an address has been selected (and temporarily stored in the *address register* of the memory system), the word stored in the addressed location can be transferred by a *read* command into the *memory register*, where it is temporarily stored. Also, when an address has been selected, a new word can be transferred to this memory location by a *write* command, using write amplifier inputs.

The locations are usually numbered or addressed from 0 to $N - 1$, where N is the total number of locations. To specify the size of a memory we have to give N as well as the word length. Commercial computers have main memories in which N ranges typically from 1024 to 524,288 (these numbers have been taken here to represent certain powers of 2, 2^{10} and 2^{19}). A typical "unit" of memory size is 1024, called K. Thus typical contemporary memory sizes are denoted 4K, 8K, 16K, etc. The *word lengths* depend on many factors. Sometimes the words are divided into *bytes* of, say, eight bits each, and this might be the smallest applicable length of a memory location. If words are treated as a whole, and if memory locations are used to store addresses of other locations, as is often the case when direct addressing is used, all addresses of the memory must be representable by one word, a binary number. Small process computers very often have 16-bit words; in scientific computers primarily intended for extensive computations a common word length is 36 bits, which is selected for a number of reasons (e.g., accurate representation of numbers). Very large computers may have 60-bit words.

Figure 1-4 shows a block scheme of the main parts of a memory system. The word lengths of data registers and the memory locations are very often the same throughout the system; they are thus a basic architectural feature of the computer.

The main memory of a computer, also called the working memory, may use ferrite cores, dots of magnetic film, or electronic circuits as cells, but it is usually of the *random-access* type, which means that any address may be selected in a fixed time. The *access time* (i.e., the time needed to transfer the word from the memory location to the memory register) of ferrite-core memories is about 1 μsec or slightly shorter. In other types of memories, 100 nsec access times are not uncommon.

A computer system usually needs the slower backing memories, which can store large amounts of data. These are usually of the *serial-access* type, in which the access time depends on the location (e.g., the position of a word on a disk). In these memories, searching of the data location is done mechanically.

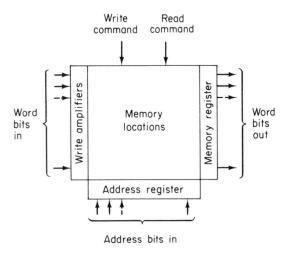

Fig. 1-4. Memory system.

Not only are numerical data stored in memories, the words can also be symbolical codes for digital operations; a collection of such codes is called a *program*. There are control circuits in the computer which can trigger a particular automatic operation at the presence of the corresponding code in a register; this is called *execution* of the operation.

Especially when memories are used to store coded programs which are fixed, it is cheaper and safer to use *read-only memories* (ROM), in which the bit values are determined during the manufacturing phase and are not alterable. Read-only memories usually have only address inputs and word outputs.

Binary Signals. The binary states in storage elements are set, information is copied, and control commands are given with the aid of messages called *binary signals*. The electric signal representing or transmitting a binary symbol is usually one of two voltage levels. We shall define later various possibilities for the assignment of binary values to these levels. For the moment, let a zero level be equivalent to 0 and a more positive level be equivalent to 1. A digital signal is an excursion of the voltage to either of these levels at a specified time. This excursion may start in synchronism with a clock and have a predetermined duration, after which the voltages are returned to zero; this is the *pulsed* signal form. Another possibility is to sustain the value of the voltage until an excursion to a new value is necessary. Such signals are called *static*; their values may be sampled at specified times.

A group of related binary signals (e.g., the bits belonging to a binary code) and the digits of a number can be distinguished by displacing them in *space* or *time*. In other words, we may have a separate wire for each bit, which then appear simultaneously; this is the *parallel* form of representation. But the bits of a code or of a binary number can also appear on a single wire at successive intervals, the least significant bit usually coming first. This is the *serial* form of representation. We have demonstrated in Fig. 1-5(a) how a binary number 110101 appears in the parallel representation. In Fig. 1-5(b) the same number is shown in serial representation by a voltage that attains a static value during successive periods. In Fig. 1-5(c) we have the corresponding pulsed serial representation. Notice that the least significant bit appears leftmost in timing diagrams (b) and (c).

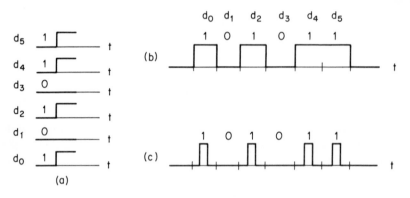

Fig. 1-5. Binary signals: (a) parallel; (b) serial, static; (c) serial, pulsed.

Parallel-to-Serial and Serial-to-Parallel Conversion of Signals. If there are N bits in a shift register and the contents are shifted N steps end-around, the original state is restored. But in right shifting, for example, if we had observed the output of the least significant bit position, we would have seen that all the bits of the original word appear in a series at this terminal. The end-around shift register thus provides a possibility to convert parallel signals, buffered as the contents of a register, into a serial digital signal.

A serial signal can be converted into parallel form using a shift register and a command pulse train generator. The serial signal is led to the input terminals of the leftmost storage cell in a right-shift register, and a shift command pulse is given during each period when the bits occur. The number of command pulses is the same as the number of bits. In this way the bit values of the serial signal are shifted into the register, from which they can be retrieved in parallel form.

1 5 Symbolic Description of Digital Computing Operations

Nature of Digital Computing. Information processing by digital systems consists of a large number of elementary arithmetic or logic operations performed in logical succession. The operands are picked up from storage locations one or a few at a time, intermediate results of operations are stored, and only the final results are written out.

Digital information is stored in registers and memory locations in the form of binary words. Thus *information transfers between registers and storage locations are the basic processes by which digital systems operate.* If arithmetic, logic, or other operations are performed on the words, they are done during the transfer by transforming the transferred signals in special logic circuits. For example, typical digital operations consist of picking up a number (operand) from a particular storage location of the main memory, adding this number with the contents of a special computing register called the *accumulator*, and transferring the result back into another specified storage location. Storage elements and registers may also be loaded with constants (i.e. with the logic values 0 and 1) and with binary words defined by switches on control panels. Digital information can also be read from the terminal bus.

It is typical in many phases of digital computation that some operations or operation sequences are automatically repeated until a given number of computations are accomplished or until specified results have been obtained. For example, if an addition must be repeated N times, a count is subtracted each time from a special register called *cycle counter*. This register was initially loaded with the number N, and the additions are automatically stopped when the contents of this register become zero.

Information transfers between all specified register pairs (between which transfer paths are provided) are conditioned by logic signals. A transfer is initiated when the corresponding function becomes 1. These functions contain clocking signals, time-dependent register contents, certain bit values of other registers, and internal or external control signals as their arguments. Several transitions may be governed by the same signal.

Symbolic Language for Internal Digital Operations. To describe the control sequences of internal digital operations of a computing system, we need a special symbolism, a kind of language in which the machine variables and the most fundamental arithmetic, logic, and transfer operations are defined. This language must include all occurring operations and yet leave freedom of choice for different electronic implementations. In this symbolism the primary variables are the contents of storage locations and registers, and the basic operations are information transfer and arithmetic or logic operations on the words, without regard to the interconnecting circuits by which these operations are implemented. For example, the same statement may be used to denote parallel and serial transfer of register contents.

In digital systems only a small portion of the stored information is operated at a time, and we shall assume that if a set of operations is explicitly indicated, the stored information not shown remains unaltered.

DEFINITION OF A SYMBOLIC LANGUAGE FOR THE GENERAL DESCRIPTION
OF DIGITAL OPERATIONS

Registers are denoted by capital letters or letter groups. The length of a
register is not usually specified, and in extreme cases a register may be
one bit long. Portions of registers may be identified by different
names.

Storage elements (flip-flops, etc.) are consequently denoted in the same way
as registers, by capital letters.

A particular bit of a register is denoted by the name of the corresponding
register indexed by a subscript showing the bit position.

The contents of a register or storage element R are denoted by parentheses
[e.g., (R) are the contents of R].

Memory location with a numerical address c is denoted by $M\langle c \rangle$. If the
address is given by the contents of the address register AR, the memory
location is denoted $M\langle (AR) \rangle$.

For brevity, a memory location is sometimes denoted by
its address only. For example, (AR) is the memory location, the
address of which is given in AR. Consequently, $((AR))$ are the contents
of this location.

Constant binary words are denoted by lowercase letters or numbers.

Transfer of information from any storage A to any storage B, or the
copying of the contents of A into B without changing A, is denoted
$B \Leftarrow (A)$.

Arithmetic, logic, or other operations during transfer are denoted with
corresponding algebraic notations; for example, $A \Leftarrow (A) + (M\langle c \rangle)$
means that the contents of the memory location with an address c
are arithmetically added to the contents of register A and the result
assigned as the new contents of A.

Register operations are operations in which the new contents of a register
are some function of the old contents, and the transformation is done
without transfer of information to another register. Examples are
counting and shifting. *Clearing* of a register (i.e., the loading of all
positions with 0) or *setting* of a register to any other particular state
are also register operations.

A logic control condition, or the condition that the subsequent operations are valid only when a logic signal f *becomes* 1, is denoted by $(f:)$ followed by the operations that are due.

Statement is a common name for all indicated operations executed at a logic control condition.

EXAMPLES OF SYMBOLIC STATEMENTS

Forward counting in a register R upon the occurrence of a counting command CP is denoted

$$CP:\ R \Leftarrow (R) + 1$$

Backward counting under similar conditions is denoted

$$CP:\ R \Leftarrow (R) - 1$$

End-around right shifting at the shifting command CP in a register A with the bits $A_n, A_{n-1}, \ldots, A_0$ (the indexes decreasing from the left to the right) is denoted

$$CP:\ A_i \Leftarrow (A_{i+1}) \text{ for } i = 0 \text{ to } n - 1, A_n \Leftarrow (A_0)$$

A special abbreviated notation may also be coined for these statements taken together, for example,

$$CP:\ A \Leftarrow g(A)$$

PROBLEMS

1-1 Convert the following binary numbers to their decimal equivalents:
(*a*) 1000000000.01
(*b*) 11011
(*c*) 0.001111

1-2 Convert the following decimal numbers to their binary equivalents:
(*a*) 59
(*b*) 0.2265625
(*c*) 110.8125

1-3 Find all the 24 possible ways to encode three variables a, b, and c with two bits.

1-4   The following numbers are expressed in 8421 code: 0001, 0111, 1001, 1100. Give all their possible representations in the following codes:
(a) 2421
(b) 5311
(c) 74(-2) (-1)   (Note negative weights.)
(d) XS3
(e) Gray

1-5   Each of the following messages, which are supposed to be Hamming codes for decimal digits, contains a single erroneous bit. Find the corrected messages by the method described in Section 1.3.
(a) 1100000
(b) 1110010
(c) 0000011

1-6   The initial contents of a shift register are 1011010111000. What are the contents of this register after (a) 6, (b) 12, and (c) 16 end-around right-shift operations when the shifting is logical?

REFERENCES

The theory of number systems is very old and has been described in several books on arithmetics and number theory. A practical representation of the conversion algorithms between different number systems can be found in

Wickes, *Logic Design with Integrated Circuits* [129]

which also serves as a collateral textbook on logic design. Coding systems are reviewed in most textbooks on computer design. An advanced discussion of coding, especially of error-detecting and correcting codes, is given in

Berlekamp, *Algebraic Coding Theory* [8].

There are more references on these subjects in the Bibliography.

chapter 2

Combinational Circuits

All binary digital operations can be described in terms of logic, and digital circuits are thus based on elementary logic gates. It is the purpose of this chapter to lay foundations for the design of logic circuits.

2.1. Logic Operations

Binary Variables and Truth Function. The *binary state* (0 or 1) of a storage element and a *binary signal* (0 or 1) transmitting the information about and implied by this state are *binary variables* which can be denoted by capital letters.

The behavior of binary digital systems obeys the rules of mathematical logic. In bivalent logic, verbal statements are called *propositions*, and, to be sensible, a proposition must be either *true* or *false*. For example, "The switch is closed" is a proposition which may be abbreviated by the symbol A. The proposition is a binary variable too, with two *truth values* denoted by T ($=$ true) and F ($=$ false).

New propositions can be formed from elementary propositions using *connectives*: the *negation operation* NOT and the connectives AND and OR. If we allow ourselves minor violations of English grammar, we may say $A =$ "closed is the switch"; another proposition, "NOT closed is the switch," which we denote NOT A, $\neg A$, or $\sim A$ has the opposite truth value. We say that NOT A is a particular *truth function* of A—the *negation* of A.

Let A and B be two propositions, without regard to their exact meaning. The proposition A AND B (or $A \wedge B$) is a truth function called the *conjunction* of A and B. It is true if and only if A and B are true.

35

The proposition A OR B (or $A \lor B$) is a truth function called the *disjunction* of A and B. It is true if and only if A or B or both are true. (In other words, it is false if and only if A and B are false.)

A function in general is a mathematical variable that is defined for all possible values of its arguments. For to define truth functions it is sufficient to tabulate the function for all possible value combinations of the arguments; abbreviating "true" by T and "false" by F we get Tables 2-1 and 2-2.

Table 2-1

A	NOT A
F	T
T	F

Table 2-2

A	B	A AND B	A OR B
F	F	F	F
F	T	F	T
T	F	F	T
T	T	T	T

Truth Tables. A proposition $P(A_1, A_2, \ldots, A_n)$ of an arbitrary number n of argument propositions which is formed by combined application of the operations NOT, AND, and OR is a truth function that can be analyzed and tabulated for all the 2^n different values of the arguments. This is shown in Table 2-3.

Table 2-3

A_1	A_2	...	A_{n-1}	A_n	$P(A_1, A_2, \ldots, A_{n-1}, A_n)$
F	F	...	F	F	$P(F, F, \ldots, F, F)$
F	F	...	F	T	$P(F, F, \ldots, F, T)$
...	...	...	...	...	...
T	T	...	T	F	$P(T, T, \ldots, T, F)$
T	T	...	T	T	$P(T, T, \ldots, T, T)$
		(all combinations)			

Switching Analogies of Truth Functions. Digital operations consist of transferring binary information from one storage to another, gating binary signals under conditions imposed by other signals, and transforming binary signals in much the same way as searching values from tables. We can infer that the essential purpose of interconnecting circuits in digital systems is to produce signals, the values of which, or the presence of which, are functions of the input signals in the sense of logic propositions. These circuits are called *logic circuits*, and together with the binary storage elements they form the building blocks of digital systems. It was first established by Shannon that the transmission function of an electrical network consisting of contacts connected in series and in parallel formed an electrical analogy for logical propositions. The transmission function Z between two points in a switching network is defined as T if the path between the points is closed and as F if the path is open. It is then obvious that a *series connection* of two switches is the *analogy for conjunction operation*, and a *parallel connection* of two switches is the *analogy for disjunction* (Fig. 2-1). Combinations of series and parallel connections are analogies of combined logic propositions formed by the connectives AND and OR.

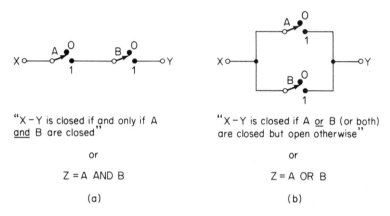

"X−Y is closed if and only if A and B are closed"

or

Z = A AND B

(a)

"X−Y is closed if A or B (or both) are closed but open otherwise"

or

Z = A OR B

(b)

Fig. 2-1. (a) Transmission-function analogy of conjunction. (b) transmission-function analogy of disjunction.

Sometimes the same argument appears in different places of the logic function. In the switching analogy, a separate switch occurs in each of these places, but the switches representing the same variable must be operated together (e.g., like all contacts of the same relay).

It would seem that the negation has no switching analogy. This is, however, a matter of definition of the state values. If in a particular switch the closed state were defined as the value F and the open state T, the transmission function would be the negation of the state.

Voltages of a Diode Circuit as Logic Variables. The transmission function of a switching circuit is not the only two-valued function that serves as an analogy for logic proposition. Using highly nonlinear components, such as diodes, we can form circuits in which input voltage levels can be put into one-to-one correspondence with logic variables entering as arguments, and the output voltage is the analogy of the logic function. Without losing too much generality, we may assume that the diodes have a zero forward resistance and an infinite backward resistance. A zero voltage is defined as the truth value *false*, whereas a voltage that is sufficiently positive corresponds to the truth value *true*. The diode circuit of Fig. 2-2(a) is clearly the analogy of the *conjunction* of *A*, *B*, and *C*, because the output is positive if and only if all input voltages are positive; the circuit of Fig. 2-2(b) is the analogy of the *disjunction* of *A*, *B*, and *C*, because a single positive input voltage makes the output positive, whereas the output is zero if all inputs are zero.

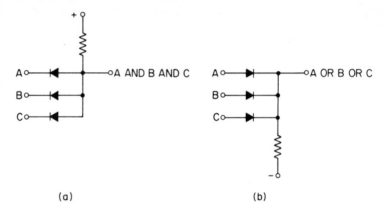

Fig. 2-2. (a) Diode-circuit analogy of conjunction; (b) diode-circuit analogy of disjunction.

With practical diodes we may be obliged to use signal voltages which are of the same order of magnitude (about 1 V) as the range of the diode voltage in which nonlinearities manifest themselves. We have, then, to abandon this highly idealized model and perform a more accurate analysis. This will be done in Chapter 6. An effect of a real diode is to deteriorate and attenuate signal voltages. Therefore, diodes alone are seldom used in logic switching circuits, and active stages must be included for the amplification and standardization of signals.

Negation in a diode circuit, however, cannot be implemented at all. Therefore, it is necessary to postulate an active circuit called an *inverter* (Fig. 2-3) whenever negations enter the logic function. An inverter is any nonlinear active circuit that produces the voltage zero when its input is positive and a positive logic voltage when the input voltage is zero.

X————○◁├——— X   Fig. 2-3. Inverter.

2.2. Logic Signals and Logic Circuits

Logic Levels. Computing devices have been based on various types of physical signals: mechanical movements, pneumatic and hydraulic pressures, etc. In the majority of computing devices, however, signals are electric, usually voltages. In the idealized picture of Section 2.1 we did not regard noise voltages, loading effects, etc. Actually there are no unique signal voltages but rather ranges of voltages, or classes of voltages which are defined as binary logic signals. Logic switching devices can be designed for positive as well as negative signal voltages, but the signal classes must be separated from each other: There is a forbidden gap between the classes. Figure 2-4(a) shows typical positive voltage levels which are used in modern integrated logic circuits; Figure 2-4(b) is an example of negative signal levels used in earlier semiconductor circuits.

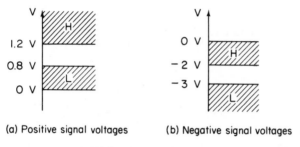

(a) Positive signal voltages   (b) Negative signal voltages

Fig. 2-4. Two examples of logic levels: (a) positive signal voltages; (b) negative signal voltages.

We are yet free to assign the truth values F and T in either order to the two classes. Let us denote the more positive voltage class in a level diagram by H (high) and the more negative by L (low). Then we may use either of the following *logic conventions*:

Positive Logic Convention. In *positive logic*, the logic levels denoted by H correspond to the truth value T, and the logic levels denoted by L correspond to the truth value F.

Negative Logic Convention. In *negative logic*, the logic levels denoted by H correspond to the truth value F, and the logic levels denoted by L correspond to the truth value T.

Comment: In logic circuits, T (= true) is denoted by 1 (also called *logical* 1) and F (= false) by 0 (also called *logical* 0).

Positive logic is more common, and unless otherwise stated we shall use positive logic in the practical circuits of this book. The reader should notice, however, that the logic convention is in no way associated with particular circuit implementation. The logic convention can be changed at any time, but then the circuit implements a different logic function. The relation between the old and new functions is expressed by the *principle of duality*[1]:

Principle of Duality. If the logic function of a circuit is defined using positive logic, the corresponding logic function of the same circuit in negative logic is obtained by replacing each conjunction by a disjunction and vice versa. The same is true if a circuit that has been defined using negative logic is redefined using positive logic.

Example 2-1. As an illustration of the principle of duality, consider a Boolean function that is defined in positive logic:

$$F = A \text{ AND } B \qquad [2\text{-}1]$$

If the variables are redefined using negative logic, the truth value of each one is inverted: NOT $F \longrightarrow F$, NOT $A \longrightarrow A$, and NOT $B \longrightarrow B$. Replacing the conjunction by disjunction yields

$$\text{NOT } F = (\text{NOT } A) \text{ OR } (\text{NOT } B) \qquad [2\text{-}2]$$

Logic Circuits. Analogies to logic operations in addition to those discussed in previous sections can be implemented with many types of active semiconductor circuits. Their operation is always based on highly nonlinear voltage-transfer characteristics of these circuits.

A *logic circuit* is an elementary circuit block having one or several inputs and one or several outputs. It is usually designed as a standardized operational unit. *The purpose of the logic circuit is to accept standardized logic signals at the inputs and to produce to the outputs other standardized logic signals the values of which are functions of the values of the input signals in the sense of propositional calculus.* The internal structure of the logic circuit may consist of different kinds of logic switches and the internal signals need not be standardized. Figure 2-5 shows the general notation for a logic circuit. The logic variables (having the values 0 or 1) are denoted by A, B, $\ldots$, X, Y, Z.

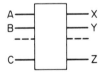

Fig. 2-5. Notation for a general logic circuit.

There are two kinds of logic circuits, *combinational* and *sequential*. In the combinational circuits, the outputs depend only on the present values

[1] This principle will be discussed in Section 2.3., too.

of input variables. Sequential circuits contain memories, and the outputs depend also on the earlier values of the input variables. The logic circuits are unilateral electronic devices, which means that the input signals affect the output signals, but a forced control of the output voltage or current within stated limits should not affect the input signals. The output can be loaded with specified currents from and to the device (also called *source* and *sink* currents of the output circuit, respectively), without changing the intended logic operation of the block.

To begin we define the basic logic gates called the AND, the OR, and the NOT circuits (Fig. 2-6).

Name	Graphic symbols	Definition
AND	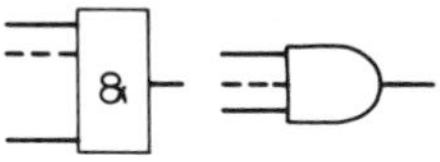 &	The output voltage of this circuit is a signal defined as the logical 1 if and only if all input voltages are signals defined as the logical 1's.
OR	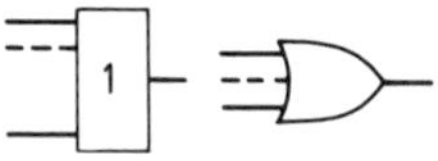 1	The output voltage of this circuit is a signal defined as the logical 1 if at least one of the input voltages is a signal defined as the logical 1.
NOT (inverter)		The output voltage of this circuit is a signal having the logical value $\bar{A}$ if the input voltage is a signal having the logical value A.

Figure 2-6

Actually the symbol for an inverter is a combination of the symbol for a noninverting amplifier (i.e., an operational block that provides current gain but leaves the logic voltage levels unaltered), a triangle, and a symbol for inversion in general, denoted by a small circle. This circle can be inserted to the output of a whole block, where it has the effect of inverting the logical value of the output signal. But the circle can be drawn to the input as well, and it then has the effect of inverting the value of the incoming signal before feeding into the logic circuit.

The AND operation followed by an inversion is called NAND ("not and") and the OR operation followed by an inversion is called NOR. These two functions, which are usually implemented as electronic operational blocks, are shown and defined in Fig. 2-7.

There is an additional basic logic gate that frequently occurs as an electronic operational block, called EXCLUSIVE OR, or EXOR. Normally it has two inputs and its definition is given in Fig. 2-8.

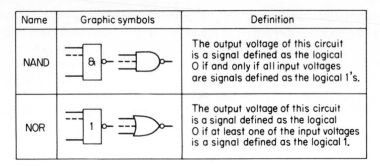

Name	Graphic symbols	Definition
NAND		The output voltage of this circuit is a signal defined as the logical 0 if and only if all input voltages are signals defined as the logical 1's.
NOR		The output voltage of this circuit is a signal defined as the logical 0 if at least one of the input voltages is a signal defined as the logical 1.

Figure 2-7

Name	Graphic symbols	Definition
EXCLUSIVE OR		The output voltage of this circuit is a signal defined as the logical 1 if and only if one of the input voltages is a signal defined as the logic 1.

Figure 2-8

2.3. Fundamental Rules of Boolean Algebra

Boolean Functions and the Definition of Boolean Algebra. The truth values of combined logical propositions can be studied with the aid of a calculus called *Boolean algebra*, which originally was established for the definition and handling of thought processes. It later became the basis for the analysis and design of binary digital systems. George Boole (1815–1864) found that the structure of truth functions can be expressed in an algebraic form in which conjunction is very similar to the ordinary multiplication of variables and disjunction corresponds to addition, with the exception that $1 + 1 = 2$ in ordinary algebra must be replaced by the rule $1 + 1 = 1$. Negation must be introduced as a separate operation. An algebra is a mathematical system that must be invariant regardless of what the variables may be, and therefore we should not confuse the propositional calculus with Boolean algebra, where we in fact are dealing with numbers. However, such terms as "truth value," "true ($= 1$)," "false ($= 0$)," "logic function," "not," "and," "or," etc., have been generally accepted in digital technology, and they may also be used intermittently in Boolean algebra.

Boolean variables can take only two values, 0 and 1. *Boolean functions* are functions in which the arguments and the functions themselves are Boolean variables. In mathematics, (general) *Boolean algebra* is defined in a way that allows its application to rather general elements. In a more restricted sense, the (special) Boolean algebra or the *switching algebra* deals with Boolean variables and Boolean functions, considered together with the operations of negation, conjunction, and disjunction defined over them. To avoid confusion, we shall speak in the following of *Boolean (switching) algebra*.

Postulates. 1. The variables (arguments and functions) of this algebra are elements of the set $\{0, 1\}$, and there are two *binary operations*, $+$ and $\cdot$, defined over them. For brevity, $A \cdot B$ will also be written AB. The rules of this calculus are

$$0 + A = A$$

$$1 + A = 1$$

$$0 \cdot A = 0$$

$$1 \cdot A = A$$

where A is an arbitrary expression of this algebra.

2. The operations $+$ and $\cdot$ are commutative.

3. Each operation is distributive over the other.

4. For every variable A there exists a variable $\bar{A}$ (*negation* or *inversion* of A) so that the following two relations are satisfied:

$$A \cdot \bar{A} = 0$$

$$A + \bar{A} = 1$$

Priority rules are the same as in ordinary algebra, and expressions can be in parentheses. We shall also call $\cdot$ the *conjunction operation* or the *logical product*, and $+$ the *disjunction operation* or the *logical sum*.

A few direct exemplifications of these postulates, in a more comprehensible form, for easy reviewing, are as follows:

$$\bar{0} = 1 \qquad\qquad\qquad\qquad\qquad \text{[2-3]}$$

$$\bar{1} = 0 \qquad\qquad\qquad\qquad\qquad \text{[2-4]}$$

$$\bar{\bar{A}} = A \qquad\qquad\qquad\qquad\qquad \text{[2-5]}$$

$$A \cdot B = B \cdot A \qquad\qquad\qquad \text{(commutative law for } \cdot \text{)} \qquad \text{[2-6]}$$

$$A + B = B + A$$

(commutative law
for $+$) [2-7]

$$A \cdot (B + C) = A \cdot B + A \cdot C$$

(distributive law
of $\cdot$ over $+$) [2-8]

$$A + (B \cdot C) = (A + B) \cdot (A + C)$$

(distributive law
of $+$ over $\cdot$) [2-9]

$$(A \cdot B) \cdot C = A \cdot (B \cdot C)$$

(associative law
for $\cdot$) [2-10]

$$(A + B) + C = A + (B + C)$$

(associative law
for $+$) [2-11]

As in ordinary algebra, the parentheses in expressions obeying associative laws may be deleted.

In the following we have some important theorems which can be proved algebraically. However, it is simplest to verify these rules by tabulating the values of both sides of these equations for all values of arguments and finding the tables identical.

$$A \cdot A = A$$ (idempotent law for $\cdot$) [2-12]

$$A + A = A$$ (idempotent law for $+$) [2-13]

$$\overline{A \cdot B} = \bar{A} + \bar{B}$$ (De Morgan's rule for $\cdot$) [2-14]

$$\overline{A + B} = \bar{A} \cdot \bar{B}$$ (De Morgan's rule for $+$) [2-15]

Besides these rather fundamental rules, we should like to mention a few theorems which are easily derived from the previous rules but can also be verified by checking for all values of the arguments:

$$A + A \cdot B = A$$ [2-16]

$$A + \bar{A} \cdot B = A + B$$ [2-17]

$$f(A, B, C, \ldots) = A \cdot f(1, B, C, \ldots) + \bar{A} \cdot f(0, B, C, \ldots)$$ [2-18]

$$f(A, B, C, \ldots) = [A + f(0, B, C, \ldots)] \cdot [\bar{A} + f(1, B, C, \ldots)]$$ [2-19]

Duality. It follows directly from Postulates 1 through 4 that an equality of two Boolean expressions remains valid if every binary operation $\cdot$ is replaced by $+$ and vice versa, and if every variable, term, and expression is inverted. This is the *principle of duality* in Boolean algebra. (See also Section 2.2.) For example, De Morgan's rules, Eqs. [2-14] and [2-15], are direct implications of duality.

2.4. Combinational Circuits

Translation of Logic Propositions into Boolean Functions. If the operation of a system can be analyzed and stated in terms of logic propositions, as is the case with all digital circuits, it is a straightforward task to translate these into expressions of Boolean algebra considering the propositions as Boolean variables and realizing that the connectives AND, OR, and NOT can be directly replaced by the binary operations $\cdot$ or the logical product, $+$ or the logical sum, and $^-$ or the negation, respectively. Usually logical deductions can be directly stated in terms of Boolean algebra, without intermediate notations of propositional calculus.

Example 2-2. The windshield-wiper motor of a car is controlled by the ignition switch, the wiper switch, and a limit contact which breaks when the wiper is in its leftmost position. All the important propositions associated with this operation are the following:

$M =$ "the motor is running"
$S =$ "the ignition switch is on"
$W =$ "the wiper switch is on"
$L =$ "the limit contact is open" or "the wiper is in its leftmost position"

The required relation reads: "The motor is running when the ignition switch is on, AND when the wiper switch is on OR if the wiper is NOT in its leftmost position," which is simply written as

$$M = S \cdot (W + \bar{L}) \qquad [2\text{-}20]$$

Notice the order of priority of various operations.

Boolean Functions of Combinational Circuits. A general *combinational circuit* is a system with n independent logic input signals denoted by X_1, $X_2, \ldots, X_n$ and m logic output signals $Y_1, Y_2, \ldots, Y_m$, related in such a way that each Y_j is an explicit Boolean function of the X_i's for $i = 1$ to n and $j = 1$ to m (Figure 2-9).

For the basic logic gates defined in Section 2.2 the Boolean expressions are listed below; denoting the inputs by $A, B, C, \ldots$, the algebraic expressions are shown in Table 2-4.

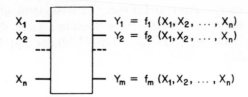

Fig. 2-9. General combinational
circuit.

Table 2-4. Boolean Expressions for the Basic Logic Gates

Name	Input Variables	Boolean Expression
AND	$A, B, C, \ldots$	$A \cdot B \cdot C \cdots$
OR	$A, B, C, \ldots$	$A + B + C + \cdots$
NOT	A	$\bar{A}$
NAND	$A, B, C, \ldots$	$\overline{A \cdot B \cdot C \cdots}$
NOR	$A, B, C, \ldots$	$\overline{A + B + C \cdots}$
EXCLUSIVE OR	A, B	$\bar{A}B + A\bar{B}$

Similarly, as switches can be combined into parallel and serial networks, so can the basic logic gates combined into networks by interconnections, and it is a straightforward task to write down the input–output transfer relations of logic signals in terms of Boolean functions. This is exemplified with two circuits.

Example 2-3. Find the input–output relations $P(A, B, C, D, E, F)$ of the network of Fig. 2-10, which is a *three-level* combinational circuit with six inputs and one output. (Note: Levels mean layers of gates here.)

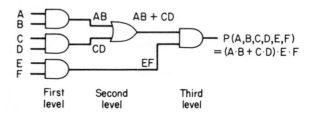

Figure 2-10

Example 2-4. Find the two Boolean functions defining the input–output relations $P_1(A, B, C, D)$ and $P_2(A, B, C, D)$ of the network of Fig. 2-11. This is a two-level combinational circuit with four inputs and two outputs.

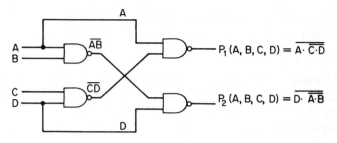

Figure 2-11

Implementation of Boolean Functions by Combinational Circuits. An inverse problem to the previous one is the synthesis of combinational circuits when their Boolean functions are given. It is obvious that every Boolean function can be implemented by using AND, OR, and NOT gates by simply following the priority rules of operations and replacing algebraic operations by their circuit equivalents.

Example 2-5. Find a logic circuit that implements the Boolean function

$$F = [(A + \bar{B}) \cdot C + \bar{D}] \cdot E \qquad [2\text{-}21]$$

If A, B, C, D, and E are primary logic variables, then B and D must first be inverted. Logic products have a higher priority than logic sums but parenthetical expressions must be evaluated first, starting with the innermost parentheses. In logic diagrams, the direction of signal flow is usually from the left to the right, and from top to bottom. The drawing of the logic diagram is therefore started on the left with the highest priority operations, proceeding to the right. The diagram of Fig. 2-12 ought to be self-explanatory.

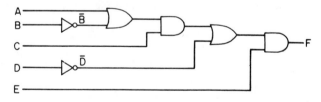

Fig. 2-12. Logic diagram for Example 2-5.

We shall see later that there are infinitely many implementations of a Boolean function, some of them simpler than the others. Therefore, a Boolean function ought never be implemented without first trying to simplify it.

The practical goal of logical design is to find the simplest network which is composed of basic logic gates and by which all given Boolean functions are implemented.

Minimal Set of Basic Logic Operations. For m independent variables, there are 2^{2^m} different truth tables (Boolean functions). It will be shown later that every truth table can be described by a Boolean expression (also called a Boolean function), using the operations $\cdot$, $+$, and $^-$. A question arises as to whether we could do with fewer basic logic operations. In fact, since De Morgan's rules allow us to express logic products in terms of logic sums and negations, or logic sums in terms of logic products and negations, we can eliminate either AND or OR gates from the set of basic gates. But NAND and NOR are regarded as basic logic gates, too, and it is possible to show that any Boolean function can be implemented using NAND's or NOR's only. In other words, any one of these gates alone represents the *minimal set of logic operations.* Very often electronic implementations of NAND and NOR are simpler than of AND and OR.

To show that any logic circuit can be implemented in terms of NAND's, it will suffice to show (according to what was told before) that NOT and OR can be implemented by NAND's. This is obvious from the following:

$$\overline{A} = \overline{A \cdot A} \qquad \text{by Eq. [2-12]}$$

$$A + B = \overline{\overline{A} \cdot \overline{B}} = \overline{\overline{A \cdot A} \cdot \overline{B \cdot B}} \qquad \text{by Eqs. [2-14] and [2-12]}$$

A similar proof can be given for NOR's.

Example 2-6. According to De Morgan's rules, a two-level network composed of AND's and OR's is directly implemented by an equal number of NAND's as shown by the example of Eq. [2-22] and Fig. 2-13:

$$AB + CD = \overline{\overline{AB}\ \overline{CD}} \qquad [2-22]$$

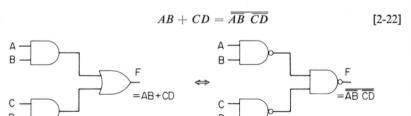

Fig. 2-13. Implementation of a two-level AND–OR circuit by NAND's.

Truth Tables for Boolean Functions. When the Boolean functions of a combinational circuit have been defined, we can tabulate the output values for all input values. For each input value combination separately, the binary values of the Boolean expressions are computed term by term and factor by factor. A *truth table* is thus formed in which all input combinations, or possible values of the independent variables, occur as entries.

Example 2-7. Construct the truth table for the Boolean function

$$F = (A + \bar{B}C)(B + C)$$

Since there are three independent variables A, B, and C, the truth table contains $2^3 = 8$ entries. Priority rules of algebraic expressions require that $\bar{B}C$ has to be computed first, then the two parenthetical factors, and finally the complete expression.

A	B	C	$\bar{B}C$	$A + \bar{B}C$	$B + C$	F
0	0	0	0	0	0	0
0	0	1	1	1	1	1
0	1	0	0	0	1	0
0	1	1	0	0	1	0
1	0	0	0	1	0	0
1	0	1	1	1	1	1
1	1	0	0	1	1	1
1	1	1	0	1	1	1

Setting up the Minimization (Simplification) Problem. Let us recall that a *Boolean function* $f(X_1, X_2, \ldots, X_n)$ of n variables is completely defined if it is tabulated for all the 2^n different value combinations of the arguments. Such tables are called *truth tables*. There are infinitely many *algebraic forms* of Boolean functions from which the values of f can be calculated. If there is no danger of confusion, the algebraic forms will be called Boolean functions in what follows.

In a search for the "simplest" Boolean function for practical purposes, we might meet restrictions: The number of allowed inputs to a gate is limited; allowed interconnections depend on particular techniques; etc. If the total cost has to be minimized, a unit price should be assigned to each gate or gate combination. Thus total costs have a rather complicated structure, and the optimization problem is most conveniently solved by a computer.

The optimization problem becomes quite different and is considerably simplified if we restrict ourselves to a particular circuit type. It will be shown in the next sections that every logic circuit can be implemented by a one- or two-level circuit in which one of the levels consists of AND's and the other of OR's (provided that all variables together with their negations are available) and which is easily transformed into a NAND network.

In the following we are searching for the simplest one- or two-level circuits which comprise a suboptimal solution to the minimization problem. This approach is justified for several reasons:

1. The cost function can be formulated in a simple and unique way; of two networks, the one with fewer gates is assumed cheaper (simpler), and of two networks with an equal number of gates, the one with the smaller total number of inputs is assumed simpler.[1] (Actually it is assumed that the cost of an input is much smaller than the cost of a gate. Sometimes this is not the case, and the total costs must be considered more carefully. The systematic reduction methods discussed in this chapter will in general reduce the numbers of gates and inputs simultaneously.)

2. Logic circuits with a small number of levels are preferable since the internal signal delays are then small.

3. There are several analytical methods available for one- or two-level problems; three-level circuits can be solved only in special cases, but there are no methods for finding the simplest expressions among networks with an arbitrary number of levels.

2.5. From Truth Table to Boolean Function

Some terminology used in this book will be defined in the following:

Independent Variable. A Boolean variable that is independent of any other occurring Boolean variables. In a network under consideration, independent variables correspond to input signals. In these definitions, independent variables are denoted by capital letters.

Literal. An independent variable (*true* independent variable) or its negation. For example, A, $\bar{B}$.

Boolean product term (product term, product). A literal, or a logical product formed of literals, in which independent variables may occur only once. For example, A, $\bar{B}$, $\bar{A}B$, $A\bar{B}\bar{C}$, etc., are Boolean product terms, whereas $AB\bar{B}$, $\overline{AB}$, and $\overline{\bar{A}\bar{B}\bar{C}}$ are not.

---

[1] We shall see later that the simplification methods discussed in Sections 2.6. and 2.7. do not introduce extra inversions. Thus the cost of an inversion need not be taken into account.

Boolean sum term (sum term, sum). A literal, or a logical sum formed of literals, in which independent variables may occur only once. For example, A, $\bar{B}$, $\bar{A} + B$, $A + \bar{B} + \bar{C}$, etc., are Boolean sum terms, whereas $A + \bar{A}$, $AB + C$, and $\overline{A + B}$ are not.

Normal product (canonical product, minterm). A Boolean product term that contains each one of the independent variables (or their negations) once and only once. For example, for two independent variables A and B, $A\bar{B}$ and $\bar{A}B$ are normal products, and for three variables A, B, and C, $\bar{A}B\bar{C}$ is a normal product.

Normal sum (canonical sum, maxterm). A Boolean sum term that contains each one of the independent variables (or their negations) once and only once. For example, for three independent variables A, B, and C, $\bar{A} + B + \bar{C}$ is a normal sum.

Disjunctive normal form (sum of normal products, first canonical form, minterm sum). A normal product, or a logical sum of normal products. For example, for three variables, A, B, and C, ABC and $\bar{A}\bar{B}C + \bar{A}B\bar{C}$ are disjunctive normal forms.

Conjunctive normal form (product of normal sums, second canonical form, maxterm product). A normal sum, or a logical product of normal sums. For example, for three variables A, B, and C, $\bar{A} + \bar{B} + \bar{C}$ and $(A + B + C)(\bar{A} + B + C)$ are conjunctive normal forms.

Two important rules are easily established:

1a. A normal product is 1 for one and only one value combination of the independent variables (and 0 otherwise).

1b. A normal sum is 0 for one and only one value combination of the independent variables (and 1 otherwise).

Proofs. If a literal in a normal product is a true independent variable, take 1 for it, and if the literal is a negation of an independent variable, take 0 for the variable (i.e., 1 for the literal). For this kind of value combination, all literals in the normal product are then 1's, giving the value 1 for the normal product. For any other value combination, at least one of the literals is 0, giving the value 0 for the normal product. If a literal in a normal sum is the true independent variable, take 0 for it, and if the literal is the negation of the independent variable, take 1 for the variable (i.e., 0 for the literal). For this kind of value combination, all literals in the normal sum are then

0's, giving the value 0 for the normal sum. For any other value combination, at least one of the literals is 1, giving the value 1 for the normal sum.

These rules are exemplified with truth tables. Let us denote the normal products by P_i, where i is equal to the binary number formed of the value combination of independent variables which gives the value 1 for this normal product. Denote the normal sums by S_j, where j is equal to the binary number formed of the value combination which gives the value 0 for this normal sum.

A	B	C	$P_0 = \bar{A}\bar{B}\bar{C}$	$P_1 = \bar{A}\bar{B}C$	$P_2 = \bar{A}B\bar{C}$	$P_3 = \bar{A}BC$	$P_4 = A\bar{B}\bar{C}$	$P_5 = A\bar{B}C$	$P_6 = AB\bar{C}$	$P_7 = ABC$	$S_0 = A+B+C$	$S_1 = A+B+\bar{C}$	$S_2 = A+\bar{B}+C$	$S_3 = A+\bar{B}+\bar{C}$	$S_4 = \bar{A}+B+C$	$S_5 = \bar{A}+B+\bar{C}$	$S_6 = \bar{A}+\bar{B}+C$	$S_7 = \bar{A}+\bar{B}+\bar{C}$
0	0	0	1	0	0	0	0	0	0	0	0	1	1	1	1	1	1	1
0	0	1	0	1	0	0	0	0	0	0	1	0	1	1	1	1	1	1
0	1	0	0	0	1	0	0	0	0	0	1	1	0	1	1	1	1	1
0	1	1	0	0	0	1	0	0	0	0	1	1	1	0	1	1	1	1
1	0	0	0	0	0	0	1	0	0	0	1	1	1	1	0	1	1	1
1	0	1	0	0	0	0	0	1	0	0	1	1	1	1	1	0	1	1
1	1	0	0	0	0	0	0	0	1	0	1	1	1	1	1	1	0	1
1	1	1	0	0	0	0	0	0	0	1	1	1	1	1	1	1	1	0

We can now reformulate two additional rules:

2a. If a Boolean function becomes 1 with one and only one value combination of the independent variables, it can be represented by a normal product in which a true independent variable occurs as a literal if in the said value combination this variable was 1, and a negation of an independent variable occurs as a literal if this variable was 0 in the said value combination.

2b. If a Boolean function becomes 0 with one and only one value combination of the independent variables, it can be represented by a normal sum in which the negation of an independent variable occurs as a literal if this variable was 1 in the said value combination, and a true independent variable occurs as a literal if this variable was 0 in the said value combination.

Implication. A Boolean expression F implies another expression G, or $F \rightarrow G$, if G is 1 every time when F is 1 (but the converse is not necessarily true). An implication is a proposition which is true or false depending on particular values of F and G. Therefore it can be understood as a Boolean variable, with *true* denoted by 1 and *false* by 0.

F G	$F \to G$
0 0	1
0 1	1
1 0	0
1 1	1

Theorem 2-1. Each term of a logical sum implies the sum, e.g.,

$$A \longrightarrow A + B \longrightarrow A + B + C$$

Theorem 2-2. A product implies all its factors, e.g.,

$$ABC \longrightarrow BC \longrightarrow C$$

These theorems follow directly from the definitions of the logical sum and the logical product.

Finding a Boolean Function for an Arbitrary Truth Table. We shall now show that every truth table can be represented by both a disjunctive and a conjunctive normal form.

To show that any truth table can be represented by a disjunctive normal form, it is necessary to notice that two different normal products have a 1 in different places of the truth table, and a sum of normal products has a 1 everywhere in the truth table where the terms of this sum have it but 0's elsewhere. It is said that the normal products imply the sum of them.

To show that any truth table can be represented by a conjunctive normal form, it is necessary to notice that two different normal sums have a 0 in different places of the truth table, and a product of normal sums has a 0 everywhere in the truth table where the factors of this product have it but 1's elsewhere.

Thus we obtain two rules to find a Boolean function for a truth table:

1. *Disjunctive normal form:* For each 1 in the truth table, take the corresponding normal product and form a sum of these products.
2. *Conjunctive normal form:* For each 0 in the truth table, take the corresponding normal sum and form the product of these sums.

Example 2-8. The following proposition: "Binary variables A and B are equivalent" is true whenever A and B have the same value but is false otherwise. The equivalence function E of A and B is thus defined by the following truth table, from which the disjunctive and the conjunctive normal forms for the Boolean function are easily derived.

A	B	E
0	0	1
0	1	0
1	0	0
1	1	1

1. $E = \bar{A}\bar{B} + AB$
2. $E = (A + \bar{B})(\bar{A} + B)$

We can also state that

$$(A + \bar{B})(\bar{A} + B) = A\bar{A} + AB + \bar{B}\bar{A} + \bar{B}B$$
$$= \bar{A}\bar{B} + AB$$

Example 2-9. Design a combinational circuit that indicates the presence of any one of the binary numbers 001 ($= 1_{10}$) and 110 ($= 6_{10}$) in a three-bit register.

We construct a truth table that has 1's for these value combinations and 0's elsewhere, and form the disjunctive normal form for this table. This function is implemented by a two-level circuit.

X_2	X_1	X_0	F
0	0	0	0
0	0	1	1
0	1	0	0
0	1	1	0
1	0	0	0
1	0	1	0
1	1	0	1
1	1	1	0

$$F = \bar{X}_2\bar{X}_1X_0 + X_2X_1\bar{X}_0$$

2.6. Quine–McCluskey Simplification

Before any Boolean functions are implemented by logic gates, it is advisable to search for the simplest forms among equivalent algebraic expressions of it. All simplification methods are based on calculational rules of Boolean algebra, but it is difficult to see "shortcut" methods which would yield the simplest expression except in the most primitive cases. We restrict ourselves to the implementation of one- and two-level circuits, which are always found starting from truth tables. The first task is to construct a truth table for the Boolean function (see Fig. 2-14).

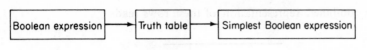

Figure 2-14

Complete Disjunctive Normal Forms. Postulate 4 of Section 2-1 states that for every Boolean variable A_i, $A_i + \bar{A}_i = 1$. Therefore we obtain the identity for an arbitrary Boolean function,

$$\prod_{i=1}^{m} (A_i + \bar{A}_i)Q = Q \qquad [2\text{-}23]$$

If the binomial product $\prod_{i=1}^{m} (A_i + \bar{A}_i)$ is expanded, we obtain a disjunctive normal form over the subset $\{A_1, \ldots, A_m\}$ of independent variables, in which all the 2^m different normal products are present. Such a form is called *complete disjunctive normal form*, and its value is identically 1; e.g.,

$$m = 1: \quad A_0 + \bar{A}_0 = 1$$

$$m = 2: \quad A_0 A_1 + A_0 \bar{A}_1 + \bar{A}_0 A_1 + \bar{A}_0 \bar{A}_1 = 1$$

$$m = 3: \quad A_0 A_1 A_2 + \cdots + \bar{A}_0 \bar{A}_1 \bar{A}_2 = 1 \qquad \text{(8 terms)}$$

If in a Boolean function expressed in a disjunctive normal form 2^m terms can be factorized into a common term Q and a complete disjunctive normal form over m *independent variables, the Boolean function can be simplified using Eq.* [2-23].

"Don't-Care" Conditions. If a Boolean function F does not depend on a particular variable A but we want to show all independent variables as arguments, we can write

$$F(A, B, C, \ldots) = F(\varnothing, B, C, \ldots) \qquad [2\text{-}24]$$

where $\varnothing$ can be read "either 0 or 1" or "don't care (of this value)". The "don't-care" notation can be used in truth tables for the simultaneous representation of several rows; assume that the truth table contains a 1 for two value combinations that differ from each other only by one bit:

A	B	C	D	F	
·	·	·			
1	0	0	1	1	$(A\bar{B}\bar{C}D \longrightarrow F)$
·	·	·			
1	1	0	1	1	$(AB\bar{C}D \longrightarrow F)$
·	·	·			

The combined representation of these rows reads

A	B	C	D	F	
.	.	.			
1	$\varnothing$	0	1	1	$(A\bar{C}D \longrightarrow F)$
.	.	.			

Notice that $A\bar{B}\bar{C}D + AB\bar{C}D = A\bar{C}D$, and the two 1's in F are also implied by the shorter expression $A\bar{C}D$, from which B has been deleted.

If a Boolean function is independent of several Boolean variables, we may write $\varnothing$ for each one of these; for example, if A and B do not occur in F, we can denote

$$F(A, B, C, \ldots) = F(\varnothing, \varnothing, C, \ldots)$$

which is true for all value combinations $(A, B) = (0, 0), (0, 1), (1, 0)$, and $(1, 1)$.

Several rows in a truth table can be represented simultaneously by using the $\varnothing$ notations, provided that F has the same value (0 or 1) for combinations which differ from each other only in a subset of variables and that *all possible value combinations occur in this subset:*

A	B	C	D	F
.	.	.		
0	0	0	1	1
.	.	.		
0	1	0	1	1
.	.	.		
1	0	0	1	1
.	.	.		
1	1	0	1	1
.	.	.		

The combined representation reads

A	B	C	D	F
.	.	.		
$\varnothing$	$\varnothing$	0	1	1
.	.	.		

Notice that

$$\bar{A}\bar{B}\bar{C}D + \bar{A}B\bar{C}D + A\bar{B}\bar{C}D + AB\bar{C}D = \bar{C}D$$

and the four 1's in the original truth table of F are also implied by the shorter expression $\bar{C}D$, from which A and B have been deleted.

In fact, representing F with a combined truth table is closely related to the factorizing of the algebraic expression of F into a common factor and a complete disjunctive normal form over a subset of variables, and deleting the latter.

Comment. Less than four value combinations of two variables (e.g., 00 and 11) cannot be represented by $\varnothing \varnothing$ because there are missing combinations (e.g., 01 and 10 here) which also belong to the same representation.

We may now state the rule by which Boolean functions can be written from combined truth tables:

If rows in a truth table can be combined, the truth table is represented by a sum of Boolean product terms in which each product term corresponds to a 1 in the combined table. These terms are otherwise formed in the same way as the normal products were formed from the original table but deleting the literals indicated by $\varnothing$'s from the normal products.

Systematic Combination of Rows.[1] It is obvious that a Boolean expression is simpler if it is written from any combined table instead of the original one. Therefore, we want to search for all possible combinations of the rows, and this is systematically done in the *Quine–McCluskey method*. Since two value combinations can be combined only when they differ in one and only one variable, we start by grouping the value combinations (rows) into sets according to the number of 1's in the combination. We are interested only in the rows in which the truth table has a 1. These rows are now rewritten into groups which contain 0, 1, 2, etc., 1's in the value combination. Notice that if any two rows can be combined, they must necessarily exist in adjacent groups.

Example 2-10. Let us discuss the truth table given as Table I, in which all rows with $F = 1$ are shown and which is already in the rearranged form. The column for F has been shown only to express the fact that this is still a truth table. In the following there is no reason to keep this column.

---

[1] In what follows, we shall only simplify disjunctive normal forms. The conjunctive normal forms are simplified in a dual way, but since this method usually does not bring about essentially simpler forms, it will not be discussed in this book.

Table I

	A	B	C	D	F
✓	0	0	0	0	1
✓	0	0	1	0	1
✓	0	1	0	0	1
✓	0	0	1	1	1
✓	0	1	0	1	1
✓	1	0	0	1	1
✓	1	0	1	0	1
✓	1	1	0	0	1
✓	0	1	1	1	1
✓	1	0	1	1	1
✓	1	1	0	1	1
✓	1	1	1	1	1

$(F = 0$ otherwise$)$

Table II

	A	B	C	D
	0	0	Ø	0
	0	Ø	0	0
✓	0	0	1	Ø
✓	Ø	0	1	0
✓	0	1	0	Ø
✓	Ø	1	0	0
✓	0	Ø	1	1
✓	Ø	0	1	1
✓	0	1	Ø	1
✓	Ø	1	0	1
✓	1	0	Ø	1
✓	1	Ø	0	1
✓	1	0	1	Ø
✓	1	1	0	Ø
✓	Ø	1	1	1
✓	1	Ø	1	1
✓	1	1	Ø	1

When two rows are combined, the original rows are deleted from the truth table and denoted by a check mark ($\checkmark$). Now *every row is compared against all rows in the next group*, and whenever a combination is possible, the combined rows are written into a new table (Table II). Because of the identity $X = X + X$, where X can represent any normal product, the same row may be added any number of times to the truth table, and this is equivalent to stating that a row can be reused several times for different combinations.

This procedure is now continued in Table II, in which rows of adjacent groups can be combined further if they have $\varnothing$'s in the same positions and differ in one bit only. There are no possible combinations between the first and second groups in this example. Combined rows are checked out and rewritten as in Table III.

Table III

A	B	C	D
$\varnothing$	0	1	$\varnothing$
~~$\varnothing$~~	~~0~~	~~1~~	~~$\varnothing$~~
$\varnothing$	1	0	$\varnothing$
~~$\varnothing$~~	~~1~~	~~0~~	~~$\varnothing$~~
~~$\varnothing$~~	~~1~~	~~0~~	~~$\varnothing$~~
$\varnothing$	$\varnothing$	1	1
~~$\varnothing$~~	~~$\varnothing$~~	~~1~~	~~1~~
$\varnothing$	1	$\varnothing$	1
~~$\varnothing$~~	~~1~~	~~$\varnothing$~~	~~1~~
1	$\varnothing$	$\varnothing$	1
~~1~~	~~$\varnothing$~~	~~$\varnothing$~~	~~1~~

This process is continued until no more combinations are possible. (Here Table III is the last one.) *Duplicate rows are deleted.* In this way all remaining (unchecked) rows in Table I, Table II, Table III, etc., taken together represent a combined truth table, and we can write an expression for F on the basis of all unchecked rows:

$$ F = \bar{A}\bar{B}\bar{D} + \bar{A}\bar{C}\bar{D} + \bar{B}C + B\bar{C} + CD + BD + AD $$

Prime Implicants. In a sum of products each term implies the sum (i.e., is an *implicant* of the sum).

The Boolean expression that is written from the combined truth table obtained by the Quine–McCluskey method is a sum of products, which, by construction, is formed of the disjunctive normal form by the elimination of terms and literals. The normal products were combined in all possible ways, and the terms that are left are called *prime implicants*.

Definition. A prime implicant is a product term P_i in a sum-of-products representation of the Boolean function F such that $P_i \longrightarrow F$, but there exists no other product term P_i in this expression such that $P_i \longrightarrow P_j \longrightarrow F$.

The relation $P_i \longrightarrow P_j$ would mean that P_j becomes 1 with the same value combination of the independent variables as P_i and perhaps for other combinations, too. Thus P_i must contain the same literals as P_j and possibly also other literals. But then P_j would be a simpler term which would have been formed of P_i and other terms in the combination process. This is in contradiction with the assumption that P_i is still there. Therefore, we conclude that the Quine–McCluskey method leaves only prime implicants in F.

A product term can be represented by a truth table which has 1's in the same places in which the normal products, of which this product term is formed, have them. For simplicity, we will say that the normal products *cover* this product term. (The 1's in the truth table of the normal products cover the 1's in the truth table of this product term.) Similarly, we say that F is covered by the corresponding normal products.

Definition. A Boolean function F is said to be covered by the terms P_i if F has 1's in the truth table at all places where the P_i's have them and nowhere else.

It should be noted that F might also be covered by several different subsets of the P_i's. *For the simplest sum-of-products expression of a Boolean function F, we have to find the simplest subset of prime implicants that still covers F.*

Prime-Implicant Table and Its Solutions. A prime-implicant table is simply a shortened notation of the original truth tables of F and its prime implicants, where only those rows are shown in which F has 1. To save writing, the 0's are not shown. (The column for F may also be deleted.) The simplest set of prime implicants which covers F is the solution of this table.

Returning now to the previous example of the Quine–McCluskey method, we have its prime-implicant table (Table 2-5; the dashed lines will be explained later). When we are deciding which subset of prime implicants should be selected for the covering of F, we first conclude that certain 1's in F are covered only by single terms, whereas the coverage of some other positions is redundantly made by two or more terms. *The prime implicants, which alone are responsible for the covering of certain 1's* (shown by vertical arrows in this example), *are called essential prime implicants* and they are

Table 2-5. Prime-Implicant Table for Example 2-10

A	B	C	D	$AB\bar{D}$	$\bar{A}CD$	$\bar{B}\bar{C}$	$B\bar{C}$	CD	BD	$A\bar{D}$	F
0	0	0	0	1	1						1
0	0	1	0	1							1
0	0	1	1					1			1
0	1	0	0			1					1
0	1	0	1						1		1
0	1	1	1					1	1		1
1	0	0	1							1	1
1	0	1	0							1	1
1	0	1	1					1		1	1
1	1	0	0				1			1	1
1	1	0	1				1		1	1	1
1	1	1	1					1	1	1	1

↑   ↑   ↑

obligatory in the final expression. Let us record these terms somewhere and then delete the corresponding columns, as well as *all rows on which these terms have* 1's (shown with dashed lines). (The 1's of *F* implied by the essential prime implicants have now been covered by the recorded terms.) There is a possibility that all rows become deleted in this way, in which case only essential prime implicants enter the final sum-of-products expression, which is now simplest and unique. Usually, however, some rows are left. We shall now see (and it can be proved) that there are at least two 1's on each remaining row. Therefore, we usually have several choices for the terms that would cover *F*. The best choice is naturally that one which can be implemented with the least cost. For this, we must consider a compromise between the least number of gates and the least total number of inputs (literals) in the final expression, and this choice is usually not unique.

    Whatever the cost function of terms and literals is, there are a few rules by which the prime-implicant table can be further simplified. The first of these says that a column (product term) can be deleted if it is *equivalent* to another column and if the cost of the latter (especially with respect to the number of literals and inversions) is not larger. Another rule states that of two columns P_i and P_j, P_j can be deleted if P_i *dominates* over P_j and if the cost of P_j is not larger.

Definition. Two columns P_i and P_j are *equivalent* ($P_i = P_j$) if they have 1's on the same rows. (This does not necessarily mean that the product terms P_i and P_j are identical, because different 1's in the original prime-implicant table might have been deleted previously.)

Definition. If a column P_i has 1's on every row where another column P_j has them, and on at least one other row, it is said that P_i *dominates* over P_j $(P_i \supset P_j)$.

When as many terms as possible are deleted by the application of the rules of equivalence and dominance, we may find that some rows are again covered only by single 1's. Therefore, we are in a position of defining the *secondary essential prime implicants*, which must be included in the final expression. Columns and rows are now deleted in the same way as in the original prime-implicant table and we may continue this way.

Alternatively, rows can be deleted on the basis of equivalence and dominance. A row R_i is equivalent to another row R_j $(R_i = R_j)$ if it has 1's in the same columns. Also, a row R_i dominates over R_j if R_i has 1's in all columns where R_j has them and in at least one other column. If $R_i = R_j$, either one of the rows can be deleted. If $R_i \supset R_j$, it is now R_i (the dominating row) that can be deleted. The proof follows from the facts that if $R_i = R_j$ or $R_i \supset R_j$, the normal product R_j is included in the same sum of product terms in which R_i is included.

When we have applied the previous rules in all possible ways, we end up with two possibilities:

1. All rows will be deleted and the solution, the sum of all recorded terms, is a unique solution for the simplest sum of products.
2. No more rows can be deleted, and there are rows left with redundant 1's. This table is now called the *cyclic prime-implicant table*. The product terms in it can be selected in several ways to cover F, and the simplest solution is found by checking all alternatives.

Believing that most practical problems which are treated by the Quine–McCluskey method will leave only so few rows in the cyclic prime-implicant table that its solution can be seen directly, we do not intend to discuss more refined systematic methods for solutions. Let it suffice to state that such methods exist and have been carried out as computer-aided design programs.

In our example we end up with a secondary prime-implicant table (Table 2-6). Obviously there are equivalent columns in it.

In order to cover the remaining 1's in F we can now choose among the following alternatives, which obviously have the same cost:

**Table 2-6.　Secondary Prime-Implicant
Table for Example 2-10**

A	B	C	D	$\bar{A}\bar{B}\bar{D}$	$\bar{A}\bar{C}\bar{D}$	CD	BD	F
0	0	0	0	1	1			1
0	1	1	1			1	1	1

$$\bar{A}\bar{B}\bar{D} \quad \text{and} \quad CD$$

$$\bar{A}\bar{B}\bar{D} \quad \text{and} \quad BD$$

$$\bar{A}\bar{C}\bar{D} \quad \text{and} \quad CD$$

$$\bar{A}\bar{C}\bar{D} \quad \text{and} \quad BD$$

Making (arbitrarily) a choice of the second alternative, we have for the complete expression of F:

$$F = \bar{A}\bar{B}\bar{D} + BD + \bar{B}C + B\bar{C} + AD$$

Simplification of Incompletely Defined Boolean Functions. In practical logic circuits it occurs rather often that some value combinations of the input signals are forbidden. This is the case, for example, if the variables are derived from binary-coded decimal registers where only 10 of the 16 possible states (of four flip-flops) are legitimate. Input variables to a combinational circuit may also be derived from other combinational circuits and the variables consequently are not independent.

If a Boolean function is not or cannot be defined over all value combinations of its arguments, we may assign arbitrary values to the truth table for those value combinations "which never occur anyway." To indicate that the choice is free, we fill the truth table with "don't-care" notations.

Example 2-11. A Boolean function F has M and N as its arguments and has to be implemented. However, M and N are derived from other combinational circuits in which A and B occur as independent variables:

A	B	M	N
0	0	0	0
0	1	1	1
1	0	1	1
1	1	0	1

We see that (M, N) can take only three value combinations: $(0, 0)$, $(0, 1)$, and $(1, 1)$.
Let us define F as follows:

M	N	F
0	0	1
0	1	0
1	0	$\varnothing$
1	1	1

The extra "don't-care" rows provide means to eliminate more terms and literals from normal products by combination than without them; i.e., *if a 1 is taken for* $\varnothing$, *the corresponding row may be combined with a row of the defined portion of the truth table.* In this way the final expression is further simplified.

The Quine–McCluskey method can be modified for incompletely defined Boolean functions: the "don't-care" rows are grouped separately in Table I, and they are also compared against all legitimate value combinations. With the aid of extra rows, legitimate rows can be "checked out" freely, and the structure of prime implicants becomes simpler. The extra rows need not be included in the prime implicants.

Example 2-11 (continued) If the truth table of F were implemented assuming 0 for $\varnothing$, F would be represented by the expression

$$F = \bar{M}\bar{N} + MN$$

We shall now show that further simplification is possible because of the existence of the forbidden combination. Let us group the rows for the Quine–McCluskey method and adjunct the "don't-care" row to a separate group below a line; the two legitimate rows do not combine mutually,

Table I			Table II	
	M	N	M	N
$\checkmark$	0	0	$\varnothing$	0
	—		1	$\varnothing$
$\checkmark$	1	1		
$\checkmark$	1	0		

but each one can be combined with the "don't-care" row and checked out (Tables I and II). In this simple example, a prime-implicant table is not needed.

$$F = M + \bar{N}$$

Example 2-12. A bit more tedious example is the simplification of a Boolean function defined by the truth table given on page 66. This example will also be discussed later with the Karnaugh map method, and the reader may skip it here. The Tables I through IV of this example are given on page 66, too. The combinations that are formed with the aid of "don't-care" rows are in parentheses and duplicate or triplicate rows are deleted, which is shown by overruling. This example is included here only to show the details of this rather mechanical calculation. The sum of prime implicants is

$$F = \bar{A} + C$$

This result is simple enough, so the prime-implicant table need not be set up. If zeros would have been put for all $\varnothing$'s in the truth table, the simplest expression would have read

$$F = \bar{A}\bar{B} + \bar{A}\bar{C} + C\bar{D}$$

In the case of incompletely defined Boolean functions, the simplification should normally end up with the solution of the prime-implicant table. To it we need not take any terms corresponding to rows of the "don't-care" group, but we should take all terms that remain when the "don't-care" rows are combined in all possible ways with the legitimate combinations.

Simultaneous Simplification of Several Boolean Functions. In a combinational two-level circuit having multiple outputs the network is in many cases simpler if common gates can be used to implement the different Boolean functions. With the simultaneous simplification of several Boolean functions, this means that if similar product terms occur in different functions, only one needs to be taken into account in the cost function since its implementation is done only once.

In the simultaneous simplification of $F_1, F_2, \ldots$ the essential problem is thus to find common terms. The common 1's in F_i and F_j ($j \neq i$) are directly indicated by the truth table of the *product* $F_i F_j$; and the common 1's in several F_i's are given by the truth table of the product of these F_i's. Hence the procedure for finding the simplest simultaneous expressions is to write down the truth tables of all F_i's as well as of all mutual products $F_i F_j$ ($j \neq i$), $F_i F_j F_k$ (i, j, k different), etc.; the augmented prime implicants of each F_i now comprise the usual prime implicants as well as the prime implicants of all products in which F_i is a factor.

Truth Table

A	B	C	D	F
0	0	0	0	1
0	0	0	1	1
0	0	1	0	1
0	0	1	1	1
0	1	0	0	1
0	1	0	1	1
0	1	1	0	1
0	1	1	1	∅
1	0	0	0	0
1	0	0	1	0
1	0	1	0	1
1	0	1	1	∅
1	1	0	0	0
1	1	0	1	0
1	1	1	0	1
1	1	1	1	∅

Table I

	A	B	C	D
✓	0	0	0	0
	—	—	—	—
✓	0	0	0	1
✓	0	0	1	0
✓	0	1	0	0
	—	—	—	—
✓	0	0	1	1
✓	0	1	0	1
✓	0	1	1	0
✓	1	0	1	0
	—	—	—	—
✓	1	1	1	0
	—	—	—	—
✓	0	1	1	1
✓	1	0	1	1
	—	—	—	—
✓	1	1	1	1

Table II

	A	B	C	D
✓	0	0	0	∅
✓	0	0	∅	0
✓	0	∅	0	0
	—	—	—	—
✓	0	0	∅	1
✓	0	∅	0	1
✓	0	0	1	∅
✓	0	∅	1	0
✓	∅	0	1	0
✓	0	1	0	∅
✓	0	1	∅	0
	—	—	—	—
✓	(0	∅	1	1)
✓	(∅	0	1	1)
✓	(0	1	∅	1)
✓	∅	1	1	0
✓	(0	1	1	∅)
✓	1	∅	1	0
✓	(1	0	1	∅)
	—	—	—	—
✓	(1	1	1	∅)
✓	(∅	1	1	1)
✓	(1	∅	1	1)

Table III

	A	B	C	D
✓	0	0	∅	∅
✓	0	∅	0	∅
	~~0~~	~~0~~	~~∅~~	~~∅~~
✓	0	∅	∅	0
	~~0~~	~~∅~~	~~0~~	~~∅~~
	~~0~~	~~∅~~	~~∅~~	~~0~~
	—	—	—	—
✓	(0	∅	∅	1)
	~~(0~~	~~∅~~	~~∅~~	~~1)~~
✓	(0	∅	1	∅)
✓	(∅	0	1	∅)
	~~(0~~	~~∅~~	~~1~~	~~∅)~~
✓	∅	∅	1	0
	~~(∅~~	~~0~~	~~1~~	~~∅)~~
	~~(∅~~	~~∅~~	~~1~~	~~0)~~
✓	(0	1	∅	∅)
	~~(0~~	~~1~~	~~∅~~	~~∅)~~
	—	—	—	—
✓	(∅	∅	1	1)
	~~(∅~~	~~∅~~	~~1~~	~~1)~~
✓	(∅	1	1	∅)
	~~(∅~~	~~1~~	~~1~~	~~∅)~~
✓	(1	∅	1	∅)
	~~(1~~	~~∅~~	~~1~~	~~∅)~~

Table IV

A	B	C	D
(0	∅	∅	∅)
~~(0~~	~~∅~~	~~∅~~	~~∅)~~
~~(0~~	~~∅~~	~~∅~~	~~∅)~~
—	—	—	—
(∅	∅	1	∅)
~~(∅~~	~~∅~~	~~1~~	~~∅)~~
~~(∅~~	~~∅~~	~~1~~	~~∅)~~

Tables for Example 2-12.

2.7. Karnaugh Map Simplification

Map Form of a Truth Table. A Karnaugh map is a special form of a truth table in which the value combinations of independent variables are shown two-dimensionally. The variables $A, B, C, \ldots$ are divided into two groups, and in the Karnaugh map we have rows for the value combinations of the first group and columns for the value combinations of the second group. Further, we want to order the rows and columns in a special way according to the value combinations (i.e., the combinations in the adjacent rows and columns shall differ from each other only by one bit). As discussed in Section 1.3, the reflected codes have this property.

Each square in the map corresponds to a normal product P_i which has the value 1 in this square. Some simple Karnaugh maps together with the positions of the P_i's are shown in Fig. 2-15.

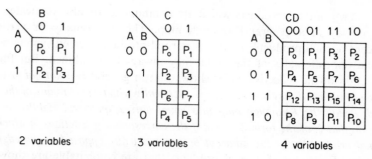

Fig. 2-15. Karnaugh maps.

The size of a Karnaugh map depends, of course, on the number of independent variables, but division into rows and columns can be accomplished in several ways.

Example 2-13. The following Karnaugh map of a function $F(A, B, C, D)$ is for four variables. (The curved lines surrounding 1's will be explained in the next paragraph.) The disjunctive normal form is obtained in the same way as from a regular table:

$$F = P_1 + P_5 + P_{12} + P_{14}$$
$$= \bar{A}\bar{B}\bar{C}D + \bar{A}B\bar{C}D + AB\bar{C}\bar{D} + ABC\bar{D}$$

	CD			
A B	00	01	11	10
0 0	0	1	0	0
0 1	0	1	0	0
1 1	1	0	0	1
1 0	0	0	0	0

Combination of Normal Products. Because of the special ordering of rows and columns, we get a simple rule for the combination of normal products in a Karnaugh map up to four variables: *All horizontally or vertically adjacent normal products, or normal products which are located symmetrically on opposite horizontal or vertical sides of the table, can be combined.* This has been shown by curved lines surrounding the 1's in Example 2-13. *Two 1's which have been combined are now equivalent to a single Boolean product term in which one of the literals has been eliminated.* In Example 2-13 the combination of P_1 and P_5 yields

$$P_1 + P_5 = \bar{A}\bar{B}\bar{C}D + \bar{A}B\bar{C}D = \bar{A}\bar{C}D$$

and the combination of P_{12} and P_{14} yields

$$P_{12} + P_{14} = AB\bar{C}\bar{D} + ABC\bar{D} = AB\bar{D}$$

Two normal products which are combinable are always located in a special way in a general Karnaugh map. The horizontally or vertically adjacent normal products, as well as normal products located symmetrically on the opposite sides of the table, may always be combined. But further inspection of the reflective codes shows that in general *two normal products can be combined if they are located symmetrically with respect to any of the lines which divide the Karnaugh map into equal halves, quarters, eighths, or any fraction that is of the form 2^{-n}, and if the corresponding fractions in which the normal products exist are adjacent or symmetrically opposed at the edges of the table.* Examples of normal products that are combinable are shown in Fig. 2-16. Combinable pairs are indicated here by the same lowercase letter.

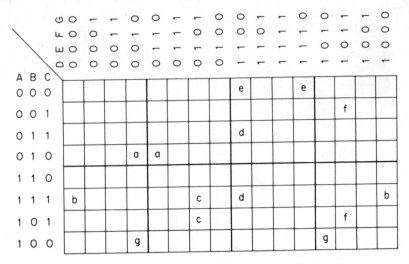

Fig. 2-16. Combinable pairs of normal products in a seven-variable Karnaugh map.

Combination of Product Terms. The previously stated adjacency and symmetry rules for the combination of normal products are valid for the combination of other types of product terms, too, represented by groups of squares in the Karnaugh map. However, we shall see in the next paragraph that these rules can also be founded by set theory.

Set-Theoretical Explanation of the Karnaugh Map. The Karnaugh map is analogous to the Venn diagram of set theory. The normal products are elements in this theory, and the truth table is equivalent to the set of all elements, or the universe. The table can be divided into subsets denoted by $\mathfrak{A}$, $\mathfrak{B}$, $\mathfrak{C}$, ... For a four-variable Karnaugh map these subsets are shown in Fig. 2-17. Complementary sets to $\mathfrak{A}$, $\mathfrak{B}$, $\mathfrak{C}$, . . . are denoted in the same

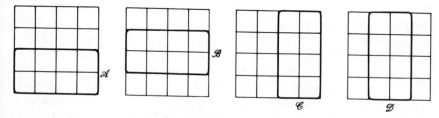

Figure 2-17

way as negations (e.g., $\bar{\mathfrak{A}}$). The subsets denoted by one letter cover one half of the truth table. Smaller areas or smaller subsets are formed by intersections; for example, the intersection of $\mathfrak{B}$ and the complementary set of $\mathfrak{A}$ (denoted by $\bar{\mathfrak{A}}$) are equivalent to the second horizontal row of the table. Obviously, the intersection of $\bar{\mathfrak{A}}$ and $\mathfrak{B}$, $\bar{\mathfrak{A}} \cap \mathfrak{B}$, is the area of the table in which a Boolean product term $\bar{A}B$ has the value 1, or $A = 0$ and $B = 1$. Similarly, intersections of several subsets are represented by areas in which corresponding Boolean product terms have the value 1 (e.g., $\mathfrak{A} \cap \bar{\mathfrak{B}} \cap \mathfrak{C}$ is equivalent to the area $A\bar{B}C = 1$).

Analogously with the set notations it is convenient to denote rows and columns of the Karnaugh map in the way shown in Fig. 2-18. The brackets are used to point out the areas in which the corresponding Boolean variable has the value 1, and in areas outside the brackets the negation of this variable is 1. It is now straightforward to construct Boolean expressions for general product terms by visualizing the areas of the truth table that are filled with 1's as subsets which are formed of larger subsets by intersections. This procedure is exemplified by the illustrations of Fig. 2-18. Maps with set-theoretical notations are also called *Veitch diagrams*.

Finding the Simplest Boolean Expression from a Karnaugh Map. Although the theory of simplification of disjunctive normal forms is the same regardless of whether a usual or a map form of truth tables is used, it is a particular advantage of the Karnaugh map representation that the

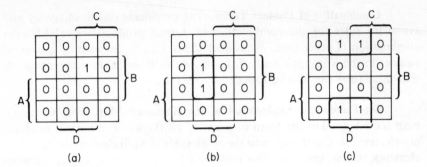

Fig. 2-18. (a) $F = 1$ in $\bar{\mathfrak{A}} \cap \mathfrak{B} \cap \mathfrak{C} \cap \mathfrak{D}$, or $F = \bar{A}BCD$; (b) $F = 1$ in $\mathfrak{B} \cap \bar{\mathfrak{C}} \cap \mathfrak{D}$, or $F = B\bar{C}D$; (c) $F = 1$ in $\bar{\mathfrak{B}} \cap \mathfrak{D}$, or $F = \bar{B}D$.

simplest terms represented by certain subsets in the table can be directly visualized. A human being is well trained to perceive two-dimensional figures (e.g., symmetrically located subsets of squares) and to recognize which of them belong together although these areas are scattered over the table.

The simplification of Boolean expressions is thus equivalent to setting up a truth table in map form and then *trying to find the minimum number of product terms which would cover the 1's in the table.* These subsets should be selected as large as possible because they are then represented by products with the smallest number of literals.

Notice that the subsets are allowed to overlap. It is also to be noted that there are often several alternative ways to cover the 1's. On the other hand, visual perception is here very effective in finding the simplest combinations. This is shown with the aid of a few examples in Fig. 2-19.

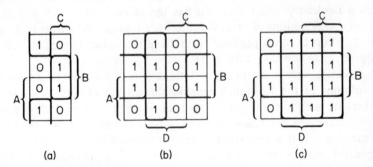

Fig. 2-19. Examples of the simplest Boolean expressions: (a) $F = \bar{B}\bar{C} + BC$; (b) $F = B\bar{D} + \bar{C}D$; (c) $F = B + C + D$.

Boolean Functions for Incompletely Defined Karnaugh Maps. Since a Karnaugh map in itself is a truth table, the forbidden value combinations of input variables are shown by "don't-care" notations in the proper positions. When looking for the simplest set of terms covering all *defined* 1's in the table, the "don't cares" can be included in subsets if they in this way complete product terms, but they can also be left outside the subsets since they need not be covered. (In this way the ⌀'s act similar to certain bivalent cards in card games.)

We shall now treat Example 2-12 to demonstrate the power of the Karnaugh map method; the truth table and the simplest subsets covering the 1's are shown in Fig. 2-20.

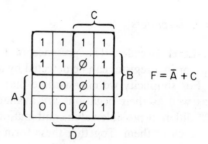

Fig. 2-20. Example 2-12 solved by the Karnaugh map method.

Note. It seems that Karnaugh map simplification can be made much easier than by using the Quine–McCluskey method, especially for incompletely defined truth tables. The Quine–McCluskey method, however, can be readily converted into a computer program due to the rather mechanical rules of this procedure. (There are also computer programs for Karnaugh maps.)

2.8.   Some Important Combinational Circuits

2.8.1.   Selector Gates

Transfer paths for digital signals can be created by means of special combinational circuits called *selector gates* or *copy gates*. Let us assume that we have control signals $T_A, T_B, T_C, \ldots$, of which only one may be simultaneously 1. When a particular one of them T_X is 1, the logic value of the respective signal X ($X = A, B, C, \ldots$) must be transferred to the output terminal, denoted by F. Obviously this is the case if the selector gate implements a Boolean function

$$F = T_A A + T_B B + T_C C + \cdots \qquad [2\text{-}25]$$

This circuit may be implemented by a two-level AND–OR combinational circuit as in Fig. 2-21, or its NAND or NOR equivalents. When a particular one of several binary words must be transferred to the output terminals in parallel, a selector gate must be provided for each bit separately. The control signals T_X of these gates are then connected in parallel.

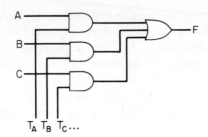

T_A T_B T_C ...

Fig. 2-21. Selector gate.

2.8.2. Decoders

One-Level Decoder. The presence of a particular binary word on parallel lines or in a register can be detected by a combinational circuit called a *decoder*. For simplicity we may assume that signals representing primary variables as well as their negations are available. For m binary variables, there are 2^m different possible value combinations. A normal product term is assigned to each of them. Together these form a combinational circuit with 2^m outputs, each giving a response (logical 1) for a particular input-value combination. Of course, only a part of all possible decoders may be utilized.

Example 2-14. Figure 2-22 shows a one-level decoder which is typical in digital systems. The presence of a binary number $X_2 X_1 X_0$ is detected with a separate decoder. Thus a decoder is a kind of "lock," and the word $X_2 X_1 X_0$ is a "key" for it.

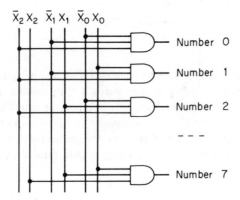

Fig. 2-22. Decoders.

Tree Decoder. An alternative for the simplest one-level decoder is the *tree decoder* or *decoding tree*, which is a multilevel combinational circuit. With n primary variables, all the normal products of $(n-1)$ primary variables are combined with the nth variable at the last level. At the next-to-last level, all normal products of $(n-2)$ primary variables are formed and combined with the $(n-1)$th variable, and so on. Figure 2-23 shows a three-

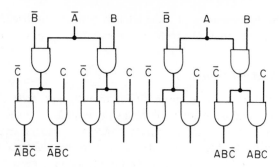

Fig. 2-23. Tree decoder.

variable tree decoder. A tree decoder has more gates but fewer inputs in total than a one-level decoder. In small circuits the one-level decoder is simpler.

Matrix Decoder. With an increasing number of primary variables, a one-level decoder becomes impractically large, because of the corresponding number of inputs to the gates. A frequently used decoding circuit is a matrix array in which all output gates have only two inputs, connected between horizontal and vertical wires in a matrix, one at each crossing. By means of smaller decoders, one horizontal wire and one vertical wire are selected. Although the number of gates is always larger than for a one-level decoder, the total number of inputs is substantially reduced for large matrices, which results in component savings. This circuit is called a *matrix decoder*.

A 3×3 matrix decoder is exemplified in Fig. 2-24, which also

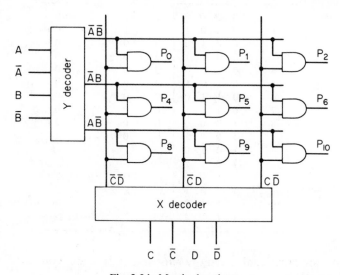

Fig. 2-24. Matrix decoder.

shows that the full capacity of a decoder may not be utilized. The outputs denoted by P_i are normal products of the variables A, B, C, and D.

2.8.3. Code Converters

A special class of combinational circuits are *code converters*, of which the most usual ones are available as integrated packages. The purpose of a code converter is to accept binary codes of numbers, letters, or other characters, and to produce new codes related to the old ones as defined by some conversion table. The most important code converters are those performing conversion between various codes for decimal digits, and those decoding numerals and other characters for displaying purposes.

Binary (8421)-to-Decimal Code Converter. This circuit has inputs for four primary variables (called here A, B, C, D) and their negations (representing the 8421 codes of decimal digits), and 10 outputs (called d_0 through d_9), of which the one corresponding to a particular digit code gives a logical 1 response. Since 6 of the 16 possible binary codes are not used, "don't-care" conditions in the truth table (Table 2-7) of this conversion are utilized to produce simpler Boolean functions.

Table 2-7. 8421-to-Decimal Conversion

Decimal Digit	A	B	C	D	d_0	d_1	$\cdots$	d_9
0	0	0	0	0	1	0		0
1	0	0	0	1	0	1		0
2	0	0	1	0	0	0		0
3	0	0	1	1	0	0		0
4	0	1	0	0	0	0		0
5	0	1	0	1	0	0		0
6	0	1	1	0	0	0		0
7	0	1	1	1	0	0		0
8	1	0	0	0	0	0		0
9	1	0	0	1	0	0		1
—	1	0	1	0	$\varnothing$	$\varnothing$		$\varnothing$
—	1	0	1	1	$\varnothing$	$\varnothing$		$\varnothing$
—	1	1	0	0	$\varnothing$	$\varnothing$		$\varnothing$
—	1	1	0	1	$\varnothing$	$\varnothing$		$\varnothing$
—	1	1	1	0	$\varnothing$	$\varnothing$		$\varnothing$
—	1	1	1	1	$\varnothing$	$\varnothing$		$\varnothing$

A combinational circuit comprising this code converter is thus defined by the Boolean functions

$$d_0 = \bar{A}\bar{B}\bar{C}\bar{D} \qquad d_5 = B\bar{C}D$$

$$d_1 = \bar{A}\bar{B}\bar{C}D \qquad d_6 = BC\bar{D}$$

$$d_2 = \bar{B}C\bar{D} \qquad d_7 = BCD \qquad\qquad [2\text{-}26]$$

$$d_3 = \bar{B}CD \qquad d_8 = A\bar{D}$$

$$d_4 = B\bar{C}\bar{D} \qquad d_9 = AD$$

If the presence of a forbidden code must be detected (e.g., to reveal operational errors in a computer), an extra circuit implementing the following Boolean function will do it:

$$F = AB + AC \qquad\qquad [2\text{-}27]$$

Gray-to-8421 Code Converter. As an example of a four-bit to four-bit code conversion, let us take the conversion of four-bit reflected (Gray) binary numbers to binary numbers of 8421 code. This is defined by Table 2-8 (see also Table 1-3). The Boolean functions E, F, G, and H defining the outputs are most conveniently implemented using EXCLUSIVE OR gates, as denoted by the corresponding algebraic symbol ($\oplus$).

Seven-Segment Decoder. For the display of decimal digits, the 8421-to-decimal converter may be used to control special digit tubes in which one

Table 2-8. Gray-to-8421 Conversion

| Gray | | | | 8 | 4 | 2 | 1 | |
A	B	C	D	E	F	G	H	
0	0	0	0	0	0	0	0	
0	0	0	1	0	0	0	1	
0	0	1	0	0	0	1	1	$E = A$
0	0	1	1	0	0	1	0	
0	1	0	0	0	1	1	1	$F = \bar{A}B + A\bar{B} = A \oplus B$
0	1	0	1	0	1	1	0	
0	1	1	0	0	1	0	0	$G = \bar{A}\bar{B}C + \bar{A}B\bar{C} + A\bar{B}\bar{C} + ABC$
0	1	1	1	0	1	0	1	$= (A \oplus B) \oplus C$
1	0	0	0	1	1	1	1	$H = \bar{A}\bar{B}\bar{C}D + \bar{A}\bar{B}C\bar{D}$
1	0	0	1	1	1	1	0	$+ \bar{A}B\bar{C}\bar{D} + \bar{A}BCD$
1	0	1	0	1	1	0	0	$+ A\bar{B}\bar{C}\bar{D} + A\bar{B}CD$
1	0	1	1	1	1	0	1	$+ ABC\bar{D} + AB\bar{C}D$
1	1	0	0	1	0	0	0	$= [(A \oplus B) \oplus C] \oplus D$
1	1	0	1	1	0	0	1	
1	1	1	0	1	0	1	1	
1	1	1	1	1	0	1	0	

of 10 numerals lights when a corresponding pin receives a signal. Cheaper indicators than those with separate numerals can be made of segment indicators which have seven lamps, each lighting up a corresponding line segment in a digit pattern (Fig. 2-25; see also Fig. 2-26). Logic control signals for the lamp drivers are defined by Table 2-9.

Table 2-9. Conversion Table for a Seven-Segment Decoder

Decimal Digit	A	B	C	D	a	b	c	d	e	f	g
0	0	0	0	0	1	1	1	1	1	1	0
1	0	0	0	1	0	1	1	0	0	0	0
2	0	0	1	0	1	1	0	1	1	0	1
3	0	0	1	1	1	1	1	1	0	0	1
4	0	1	0	0	0	1	1	0	0	1	1
5	0	1	0	1	1	0	1	1	0	1	1
6	0	1	1	0	0	0	1	1	1	1	1
7	0	1	1	1	1	1	1	0	0	0	0
8	1	0	0	0	1	1	1	1	1	1	1
9	1	0	0	1	1	1	1	0	0	1	1
—	1	0	1	0	$\varnothing$	$\varnothing$	$\varnothing$	$\varnothing$	$\varnothing$	$\varnothing$	$\varnothing$
—	1	0	1	1	$\varnothing$	$\varnothing$	$\varnothing$	$\varnothing$	$\varnothing$	$\varnothing$	$\varnothing$
—	1	1	0	0	$\varnothing$	$\varnothing$	$\varnothing$	$\varnothing$	$\varnothing$	$\varnothing$	$\varnothing$
—	1	1	0	1	$\varnothing$	$\varnothing$	$\varnothing$	$\varnothing$	$\varnothing$	$\varnothing$	$\varnothing$
—	1	1	1	0	$\varnothing$	$\varnothing$	$\varnothing$	$\varnothing$	$\varnothing$	$\varnothing$	$\varnothing$
—	1	1	1	1	$\varnothing$	$\varnothing$	$\varnothing$	$\varnothing$	$\varnothing$	$\varnothing$	$\varnothing$

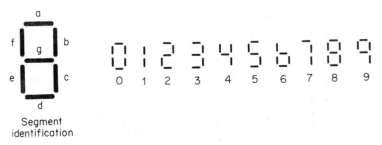

Segment identification

Fig. 2-25. Seven-segment display indicators of arabic numerals.

Boolean functions for the segment decoders which may or may not be the simplest ones, but which, owing to several common terms, are implemented by only 17 gates, are defined in Eqs. [2-28]:

Fig. 2-26. Seven-segment light-emitting diode readouts (left) driven
by integrated decoder/driver circuit packages (right). (*Courtesy of
Signetics Corp.*)

$$a = A + \bar{B}\bar{D} + BD + CD$$

$$b = A + \bar{B} + \bar{C}\bar{D} + CD$$

$$c = A + B + \bar{C} + D$$

$$d = A\bar{D} + \bar{B}C + \bar{B}\bar{D} + C\bar{D} + \bar{A}B\bar{C}D \qquad\qquad [2\text{-}28]$$

$$e = A\bar{D} + \bar{B}\bar{D} + C\bar{D}$$

$$f = B\bar{C} + B\bar{D} + \bar{C}\bar{D}$$

$$g = \bar{B}C + B\bar{C} + C\bar{D}$$

PROBLEMS

2-1   In Fig. P2-1 the relay contacts denoted by m make a contact while
the one denoted by b breaks when the corresponding solenoid is energized by a logic 1 voltage ($=$ H). Derive a Boolean function for
X in terms of A and B.

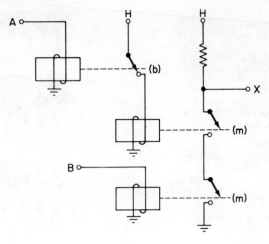

Figure P2-1

2-2   Tabulate the following Boolean functions F and G for all values of the arguments A, B, C, and D.

(a) $F = (\overline{\overline{A \cdot B}}) + (\overline{C \cdot D})$

(b) $G = A \cdot \bar{B} \cdot C \cdot \bar{D} + A \cdot \bar{C} \cdot D + B \cdot C + \bar{C} \cdot D$

2-3   (a) Write the truth table for the Boolean function

$$X = (\overline{ABC + D}) + (\overline{\overline{ACD + B}})$$

(b) Write X in disjunctive normal form.

(c) Write X in conjunctive normal form.

(d) Reduce the above function by De Morgan's rules.

2-4   Which of the following expressions are, and which can be, algebraically reduced to (1) Boolean product terms and (2) Boolean sum terms?

(a) A

(b) $X + A$

(c) $\bar{D} \cdot \bar{E} \cdot F$

(d) $D \cdot E \cdot \bar{E}$

(e) $\overline{\bar{A} \cdot \bar{B}}$

2-5   Which of the following terms are, and which can be, algebraically reduced to (1) normal products and (2) normal sums? The independent variables are A, B, C, and D.

(a) $A \cdot B \cdot C$

(b) $\bar{A} \cdot \bar{B} \cdot C \cdot \bar{D}$

(c) $\overline{A \cdot B \cdot C \cdot D}$

(d) $\bar{A} + \bar{B}\bar{C}\bar{D}$

(e) $A \cdot \bar{B} \cdot C \cdot \bar{D}$

(f) $A + B$

2-6   Find all prime implicants for the expression

$$F = (A + \bar{B} + \bar{C}) \cdot (A + B + \bar{C}) \cdot (\bar{A} + B + C)$$

2-7   The prime implicants of a five-variable Boolean function are $AB\bar{C}D$, $AB\bar{C}\bar{E}$, $ABDE$, $\bar{A}C\bar{D}E$, $ACDE$, $\bar{B}C\bar{D}E$, $\bar{B}\bar{C}DE$, $BC\bar{D}\bar{E}$, $A\bar{B}C$, $AC\bar{D}$, $BC\bar{D}$, and $\bar{C}\bar{D}\bar{E}$.

(a) Set up and solve the prime-implicant table.

(b) Write down the corresponding Boolean function.

2-8   Reduce the Boolean function

$$F = \bar{A} \cdot \bar{B} \cdot C \cdot D + \bar{A} \cdot B \cdot \bar{C} \cdot \bar{D} + A \cdot B \cdot C \cdot D + \bar{A} \cdot B \cdot C \cdot D$$
$$+ A \cdot \bar{B} \cdot \bar{C} \cdot D + A \cdot \bar{B} \cdot C \cdot D + A \cdot \bar{B} \cdot C \cdot D$$

to the simplest sum of products by the

(a) Karnaugh method.

(b) Quine–McCluskey method.

2-9   Find the simplest Boolean functions in sum-of-products form for the Karnaugh maps shown in Fig. P2-9.

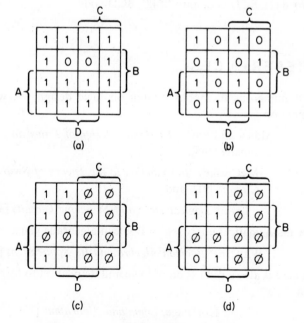

Figure P2-9

2-10   (*a*) Find the simplest Boolean sum of products for the Karnaugh map shown in Fig. P2-10.

(*b*) How many rows would you have in a truth table?

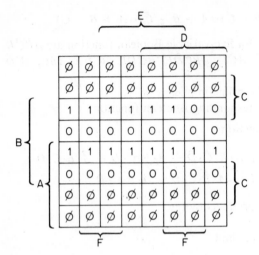

Figure P2-10

2-11   Design a combinational circuit that performs the conversion of

(*a*) 2*421 BCD code into 8421 BCD code.

(*b*) 8421 BCD code into 2*421 BCD code.

REFERENCES

Of numerous textbooks written on logic design, we would like to mention

Maley and Earle, *The Logic Design of Transistor Digital Computers* [88],

McCluskey, *Introduction to the Theory of Switching Circuits* [91], and

Lewin, *Logical Design of Switching Circuits* [80].

A mathematical approach to Boolean algebra is given in

Whitesitt, *Boolean Algebra and Its Applications* [128].

There is also a discussion of logic and its relation to formal languages in

Korfhage, *Logic and Algorithms* [75].

(See also the Bibliography.)

chapter 3

Sequential Circuits

3.1. Sequential Operations

There is no *memory* in combinational circuits; a particular output signal combination is present only as long as the corresponding inputs are activated. Digital computing, however, is usually carried out in several successive steps, which means that the logic values of signals must be temporarily stored. There are also many *serial logic operations* in which the order of occurring signals is important and must be recognized. Also, *counting* of single events calls for storage of digital signals. Such features are provided by *sequential circuits*. The simplest sequential circuit is a *flip-flop*, which has two states. A set of n flip-flops may be regarded as a system that has 2^n state combinations, called *states* of the system for short.

We shall soon see that it is possible to construct general circuits that can have several stable states; such circuits are called *polystable*. This type of circuit can jump to another state by applied input signals. The circuit thus has a *memory*, which can recall input signal values. It is the purpose of this chapter to review the fundamental properties of these memorizing or sequential circuits.

There are two basic kinds of sequential circuits and operations: *asynchronous* and *synchronous*. In asynchronous operations, the circuit is usually polystable and makes a transition from one state to another at a change in the logic values of input signals. This transition follows the signals immediately. In synchronous operations, the previous state of the system (usually the state of flip-flops) is sampled and a new state loaded in synchronism with *clocking signals* or *clocking pulses* delivered simultaneously all over the system.

81

Let us recall that digital computing systems operate on machine variables which usually are binary states of storage elements. During computing the previous states are read and transformed by combinational circuits. The outputs of combinational circuits are then used to load new contents into the storage. This is also what we would call a synchronous sequential operation. From a little different point of view we can describe the synchronous sequential operation as the transfer of information from certain registers to other registers, possibly to the original ones, whereby operational gates in the transfer path implement one step of computing operations. In Chapter 4 we shall be dealing with the detailed transformation of information during computing; in this chapter we shall establish general methods for the description and design of functional circuits by which arbitrary *state sequences* (or briefly, *sequences*) can be implemented.

Besides computing operations, implementation and recognition of state sequences are also needed in the automatic control of digital systems.

Asynchronous sequential circuits are used less often than synchronous circuits in digital systems unless we take into account that every flip-flop is itself an asynchronous circuit. The purpose of asynchronous circuits in most cases is to implement a given *output signal sequence* upon reception of input signals in a given order. The internal state of the circuit is of no primary importance. This seemingly leaves more degrees of freedom for design; however, systematic optimization of asynchronous circuits becomes rather difficult. We shall start this discussion with asynchronous sequential circuits because they provide the fundamental idea of *state transition* laws. We would like to emphasize, however, that even the most complicated processors of digital information can be implemented using synchronous sequential circuits only. In doing this, however, we are in fact always making use of asynchronous building blocks, the flip-flops.

It is necessary to analyze transitions in asynchronous circuits, at least to accentuate the existence of their usual two pitfalls—the *races* and the *hazards*—which, however, are not encountered in synchronous circuits.

3.2.   Some Basic Sequential Circuits

There are innumerable ways of making different kinds of sequential circuits. All of these can in principle be implemented by means of elementary logic gates, as will be seen. However, it is convenient to have available building blocks which themselves are elementary sequential circuits and by which more complicated sequential cirucuits can be assembled. The most important of them are the various classes of *flip-flops*. Notice also that *counters* and *shift registers* are sequential circuits.

Flip-Flops. Storage elements are elementary sequential circuits since they are intended for the "remembering" of binary signals. More general sequential circuits may be implemented by the interconnection of storage elements and combinational circuits. Electronic binary storage elements that are directly controlled by binary logic signals are called *flip-flops*.

Some flip-flop types have gained fundamental importance in digital techniques. These flip-flops are identified by the way in which their states and the transitions to new states are determined by control signals. First, we have to realize that there are two different basic kinds of control: *static* and *dynamic*. With flip-flops this means that if a control is static, a particular logic value of the control signal, together with particular logic values of some other control signals, define the state of the flip-flop. A dynamic control, also called *clocking* or *triggering*, means that the control signal may have an effect on the output of a flip-flop (perhaps conditioned by some static signals) only at the moment when the signal jumps from one binary value to another.

The most important flip-flop types are the *direct R-S flip-flop*, which has only static control inputs, and the *clocked* flip-flops, which are termed *D*, *R-S*, and *J-K flip-flops*.

The flip-flops have an output, called Q, which indicates the state of the flip-flop. Another output, $\bar{Q}$, is usually provided also. The direct $R\text{-}S$ flip-flop is defined in Fig. 3-1.

Name	Symbol	Definition
Direct R-S flip-flop	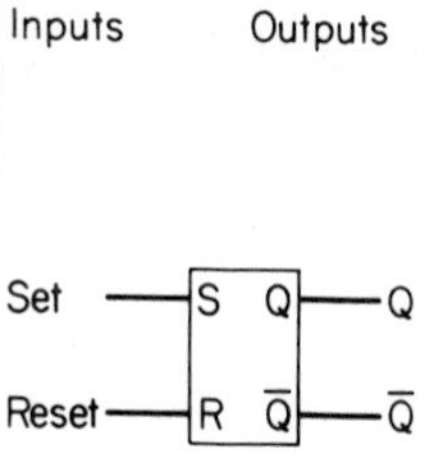 Inputs Outputs Set ——[S Q]—— Q Reset ——[R $\bar{Q}$]—— $\bar{Q}$	The outputs Q and $\bar{Q}$ are normally complements of each other. Q may attain either of the values 0 or 1, and as long as both inputs are 0, there is no change in the output values. When Set is made 1, Q becomes 1, and sustains this value if Set is returned to 0. If Reset is made 1, Q becomes 0, and sustains this value if Reset is returned to zero. The operation is undefined if both inputs are 1, and this situation should be avoided.

Figure 3-1

In clocked flip-flops the transition of Q occurs either at the leading or the trailing edge of the *clocking signal*, but to complete the cycle and to facilitate a new transition, the clocking signal must be returned to the original value. However, the ideal construction of clocked flip-flops should be such that the duration of the clocking signal should have no effect on the value of the next state. Let us denote the old values of the control signals and Q before and at the triggering instant by the subscript n and the new state by the subscript $n + 1$. In the following we define the various clocked flip-flop types (See Fig. 3-2).

Name	Symbol	Definition
D flip-flop	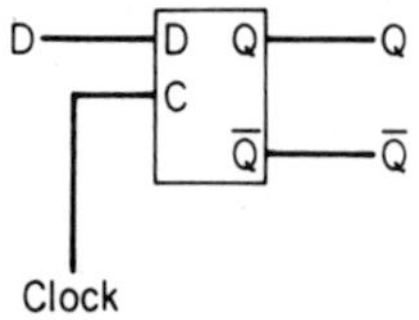	The outputs Q and $\bar{Q}$ are normally complements of each other. Q may attain either of the values 0 or 1, and as long as there are no changes in the clocking signals, the output values remain unchanged. The control inputs D, S, R, J, and K, called <u>preparatory inputs</u>, are not able to change the state of the flip-flops without a clocking signal. Upon a clocking signal, the output Q is triggered from the value Q^n to the value Q^{n+1} according to the conditions defined by the following tables:
R-S flip-flop	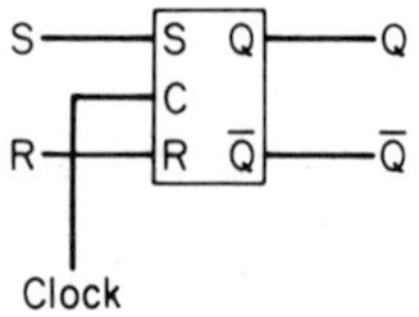	
J-K flip-flop	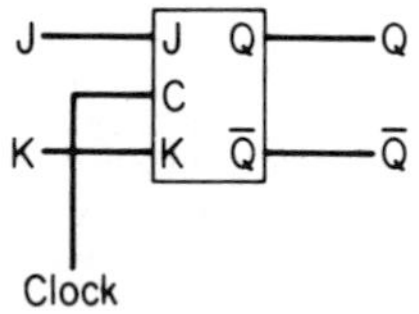	

D flip-flop:

D^n	Q^{n+1}
0	0
1	1

R-S flip-flop:

S^n	R^n	Q^{n+1}
0	0	Q^n
0	1	0
1	0	1
1	1	undefined

J-K flip-flop:

J^n	K^n	Q^{n+1}
0	0	Q^n
0	1	0
1	0	1
1	1	$\bar{Q}^n$

Figure 3-2

In the *D flip-flop*, the new state is always independent of the old state and the static control signal D is, in fact, sampled at the triggering instant and held after it. (D comes from "delay.") In the *R-S flip-flop*, the old value of Q is sustained if S and R are 0. The undefined operation with S and R simultaneously 1 at the triggering should be avoided. It is not forbidden, however, to have the static input signals at 1 between the triggering instants. If S and R are complements of each other, the output Q will be set to the value defined by S at the triggering. In the third flip-flop, the *J-K flip-flop*, which is the most versatile of all, there are no forbidden control signal combinations. If J and K are complements of each other, the output will be set to the value defined by J at the triggering instant, independent of the old value of Q. The old value is sustained if J and K are 0, and the old value is inverted if J and K are 1 (*toggle operation*).

Clock-Pulse Polarity. In practical circuits, the triggering of a clocked flip-flop occurs either at the leading edge ($0 \rightarrow 1$ transition) or the trailing edge ($1 \rightarrow 0$ transition) of the clocking signal. In order to specify the triggering edge, we shall use the convention that the C input is drawn in the normal way when transitions occur at the leading edge [Fig. 3-3(a)], whereas a symbol of inversion is used at this input if the trailing edge is used for triggering [Fig. 3-3(b)].

Fig. 3-3. (a) Triggering at the leading edge of a clocking signal; (b) triggering at the trailing edge of a clocking signal.

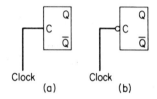

Toggles. A special form of flip-flop, called the *toggle*, has the property that the output Q will always be inverted at the triggering signal:

$$Q_{n+1} = \bar{Q}_n \qquad [3\text{-}1]$$

Special toggles can be found in integrated systems, but usually they are implemented by various circuit connections from other clocked flip-flops. The reader should convince himself that all the circuits in Fig. 3-4 are toggles. The input signals denoted by 1 are voltages with a constant logical value 1.

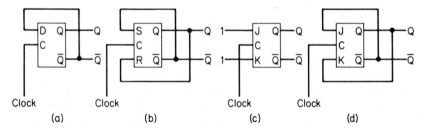

Fig. 3-4. Toggles.

Next we shall show how a simple binary counter and a shift register can be implemented using clocked flip-flops.

Binary Counter. Let us assume that the triggering of a clocked flip-flop occurs at the trailing edge of the clocking signal (i.e., at the transition $1 \rightarrow 0$) and that the clocking signal has the same specifications as the logic signals. From the circuit diagram of Fig. 3-5(a) and its timing diagram [Fig. 3-5(b)] we see that a triggering of the next stage on the left occurs when the output Q of the former stage changes from 1 to 0. Every stage thus divides the frequency of clocking signals by 2. If the weights $2^0, 2^1, 2^2, \ldots,$ are assigned to the outputs $Q_0, Q_1, Q_2, \ldots,$ respectively, and the Q's are 0 at the beginning, we can see that immediately after N clocking signals received by the rightmost stage, the binary number

$$Q_n Q_{n-1} \cdots Q_1 Q_0$$

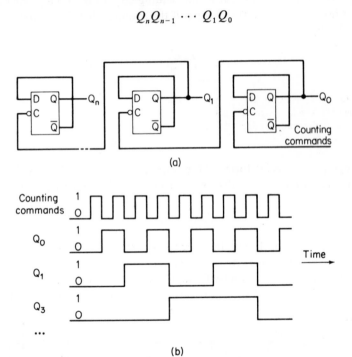

(a)

(b)

Fig. 3-5. Binary ripple counter.

represents the number of counted clocking signals. This system of $n + 1$ flip-flops is also referred to as a *ripple counter*, which is characterized by the fact that the carry bit to the next stage is given by the output of the former stage. Other types of counters are discussed in Chapter 4. A ripple counter may be built of any of the types of toggles shown in Fig. 3-4.

Shift Register. The binary value of one flip-flop may be copied onto another by any of the interconnections shown in Fig. 3-6. Every flip-flop in the chain must receive a new value from the left at the same time that it is transmitting its old value to the next flip-flop on the right. With clocked

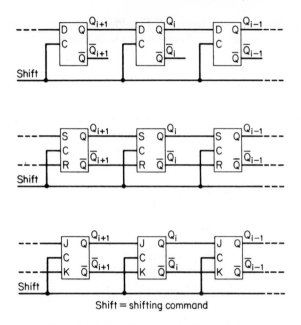

Shift = shifting command

Fig. 3-6. Right-shift register implementations.

flip-flops this is no problem if a common clocking signal is used for all stages, and if it is assumed that the triggering effect on all flip-flops is simultaneous, at least within limits that differ less from each other than the internal signal delays in each flip-flop.

In a left-shift register the preparatory inputs of the ith flip-flop are connected to the outputs of the $(i - 1)$th flip-flop.

3.2.1. Design of Elementary Sequential Circuits

Logical Feedback. Output signals in combinational circuits are single-valued functions of steady-state input signals.

We can now raise the question of what happens if some outputs are interconnected with inputs. In linear circuits this operation would be called signal feedback; here we shall call it *logical feedback.*

In terms of mathematical concepts, *feedback implements a recursive relation: the value of a function depends implicitly on itself.*

We shall now show that logical feedback can change a combinational circuit to a polystable system; in other words, the circuit becomes sequential and can be used as a digital memory. Bistability in a logic circuit is a special case and is implemented by the feedback of a single output signal, as will be seen.

Implementation of a Direct R-S Flip-Flop. Let us try to build a logic circuit that would implement a Boolean function X defined by the recursive relation

$$X = S + \bar{R}X \qquad\qquad [3\text{-}2]$$

where S and R are parameters (Boolean variables with fixed values) and X is an unknown Boolean variable. A straightforward network analogy for it is shown in Fig. 3-7. Equation [3-2] is, in fact, what we would call a *Boolean*

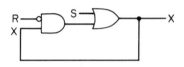

Fig. 3-7. Network analogy for Eq. [3-2].

equation for X; we are looking for solutions for X which make both sides of Eq. [3-2] equal. Since there are only three Boolean variables with $2^3 = 8$ different value combinations, the equality of both sides is readily inspected for all of these. The results are represented in Table 3-1, from which we can see that in this case there exist solutions for all value combinations of the parameters S and R, and for $S = R = 0$ there are *two* solutions for X.

Table 3-1

S	R	X
0	0	0, 1
0	1	0
1	0	1
1	1	1

The values of X tabulated in Table 3-1 are possible steady-state signal values, and for $S = R = 0$ the system is *bistable*. For all other values of S and R the variable X is uniquely defined. This system, in fact, constitutes a static *direct R-S flip-flop* which satisfies the definition given in Chapter 1.

Systems with several logical feedback paths can also be described by Boolean equations. Depending on particular values of parameters (such as values of independent input signals), the Boolean equations may have none, one, or several solutions. To deduce whether the system is polystable, it is only necessary to solve the equations, which is done by checking the equality of both sides for all value combinations and accepting those values for which the equality holds.

Dynamic Analysis of the Direct R-S Flip-Flop. There is always a small nonzero signal delay from the inputs to the outputs in a combinational circuit. For a particular type of logic gate, this delay depends primarily on the number of logic levels along a signal path. If the largest delay through the system is denoted by τ_{max}, and the input signals are constant from time t on, we must specify the algebraic relations between the input and output signals in the following way:

$$Y_j(t + \tau) = f_j[X_1(t), X_2(t), \ldots, X_n(t)] \quad \text{for all } j \text{ and for } \tau \geq \tau_{max} \quad [3\text{-}3]$$

The behavior of the system is thus completely undefined for $\tau < \tau_{max}$.

With a delay in the logical function, the flip-flop equation should, in fact, read

$$X(t + \tau) = S(t) + \bar{R}(t)X(t) \quad (\tau \geq \tau_{max}) \quad [3\text{-}4]$$

and we can follow the dynamic behavior of this system by handling all variables on the right as input variables and the variable on the left as the output. For this purpose we have to write down (as Table 3-2) the complete

Table 3-2

$S(t)$	$R(t)$	$X(t)$	$X(t + \tau)$
0	0	0	0
0	0	1	1
0	1	0	0
0	1	1	0
1	0	0	1
1	0	1	1
1	1	0	1
1	1	1	1

truth table for $S(t)$, $R(t)$, and $X(t)$. If we restrict ourselves to the conditions allowed for direct R-S flip-flops (i.e., with S and R never simultaneously equal to 1), we can ignore the two last rows in Table 3-2. Assume now that the logic gates have delays τ_1 and τ_2, respectively, whereby $\tau_{max} = \tau_1 + \tau_2$. Then changes in $S(t)$ and $R(t)$ are reflected in $X(t)$ according to Fig. 3-8. It is further assumed that the time between any successive changes in S and R is larger than τ. The waveforms of Fig. 3-8 are readily established for all times using Table 3-2. Thus the system defined by Eq. [3-2] or Fig. 3-8 indeed is a memory for a digital signal in the sense of the definition of the direct R-S flip-flop.

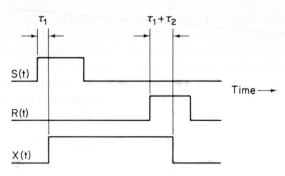

Figure 3-8

Other Forms of Direct R-S Flip-Flops. Another common type of feedback system is described by the equation

$$X = (S + X)\bar{R}$$ [3-5]

Normally R and S are never made simultaneously 1. If this is done, however, Eq. [3-2] has the solution $X = 1$, whereas the solution of Eq. [3-5] is $X = 0$. In the future we shall call the former circuit the *S-override direct R-S flip-flop* and the latter the *R-override direct R-S flip-flop*.

Using De Morgan's formulas, Eqs. [3-2] and [3-5] can be expressed in terms of NAND functions as

$$X = \overline{\bar{S}\overline{\bar{R}X}}$$ [3-6]

$$\bar{X} = \overline{\bar{S}\overline{\bar{X}\bar{R}}}$$ [3-7]

respectively, with the corresponding circuit configurations shown in Fig. 3-9.

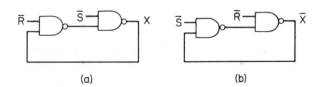

(a)   (b)

Fig. 3-9. (a) *S*-override direct *R*-*S* flip-flop; (b) *R*-override direct *R*-*S* flip-flop.

There is a similarity between these two, which can be seen from the NAND implementations; the *R*-override solution is obtained from the *S*-override solution by interchanging $\bar{S}$ with $\bar{R}$ and X with $\bar{X}$.

Very often these circuits are drawn in a symmetrical way. The *S*-override type is shown in Fig. 3-10. The other output, X', is seen to be an

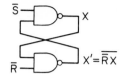

Fig. 3-10. S-override direct R-S flip-flop redrawn.

intermediate variable not explicitly occurring in Eqs. [3-6] and [3-7]. Normally $X' = \bar{X}$ except that for $S = R = 1$, we have $X' = X = 1$.

Comment. The setting and clearing in the circuits of Figs. 3-9 and 3-10 are, in fact, done by setting $\bar{S}$ and $\bar{R}$, respectively, to 0; in the normal state, these inputs are 1.

Bistable Latch. The output Q of a bistable latch will follow the logic value of a logic input variable I when a clocking signal C attains the value 1, and sustain this value when $C = 0$ again. If there are changes in I during this time, the last value of I during $C = 1$ will remain as the state of Q.

The bistable latch is described by the Boolean equation

$$Q = CI + \overline{CI}Q$$

$$= \overline{CI\,\overline{CI}Q} \qquad\qquad [3\text{-}8]$$

Its circuit implementation is shown in Fig. 3-11.

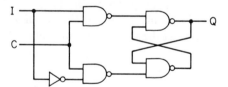

Fig. 3-11. Bistable latch.

This circuit can be transformed to a simpler equivalent form by noticing that Eq. [3-8] can be replaced by an equivalent expression,

$$Q = \overline{CI\,\overline{CCI}Q} \qquad\qquad [3\text{-}9]$$

which is readily checked for all value combinations. Thus we can save an inverter and implement the latch by four NAND gates, as shown in Fig. 3-12.

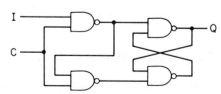

Fig. 3-12. Another   form   of bistable latch.

3.2.2. Example of Sequential Circuits: the Annunciator

Industrial or other plants are often monitored with the aid of *annunciators*, which are alarm-detecting-and-recording devices. When a sensor indicates an abnormal condition, a corresponding logic signal $A = 1$ (alarm true) is produced which lasts as long as the abnormal situation is valid. A standard procedure is to indicate and continually remind the staff of the occurrence of the alarm condition by means of a light flashing on the control panel with a frequency of 2 Hz; a siren might also sound. The condition for the flashing light and the siren may be simultaneously represented by a single logic signal F. The operations represented by F are reset when a button on the control panel is depressed. After acknowledgment of the alarm, the siren is silenced, too. However, an indicator lamp must be turned on as a sign for acknowledgment of the alarm, and the light is now changed to steady. The lamp is automatically turned off when the alarm condition has been removed ($A = 0$). The condition for the steady light is represented by a logic signal L. The signals F and L are not simultaneously 1.

To implement the foregoing sequential operations, we must have two memories: one for the alarm and one for the acknowledgment. If the condition for the depression of the reset button is denoted by $R = 1$, we may describe the "acknowledgment memory" by

$$L = R + AL \qquad [3\text{-}10]$$

which shows that the memory can be set only when the alarm is true; otherwise the lamp will light only as long as the button is depressed. The signal L will then be used to reset the annunciator memory F described by the equation

$$F = (A + F)\bar{L} \qquad [3\text{-}11]$$

The annunciator function F becomes 1 when the alarm occurs unless L is 1 (button depressed). The reset operation always cancels the annunciator function. Notice that the acknowledgment memory will be automatically reset if the alarm has been acknowledged and the abnormal situation is over ($A = 0$). The annunciator is now ready for a new operation. The acknowledgment memory cannot be reset in any other way. The annunciator circuit, with a push-button control and a closing alarm contact, is shown in Fig. 3-13.

The reader should analyze the sequential operation of this circuit on the basis of Eqs. [3-10] and [3-11] and Fig. 3-13, with various signal-status combinations, to become convinced of the correct operation.

3.3. General Asynchronous Sequential Circuits

Representation of General Asynchronous Sequential Circuits. The general asynchronous sequential circuit is shown in Fig. 3-14. The signals

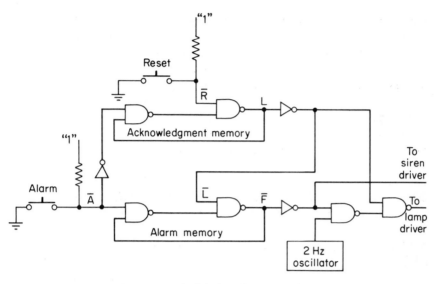

Fig. 3-13. Logical design of an annunciator.

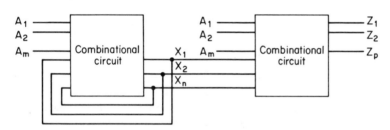

Figure 3-14. General asynchronous sequential circuit.

denoted by $A_1 \cdots A_m$ are *inputs* to the circuit. *The set of feedback signals* $\{X_1 \cdots X_n\}$ *is called the state of the circuit.* Feedback signals are not always used as the outputs of the system; in a more general case, the X_i signals together with the input signals are transformed into another combinational circuit, the outputs $Z_1 \cdots Z_p$ of which are called *system outputs*, or simply *outputs*.

It is sometimes convenient to call the elements of the signal sets $\{A_1, \ldots, A_m\}$, $\{X_1, \ldots, X_n\}$, and $\{Z_1, \ldots, Z_p\}$ the components of the *input*, *state*, and *output vectors*, respectively.

Excitation Table. There are internal delays in the first combinational circuit of Fig. 3-14 and the X_i's change immediately when signals are propagated through the circuit.

Asynchronous sequential circuits are more difficult to analyze and design than synchronous circuits—first, because we cannot state exactly in which order the feedback signals change. Therefore, we must change the input signals one at a time and allow the system to settle down before a new change is made. Assume now that after a change in one external input signal the state (i.e., the set of output signals) is changed. We shall regard the circuit as *well behaved* if the new state differs from the old one in only one signal. The feedback signals will now be changed and, as a result, a new set of output signals is obtained after a certain propagation delay, etc. Should two output signals change simultaneously at any time, however, we are no longer sure what changes this will cause in the system, and the operation becomes ambiguous. To reveal this kind of unwanted behavior, we shall first set up an *excitation table*. This is a kind of Karnaugh map which is set up for the *future* outputs $\{X_1(t + \tau), X_2(t + \tau), \ldots, X_n(t + \tau)\}$ expressed as Boolean functions of the *present* feedback inputs $\{X_1(t), X_2(t), \ldots, X_n(t)\}$ and of all external inputs. In other words, we set up the truth tables for the first combinational circuit. In order to save writing, these truth tables are combined so that we show the whole *successor state* in every element of the table.

We shall delete the time variables from the excitation table because they are clear from the context.

Example 3-1. Let the combinational circuit be described by the Boolean functions

$$X_1 = (A_1 + \bar{A}_2)X_1 + \bar{A}_1 X_2 + A_1 \bar{X}_2 + A_1 \bar{A}_2$$

$$X_2 = \bar{X}_1 + A_1 \qquad\qquad [3\text{-}12]$$

These equations are represented by an excitation table (Table 3-3). In the table pairs represent outputs of the combinational circuit (i.e., the successor state).

Table 3-3. Excitation Table

Present State $X_1\ X_2$	$A_1 A_2$				
	0 0	0 1	1 1	1 0	
0 0	0 1	0 1	1 1	1 1	Successor
0 1	1 1	1 1	0 1	1 1	State
1 1	1 0	1 0	1 1	1 1	$X_1\ X_2$
1 0	1 0	0 0	1 1	1 1	

Transition Types. In Example 3-1 we have four alternatives for $A_1 A_2$. Starting with the first column and the present state $X_1 X_2 = 00$, the successor state will be found from the table as the matrix element 01. Looking for the state following 01, we find 11, and the state following 11 is 10. *Now*

10 *is a stable state since the old and new states are equal.* We say that the system, starting from the state 00, has made a *multiple transition* to 10 through *intermediate states.*

To visualize different types of transition events, let us redraw the same excitation table without matrix elements and, starting with the state 00, indicate the subsequent matrix elements by arrows (with fixed A_1A_2) (see Table 3-4). Denote the stable states by circles. We see that there are no stable states in the second column because the old and new states are never equal. The system is left in an astable condition, *oscillation.*

Table 3-4

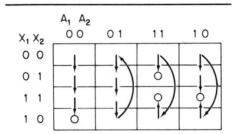

Races. In the third column we meet a condition that is very common in sequential circuits; it is called a *race.* Starting with the state 00, the new state as defined by Eqs. [3-12] or by the excitation table is 11. This would necessarily require simultaneous change in two feedback variables; in reality, as a result of indeterminate signal-path delays, the transitions begin either in the order $00 \rightarrow 01$ or $00 \rightarrow 10$. But these lead to two different sequences, as seen from the diagram. Accordingly, the system can end up in either the stable state 01 or 11.

Races are usually not allowed because the behavior of the system cannot be predicted. We call the case occurring in the third column a *critical race*, because the final state is ambiguous. In the fourth column we have what is called a *noncritical race*, and, although the intermediate transitions are ambiguous, they lead to a single well-defined state. Noncritical races may sometimes be allowed if intermediate states are not relevant.

Hazards.[1] Assume that a Boolean function, let us say F, can be expressed as a sum of two terms, F_1 and F_2. Both these terms might be functions of the same logical variable, A, the value of which is changing. Let the value of A before the change be denoted by A and, after the change, by $\bar{A}$.

---

[1] What we discuss in this book is actually called *static hazard.* For other forms, called *dynamic* and *essential hazards*, see, for example, the book of *McCluskey* [91].

Consider the expressions

$$F(A) = F_1(A) + F_2(A)$$
$$F(\bar{A}) = F_1(\bar{A}) + F_2(\bar{A})$$

[3-13]

A *hazard* occurs if $F_1(A) = 0$ and $F_2(A) = 1$, but $F_1(\bar{A}) = 1$ and $F_2(\bar{A}) = 0$. In other words, if the value of the Boolean expression F before and after the change is 1, but if the two terms that share the responsibility for the 1's are changing simultaneously, a risk is taken. There are always variations in the mutual delays of the signals, and it might well happen that both F_1 and F_2, generated by different logic circuits, for a short while are simultaneously 0. Even a short false signal value can cause unwanted transitions in sequential circuits. A check for possible consequences of hazards must always be made.

Indication and Elimination of Hazards. There is a very simple rule for the indication of hazards based on the study of the Karnaugh map: *A hazard exists if a change in the input variables causes such a transition in the Karnaugh map that if there are 1's in the corresponding variables of the initial and final states, they belong to different product terms.*

Example 3-2. In the Karnaugh map of Table 3-5(a), which forms the excitation table of X, a single feedback variable, the simplest Boolean expression for the combinational circuit would be represented by the two indicated product terms. If the initial state is $A = 0$, $B = X = 1$, and the final state is $A = B = X = 1$, a hazard exists. It can be eliminated by bridging the two terms by a third one, as shown in Table 3-5(b), and this term will be 1 during the transition.

Table 3-5

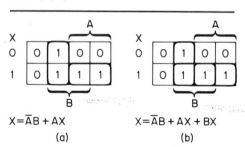

$X = \bar{A}B + AX$ $X = \bar{A}B + AX + BX$

(a)   (b)

Notice that the Boolean function obtained from Table 3-5(b) is no longer the simplest one. This is often the tradeoff when hazards are eliminated in sequential circuits.

3.4. Design of Asynchronous Sequential Circuits

An inverse problem to analysis is the *synthesis* of a circuit which implements a given sequence of events.

With asynchronous circuits, we are usually not primarily interested in the internal structure or the states of the network but on the *sequence of output signals*, which must depend on a given sequence of input signals. All systems that implement the given input–output relations are equivalent in this sense. The purpose of design is to find out the simplest one of all such equivalent circuits.

Again we are confronted with the question of what type of circuit is really the simplest. From the point of view of fabrication, it is the cost of components which is usually most decisive. On the other side, we could also try to reduce the number of feedback paths, because problems with races and hazards are expected to be simplest with fewest feedback loops. This is the approach that we make in the following.

Primitive Flow Table. Let us discuss a circuit example of the type depicted in Fig. 3-15(a). Its behavior can be defined with the aid of a *timing diagram*, Fig. 3-15(b), which relates the output signal sequences to different

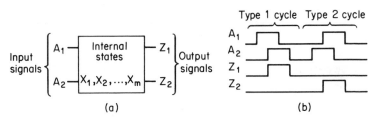

(a)   (b)

Fig. 3-15. Asynchronous sequential circuit: (a) model; (b) timing diagram.

sequences of input signals at any time they occur. In this particular example, the wanted operation is the following: The input signals A_1 and A_2 are obtained from, e.g., photoelectric sensors which detect the passage of a large object. We know that the object during some interval covers both sensors. The order of sensor signals must be detected, and the only waveforms that need to be considered are defined by Fig. 3-15(b). Since we have a finite number of inputs and outputs, the system can undergo only a finite number of different cycles of events. These are usually defined when the problem is set up, when all cases are considered carefully. All *wanted* input–output signal combinations are thus represented in the timing diagram; the intervals between signals are of no importance. It may be noted that of all *possible* sequences, some may not occur in practice; the behavior of the system remains undefined for such events.

We have previously defined the *state* of the circuit as the set of all feedback signals. More accurately, this set of signals is often called the *internal state* of the system.

It is also reasonable to assume that the system has only a finite number of internal states. It is not necessary to assume more states than there are different signal combinations in the timing diagram. Notice that if the circuit is in a given state for some input signal combinations, the next states are unambiguously given by the timing diagram for any allowed change in the input signal values. In other words, the behavior of the system is *deterministic*. To begin with, we assume that there is a stable internal state for every stationary signal combination. In Fig. 3-16 these states are labeled from left to right by encircled numbers, as shown. During changes in input or output signals, the system is in an unstable state, which is shown numbered (uncircled) according to the *successor* state.

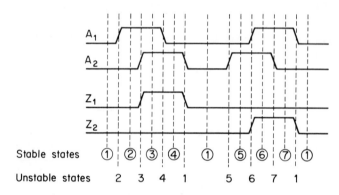

Fig. 3-16. Assignment of states to the timing diagram.

We proceed by writing down the *primitive flow table* (Table 3-6), which is supposed to represent a kind of excitation table of the system. Since we do not yet know the values of feedback signals, the order of rows in the primitive flow table is still arbitrary. Nor do we know yet the number of feedback loops or the state variables. The rows are numbered according to the stable states, and the columns are labeled by the values of input signals. The stable and unstable states are easily transferred to this table by examining present signal values and successor states in the timing diagram. In addition, the values of output signals corresponding to stable states are transferred to separate columns. Since the operation is undefined for some signal values, and only one input may be changed at a time, some elements of the table remain undefined. These places are filled by dashes, which are a kind of "don't-care" sign. In a later phase of design, such dashes serve in finding the simplest Boolean functions. Before doing so, however, we have to find out whether the same output sequences could be implemented by a simpler excitation table than Table 3-6.

Table 3-6. Primitive Flow Table for the Circuit Example of Fig. 3-15

Present State	A_1A_2				Outputs	
	0 0	0 1	1 1	1 0	Z_1	Z_2
①	①	5	—	2	0	0
②	—	—	3	②	0	0
③	—	4	③	—	1	0
④	1	④	—	—	1	0
⑤	—	⑤	6	—	0	0
⑥	—	—	⑥	7	0	1
⑦	1	—	—	⑦	0	1

Merging of States in the Primitive Flow Table. An output signal, according to Fig. 3-14, is a Boolean function of input and feedback signals. It occurs often that with two different sequences of internal state (e.g., sequences of feedback signals), the output sequences are identical. If this is the case, one state sequence could be replaced by the other one without affecting the intended operation of the system in any way. States can thus sometimes be absorbed from the system by *merging* them (i.e., assigning the same labels to both of them), which means that the internal states in the two cases are put identical.

In this book we shall discuss only one method for the merging of states. It can be applied to sequential circuits in which we can use input signals for the formation of outputs. If the input signal values for merged states are different, we can still identify the original output sequences using input signals in connection with merged state variables. There is another method, called the *Huffman–Mealy simplification* (discussed in Phister [104]), where output signals are formed of the state variables (feedback signals) only. In this method, obviously only those states can be merged for which output signals are identical. There are less degrees of freedom for the design of output circuits in the Huffman–Mealy method, but, on the other hand, the design procedure is more systematic and the output circuit usually becomes simpler.

Merging Rule. Two rows in the primitive flow table can be merged (combined) if in corresponding columns, all *defined* state labels are equal.

For example, we can merge the fifth and sixth row in the example since in the column $A_1A_2 = 11$ we have 6 in both rows, and the "don't-care dashes" can be replaced by corresponding numbers in the other row.

Merger Diagram. We shall now set up a diagram in which all combinable pairs of states are indicated by interconnecting lines (Fig. 3-17). This graph is called a *merger diagram*. Notice that if three states form a triangle, they can be merged. In order to merge four states, they must have all six interconnections (like edges in a tetrahedron), etc. The output signals are indicated for every state.

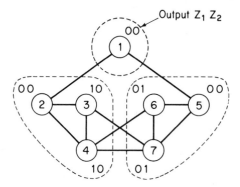

Fig. 3-17. Merger diagram.

There are several alternatives for combinable sets, and the simplest combination is usually found by inspection. In doing this we may take into account the fact that simplification of the output circuit can be achieved if states with the same output values are merged. However, in our example, the choice indicated with dashed lines seems to be a reasonable alternative which leads to three states. For this choice, output circuits must be designed separately.

Merged Flow Table. Now we write down a new flow table (Table 3-7) in which merged states have been denoted by labels of the stable states. For

Table 3-7. Merged Flow Table

Present State	A_1A_2			
	0 0	0 1	1 1	1 0
a	①	5	—	2
b	1	④	③	②
c	1	⑤	⑥	⑦

the merged states we use new symbols, $a, b,$ and $c,$ in the present-state column.

Secondary Assignment of States. With three rows, we obviously need two state variables or feedback signals. For the *secondary assignment* of proper value combinations for the labeled states, let us first inspect the allowed transitions between different states. For this purpose we draw a *transition diagram*, shown in Fig. 3-18 for this example, in which legal transitions are indicated with arrows. To avoid races in the transition table, we ought to strive for such an assignment of labels for the states that only one feedback variable would be changed at a time. One such alternative is shown in Fig. 3-18.

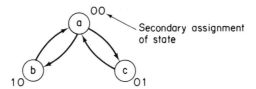

Fig. 3-18. Transition diagram.

If races in the transition diagram cannot be avoided, it is advisable to aim at such a secondary assignment that the races become noncritical. Also, a new merging scheme could be tried. If all alternatives fail, the last possibility to avoid erroneous operation is to use delay elements in the feedback paths to define the order of changes in signals.

Boolean Functions for the Feedback System. Once the secondary state assignment has been made, the main features of the excitation table have been defined, and it is a straightforward task to determine the simplest Boolean functions for it. Notice that in our example we have only three defined rows; the fourth row can be used for simplification. The defined parts of the excitation table for Table 3-7 are easily filled out, and undefined elements are equivalent to "don't cares" (Table 3-8). Notice that the order

Table 3-8. Excitation Table

	$X_1 X_2$	$A_1 A_2$ 0 0	0 1	1 1	1 0
①	0 0	⓪⓪	0 1	—	1 0
⑤	0 1	0 0	⓪①	⓪①	⓪①
—	1 1	—	—	—	—
②	1 0	0 0	①⓪	①⓪	①⓪

of rows has been changed. From Table 3-8, the Karnaugh maps and the Boolean functions, which always must be checked for hazards, are easily derived as follows:

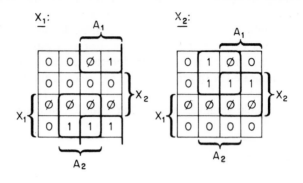

$$X_1 = A_2 X_1 + A_1 \bar{X}_2$$

$$X_2 = A_2 \bar{X}_1 + A_1 X_2$$

[3-14]

Output Equations. The outputs were orginally defined in the primitive flow table. For example, Z_1 was 1 in states ③ and ④, whereas after merging these states are represented by state a. However, in Table 3-7, the three stable states on the second row can still be identified by their original labels. Thus A_2 was 1 for states ③ and ④ and A_2 was 0 for state ②. Similarly, states ⑥ and ⑦ are identified by $A_1 = 1$. This would yield the *output equations*.

In fact, the output equations are normally written from *output tables*, which are obtained from the excitation tables by inserting defined output values for all stable states. There are also *unstable states* in the excitation table, through which the system makes transitions. Since these states last for a very short time, the output values in them usually do not matter. Usually "don't care's" can be put in those places, too, but there are cases when this is not possible. If in a transition between two stable states the output ought to remain constant, it is advisable to select those output values for unstable states that are identical with those in the corresponding stable states, to avoid hazard-type "spikes" in the output signals during transition. (See Table 3-9.)

For *undefined* elements, "don't cares" may be used without restrictions. Notice that the choice for the values of $\varnothing$ need not comply with the choice made in the Karnaugh maps of feedback variables.

Table 3-9 presents the output tables for our circuit example. (The zeros set in boldface type must be selected here in order to avoid signal "spikes" during transition, due to hazards.)

Table 3-9. Output Tables

Z_1: Z_2:

	A_1			
0	0	$\emptyset$	0	
0	0	0	0	X_2
$\emptyset$	$\emptyset$	$\emptyset$	$\emptyset$	
0	1	1	0	

X_1 spans rows; A_2 labels bottom.

	A_1			
0	0	$\emptyset$	0	
0	0	1	1	X_2
$\emptyset$	$\emptyset$	$\emptyset$	$\emptyset$	
0	0	0	0	

The simplified output equations read

$$Z_1 = A_2 X_1$$
$$Z_2 = A_1 X_2$$

[3-15]

Thus Eqs. [3-14] and [3-15] together define the wanted circuit, which is shown in Fig. 3-19. Notice that the circuit of Fig. 3-19 is only one of numerous alternatives and does not necessary constitute the absolutely simplest one, although it is probably one of the simplest.

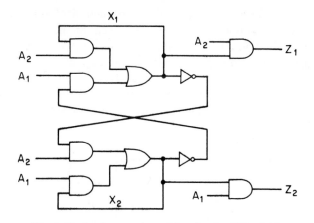

Fig. 3-19. Implementation of the circuit of Fig. 3-15.

3.5.   Clocked Flip-Flops

By the previously discussed design methods it is always possible to implement arbitrary asynchronous or synchronous sequential circuits using discrete logic gates. However, it is usually desirable to have ready-made *standardized building blocks which in themselves are elementary sequential circuits and by which arbitrary sequential circuits can be implemented.* These blocks can usually store one bit of information. The direct *R-S* flip-flop is one, although not a sufficiently general example. In this section we shall discuss *clocked flip-flops.* The *toggle,* which inverts its state upon receipt of one clocking pulse, is the simplest of these devices. However, having additional preparatory inputs as found in *D*, *R-S*, and *J-K* flip-flops as defined in Section 2-5 is a desirable feature in design because the transition to the next state of the flip-flop can be defined by logic signals applied at the preparatory inputs at the clocking instant.

Another desirable feature is that arbitrary state sequences, (i.e., sequences in which an arbitrary number of signals is changed simultaneously) can be implemented. For this to be possible, the values of preparatory inputs must depend on the previous state of the flip-flops. This seems to lead us to a logical paradox: The flip-flops must receive new information while emitting old information. All clocked flip-flops used for sequential circuits are able to handle this problem, which means that the preparatory inputs must be disabled when the clocking begins. A simple bistable latch does not have this property.

3.5.1. The Toggle

The toggle, as mentioned in the introduction of this section, shall invert its previous output value Q at every cycle of the clocking signal C (e.g., during an excursion $0 \rightarrow 1 \rightarrow 0$ or $1 \rightarrow 0 \rightarrow 1$ of it). The output signal usually changes at the trailing or negative-going edge of C, although in some circuits the change may occur at the leading edge. For our discussion we take the former case, and so the operation of the toggle is defined by the timing diagram of Fig. 3-20, in which, for simplicity, unstable states are not shown.

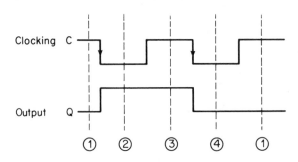

Fig. 3-20. Timing diagram of the toggle.

The synthesis of this circuit is done by standard methods. We shall see, however, that the primitive flow table shown in Table 3-10 cannot be merged at all. Therefore, the next task is the secondary assignment of states, which results, for example, in the excitation table of Table 3-11. From Table 3-11 the Karnaugh maps are derived as shown in Table 3-12. Notice that to

Table 3-10. Primitive Flow Table of the Toggle

	C		
	0	1	Q
①	2	①	0
②	②	3	1
③	4	③	1
④	④	1	0

Table 3-11. Excitation Table of the Toggle

	C	
$X_1 X_2$	0	1
0 0	01	⓪⓪
0 1	⓪①	1 1
1 1	1 0	①①
1 0	①⓪	00

$Q = X_2$

Table 3-12. Karnaugh Maps for the Toggle

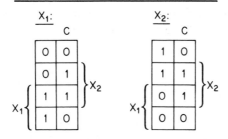

avoid hazards the simplest product terms must be bridged by third terms. From these diagrams we obtain the Boolean functions

$$X_1 = \bar{C}X_1 + CX_2 + X_1X_2$$

$$X_2 = \bar{C}\bar{X}_1 + CX_2 + \bar{X}_1X_2$$

[3-16]

Tle corresponding circuit implementation (transformed for NAND gates) is shown in Fig. 3-21.

Usually this flip-flop is not available as a building block because it can be obtained from, say, a more general D flip-flop by an external feedback.

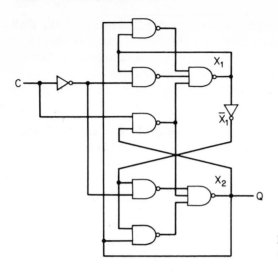

Fig. 3-21. Circuit implementation of the toggle.

3.5.2. D Flip-Flop

Systematic Design Approach. The operation of a D flip-flop in which the value of the D signal is transferred to the output Q and memorized at every trailing edge of the clocking signal C is completely defined by the timing diagram of Fig. 3-22. This diagram can be constructed by considering that

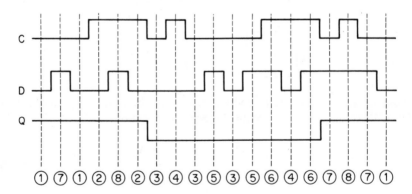

Fig. 3-22. Timing diagram of D flip-flop.

every stable state has two successor states due to the fact that either one of the input signals may change first. The synthesis is carried out in the conventional manner. The primitive flow table is set up first:

	CD				
	0 0	0 1	1 1	1 0	Q
①	①	7	—	2	1
②	3	—	8	②	1
③	③	5	—	4	0
④	3	—	6	④	0
⑤	3	⑤	6	—	0
⑥	—	7	⑥	4	0
⑦	1	⑦	8	—	1
⑧	—	7	⑧	2	1

There are many possibilities for the merging of states. Of these we select the one with four state pairs (1, 7), (2, 8), (3, 5), and (4, 6). For this choice, no combinational circuits are needed in the output. Table 3-13 is the merged flow table. The state transition diagram of Fig. 3-23 shows that

Table 3-13. Merged Flow Table of the *D* Flip-Flop

	CD				
	0 0	0 1	1 1	1 0	Q
a	①	⑦	8	2	1
b	3	7	⑧	②	1
c	③	⑤	6	4	0
d	3	7	⑥	④	0

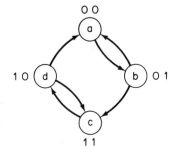

Fig. 3-23. Transition diagram of the *D* flip-flop.

reflected codes can be assigned to all four merged states. A secondary assignment of states leads to an excitation table (Table 3-14), from which the Karnaugh maps (Table 3-15) are directly written out.

Table 3-14. Excitation Table of the D Flip-Flop

	CD			
X_1X_2	0 0	0 1	1 1	1 0
0 0	0 0	0 0	0 1	0 1
0 1	1 1	0 0	0 1	0 1
1 1	1 1	1 1	1 0	1 0
1 0	1 1	0 0	1 0	1 0

Table 3-15. Karnaugh Maps for the D Flip-Flop

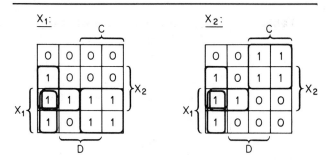

In the selection for the simplest set of product terms, the following equations are believed to belong to the simplest ones. A common term F is used.

$$X_1 = F + CX_1$$

$$X_2 = F + C\bar{X}_1$$

[3-17]

where

$$F = \bar{C}(\bar{D}X_1 + \bar{D}X_2 + X_1X_2)$$

However, there are hazards in this design. (Where?) How would you avoid them?

Commercial Version of the _D_ Flip-Flop. The following circuit, shown in Fig. 3-24, is a _D_ flip-flop which is found in many commercial integrated circuit families. It has the following advantages over the previous one:

1. It has fewer gates.
2. The output circuit is a direct _R-S_ flip-flop to which additional external control inputs (e.g., preset and preclear) can be connected.

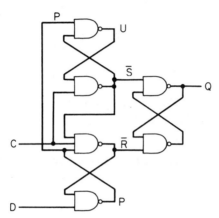

Fig. 3-24. _D_ flip-flop.

We can easily transform this circuit into a general form, in which it will be seen to have _three_ feedback paths, resulting from another choice for the merged flow table. However, it is very difficult to devise a design procedure by which we could end up with this circuit; it may be found after an extensive systematic study of all alternatives. This is also due to the fact that the simplest combinational circuit part is not a two-level circuit but an even simpler four-level circuit. Let us restrict ourselves to the analysis of the sequential behavior of this circuit.

We start by writing down all logic equations of the circuit. For the $\bar{S}$ and $\bar{R}$ variables we have

$$\bar{S} = \overline{CU}$$

$$\bar{R} = \overline{C\bar{S}P}$$

[3-18]

where

$$U = \overline{P\bar{S}}$$

$$P = \overline{\bar{R}D}$$

The direct R-S flip-flop part of this system is described by

$$Q = S + \bar{R}Q \qquad [3\text{-}19]$$

By substitutions, these equations are easily derived to an equivalent form:

$$S = D\bar{R}C + SC$$

$$R = \bar{D}\bar{S}C + R\bar{S}C \qquad [3\text{-}20]$$

$$Q = S + \bar{R}Q$$

from which we can see that the state variables, or the feedback variables, are S, R, and Q.

The excitation table for Eqs. [3-21] is shown in Table 3-16, in which the stable states have been encircled.

Table 3-16. Excitation Table for Eqs. [3-21]

| | CD | | | |
S R Q	0 0	0 1	1 1	1 0
0 0 0	⓪ⓞⓞ	⓪ⓞⓞ	1 0 0	0 1 0
0 0 1	⓪ⓞⓛ	⓪ⓞⓛ	1 0 1	0 1 1
0 1 1	0 0 0	0 0 0	0 1 0	0 1 0
0 1 0	0 0 0	0 0 0	⓪ⓛⓞ	⓪ⓛⓞ
1 1 0	0 0 1	0 0 1	1 0 1	1 0 1
1 1 1	0 0 1	0 0 1	1 0 1	1 0 1
1 0 1	0 0 1	0 0 1	ⓛⓞⓛ	ⓛⓞⓛ
1 0 0	0 0 1	0 0 1	1 0 1	1 0 1

3.5.3. Master–Slave Flip-Flops

Sequential circuits for the R-S and J-K flip-flops would be much more complicated than the previous examples, because the number of variables is increased. Therefore, another method for their implementation has been devised. This is the *master–slave* principle (Fig. 3-25), which is usually applied to J-K flip-flops. In this section we do not discuss R-S flip-flops, because their operation is included in the operation of J-K flip-flops. For special purposes, (e.g., low-power applications) special, simpler R-S flip-flops can be made using dynamic circuits.

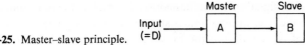

Fig. 3-25. Master–slave principle.

The master–slave flip-flop in its basic form is a kind of a *lock* (flood-gate). There are two memory cells, *A* and *B*, similar to the chambers of a lock, and we shall open the gates to these cells in a sequence; the input information is first copied into *A*, and the gate to *B* is closed. In a separate operation *A* is copied into *B*, and during this time the input gate of *A* is closed. For this purpose we need two clock phases, called C_1 and C_2; these must never be simultaneously 1.

The memory cell *A* is called the *master* and *B* the *slave*. Together the master and slave form a circuit called a *master–slave flip-flop*. It is usually the output of the slave that is used to indicate the stored information. The slave *B* is the proper bit storage and *A* is used as an intermediate storage. If the two clock phases C_1 and C_2 together were regarded as a single clocking event, the master–slave flip-flop would form a kind of *D* flip-flop. Such a system is described by Boolean equations,

$$A = DC_1 + \overline{D}C_1\ A$$

$$B = AC_2 + \overline{A}C_2\ B \tag{3-21}$$

Two Clock Phases Out of One Signal. Although it is quite possible to produce two clock phases at all places where they are needed and every time clocking is due, we would appreciate having only one primary clocking signal. The simplest method is to form C_2 out of C_1 by negation: if C_1 is normally kept at 0 and during the counting it makes an excursion $0 \rightarrow 1 \rightarrow 0$, then C_2 would normally be 1 and make an excursion $1 \rightarrow 0 \rightarrow 1$ during the clocking cycle. Thus in the normal condition the state of *A* is continuously copied into *B*, and the input gate of *A* is closed. When the clocking begins, the connection between *A* and *B* is broken and during $C_1 = 1$, new contents $(= D)$ are copied to *A*. When C_1 again returns to 0, the input gate is closed and its last contents are transferred to *B*. However, we have to point out that there are no ideal inverters; each one has an inherent delay, so the real C_1 and C_2 may overlap, as shown in Fig. 3-26. It is to be checked what happens

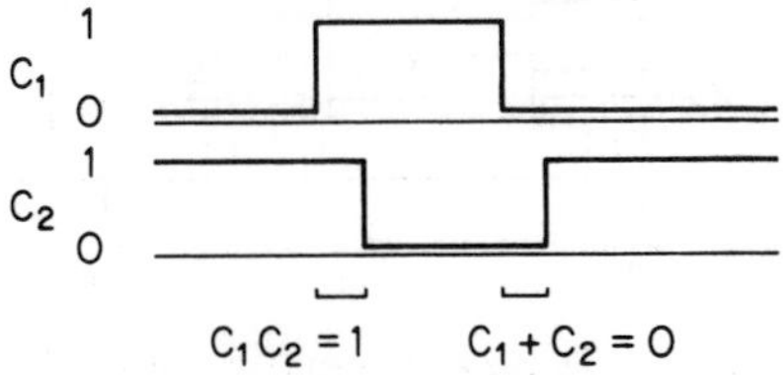

Fig. 3-26. Making C_2 out of C_1 might cause false signal combinations.

when both clocking signals are simultaneously 1, which is the most dangerous case. In practice the overlapping time is shorter than the signal delay through the logic gates of A and B, and the most critical case, which is closing the input gates of B after new contents are copied into A, is avoided by a special electronic design; in Fig. 3-27, which is the circuit implementation of Eqs. [3-22], the three shaded gates are in fact replaced by a transistor pair shown in Fig. 3-28. This pair acts as a usual logic-gate system but is much faster than the other gates.

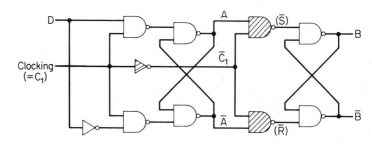

Fig. 3-27. Master–slave D flip-flop.

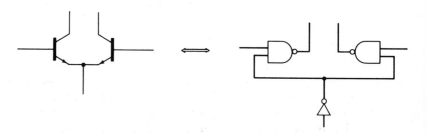

Fig. 3-28. Fast-gate part of Fig. 3-27.

Master–Slave J-K Flip-Flop. The master–slave combination is commonplace in J-K flip-flops. In order to toggle the flip-flop with $J = K = 1$, a sort of twisted feedback from the slave to the master is needed. We shall study the operation of J-K flip-flops with the aid of Fig. 3-29, where two clock

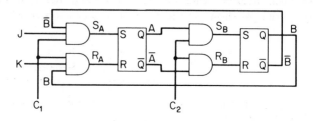

Fig. 3-29. Two-phase master–slave J-K flip-flop.

phases C_1 and C_2 are used. These two phases are usually made of the same signal, as discussed in the previous section.

Let us denote by $A^{(n)}$ the information in A before the application of the nth clocking signal C_1 and by $A^{(n+1)}$ the information at and immediately after C_1. Similarly, $B^{(n)}$ is the old information in B before the nth signal C_2 and $B^{(n+1)}$ is the new value at and after C_2.

From the circuit diagram we see that

$$S_A = J\bar{B}C_1$$

$$R_A = KBC_1$$

[3-22]

The direct R-S flip-flops A and B are assumed to be of the S-override type, where

$$A(t + \tau) = S_A(t) + \bar{R}_A(t)A(t)$$

[3-23]

After the substitution of S_A and R_A from Eq. [3-22] and observation that A will have the same value when $C_1 = 1$ and immediately afterward, we can write

$$A^{(n+1)} = J^{(n)}\bar{B}^{(n)} + \overline{K^{(n)}B^{(n)}}A^{(n)}$$

[3-24]

Taking explicit values for $J^{(n)}$ and $K^{(n)}$, we get the first transition table (for the master flip-flop), Table 3-17.

Table 3-17

$J^{(n)}$	$K^{(n)}$	$A^{(n+1)}$
0	0	$A^{(n)}$
0	1	$A^{(n)}\bar{B}^{(n)}$
1	0	$A^{(n)} + \bar{B}^{(n)}$
1	1	$\bar{B}^{(n)}$

The operation of the slave flip-flop is simpler:

$$S_B = AC_2$$

$$R_B = \bar{A}C_2$$

[3-25]

and thus the transition table for the respective values before and after C_2 is as shown in Table 3-18.

Table 3-18

$A^{(n)}$	$B^{(n)}$
0	0
1	1

After both clocking phases the final state transition in the master–slave system will be completed. Therefore, we obtain the new value $A^{(n+1)}$ by making a substitution from Table 3-18 to Table 3-17 and back (Table 3-19). This is indeed the state transition table of a *J-K* flip-flop.

Table 3-19

$J^{(n)}$	$K^{(n)}$	$A^{(n+1)}$	$B^{(n+1)}$
0	0	$A^{(n)}$	$B^{(n)}$
0	1	$A^{(n)}\bar{A}^{(n)} = 0$	0
1	0	$A^{(n)} + \bar{A}^{(n)} = 1$	1
1	1	$\bar{A}^{(n)}$	$\bar{B}^{(n)}$

3.6. Synchronous Sequential Circuits

State Graphs for Synchronous Sequential Circuits. The state of a logic circuit containing binary memory elements is defined as the combination of the states of these elements. We can also regard the set of all memory elements as a binary register called a *state counter*, the contents of which constitute the state. In synchronous sequential circuits all memory elements change their states simultaneously (synchronously) at discrete times called clocking, and the next state of the circuit, called the *successor state*, is a function of the present state as well as possibly of some external input signals. The state sequence of a synchronous sequential circuit is usually defined by a *state graph*, which comprises the primary information about the intended operation. Since the purpose of a synchronous sequential circuit is to produce a given state sequence, the associated output circuits do not play such a role as with asynchronous circuits and will be neglected in this discussion.

All different states are denoted in a state graph by separate circles which are labeled by the binary code representing the states. For brevity, the states may also be identified by decimal numbers which are in a one-to-one correspondence to the codes. If a state is a successor to another state, the corresponding transition at the clocking signal is denoted by an arrow pointing in the direction of the transition. A state may remain unaltered at the clocking and, since it is then a successor of itself, the arrow indicating the next transition points back to the circle from which it emanated. A state may also have several successor states, in which case, since the operation of the circuit will always be unambiguously defined, we must have a logic condition (e.g., a Boolean function of signals or a combination of Boolean functions which distinguishes one of these successor states). Every arrow is then labeled by the particular logic condition according to which the corresponding transition is selected.

Example 3-3. Two storage cells with the outputs Q_1 and Q_0 form the memory elements of a synchronous sequential circuit. The operation is defined by the following graph, in which the binary codes $Q_1 Q_0$ are used to label the four states. Two external input signals F and G are used to control the operation.

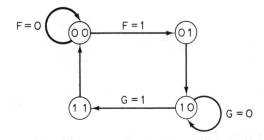

There are two unconditional transfers, $01 \rightarrow 10$ and $11 \rightarrow 00$, but the system halts in the states 10 and 00 until G and F, respectively, are made 1.

Implementation of Synchronous Sequential Circuits. The clocked flip-flops are suitable memory elements for synchronous sequential circuits since they change their states, if at all, only upon the occurrence of clocking pulses, which are common for all flip-flops in the system. The new states are defined by the logic values of the preparatory inputs, which are termed D, (S, R), or (J, K), according to the flip-flops. The preparatory input signals are taken from a combinational circuit to which the output signals of the flip-flops are fed back and which may receive external conditioning input signals, as visualized in Fig. 3-30.

A synchronous sequential circuit for any state graph can always be designed by straightforward methods. Another question is: *What is really the simplest state graph?* In practical problems we most often can assume the configuration of the state graph as known and the only question is whether there exists an optimal way for the labeling of the states.

The main phases of the logical design of synchronous sequential circuits are shown in Fig. 3-31. Let us now discuss each phase separately.

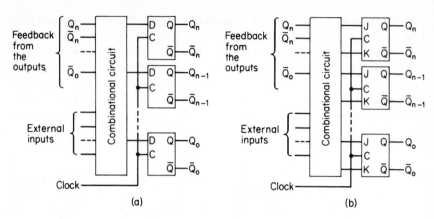

Fig. 3-30. Synchronous sequential circuits. (*R-S* flip-flops are used in the same way.)

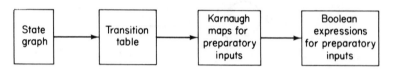

Fig. 3-31. Phases of the logical design of synchronous sequential circuits.

Transition Table. Assume that the state graph and the labeling are fixed. We have now to find the preparatory input signals for each flip-flop in Fig. 3-30, as the outputs of the combinational circuits. We start the design procedure by expressing the state graph in the form of an equivalent *transition table* in which the successor states for all present states and combinations of conditioning signals are shown (Table 3-20). The present state codes, written in the same order as reflected codes, are used to label rows of the

Table 3-20. Transition Table for Example 3-3

$Q_1^{(n)}Q_0^{(n)}$		$Q_1^{(n+1)}Q_0^{(n+1)}$						F	
0	0	0	0	0	0	0	1	0	1
0	1	1	0	1	0	1	0	1	0
1	1	0	0	0	0	0	0	0	0
1	0	1	0	1	1	1	1	1	0

G

map. If there had been forbidden states, rows provided in the present state
column would have bee regarded as "don't cares" and treated in the same
way as forbidden combinations in an incompletely defined Karnaugh map.
The columns are identified according to the conditioning signals.

Control Conditions for Preparatory Inputs. Let us recall the transition tables of D, R-S, and J-K flip-flops:

$D^{(n)}$	$Q^{(n+1)}$		$S^{(n)}$	$R^{(n)}$	$Q^{(n+1)}$		$J^{(n)}$	$K^{(n)}$	$Q^{(n+1)}$
0	0		0	0	$Q^{(n)}$		0	0	$Q^{(n)}$
1	1		0	1	0		0	1	0
			1	0	1		1	0	1
			1	1	undefined		1	1	$\bar{Q}^{(n)}$

These tables are now read in reverse order: We are searching for all entries
that produce a particular transition $Q^{(n)} \longrightarrow Q^{(n+1)}$. Obviously, $D^{(n)}$ depends
merely on $Q^{(n+1)}$ and is independent of $Q^{(n)}$. This is not the case with the
other two tables; assume, for example, that $Q^{(n)} = 0$ and $Q^{(n+1)} = 0$. The
latter value in R-S and J-K flip-flops can be obtained in two ways: (1) halting
the flip-flop by $S^{(n)} = R^{(n)} = 0$ or $J^{(n)} = K^{(n)} = 0$, respectively; (2) defining
the new state by $S^{(n)} = 0$, $R^{(n)} = 1$ or $J^{(n)} = 0$, $K^{(n)} = 1$, respectively. Thus
cases (1) and (2) combined read $S^{(n)} = 0$, $R^{(n)} = \varnothing$, or $J^{(n)} = 0$, $K^{(n)} = \varnothing$,
respectively. For a transition $Q^{(n)} = 0$, $Q^{(n+1)} = 1$ we have in R-S flip-flops
only one possible way of controlling: $S^{(n)} = 1$, $R^{(n)} = 0$. For J-K flip-flops
the said transition can be induced in two ways: (1) with $J^{(n)} = 1$,
$K^{(n)} = 0$ or (2) by toggling, $J^{(n)} = 1$, $K^{(n)} = 1$. Cases (1) and (2) combined
now read $J^{(n)} = 1$, $K^{(n)} = \varnothing$.

The possible values of preparatory inputs are now expressed as what
we call *control conditions*, for all combinations $(Q^{(n)}, Q^{(n+1)})$; these are
calculated once and for all for the various flip-flops (Table 3-21).

Table 3-21. Control Conditions

$Q^{(n)}$	$Q^{(n+1)}$	$D^{(n)}$	$S^{(n)}$	$R^{(n)}$	$J^{(n)}$	$K^{(n)}$
0	0	0	0	$\varnothing$	0	$\varnothing$
0	1	1	1	0	1	$\varnothing$
1	0	0	0	1	$\varnothing$	1
1	1	1	$\varnothing$	0	$\varnothing$	0

Karnaugh Maps. The next task is to scan all elements and all variables
Q_i of the transition table to find the corresponding transitions $Q_i^{(n)} \longrightarrow Q_i^{(n+1)}$.

According to the control conditions expressed in Table 3-21, we can now find the corresponding values of preparatory signals $D_i^{(n)}$, $(S_i^{(n)}, R_i^{(n)})$, or $(J_i^{(n)}, K_i^{(n)})$ for which Karnaugh maps are now written as illustrated in the following example.

Example 3-4. Let us start with the transition table, Table 3-20.

$$Q_1^{(n+1)} Q_0^{(n+1)}$$

$Q_1^{(n)} Q_0^{(n)}$	F		G	
0 0	0 0	0 0	0 1	0 1
0 1	1 0	1 0	1 0	1 0
1 1	0 0	0 0	0 0	0 0
1 0	1 0	1 1	1 1	1 0

Using control conditions we can now find the Karnaugh maps for the preparatory inputs $D_1^{(n)}$, $(S_1^{(n)}, R_1^{(n)})$, or $(J_1^{(n)}, K_1^{(n)})$. This procedure is repeated for $D_0^{(n)}$, $(S_0^{(n)}, R_0^{(n)})$, or $(J_0^{(n)}, K_0^{(n)})$.

FOR D FLIP-FLOPS

$D_1^{(n)}$:

$Q_1^{(n)} Q_0^{(n)}$	F		G	
0 0	0	0	0	0
0 1	1	1	1	1
1 1	0	0	0	0
1 0	1	1	1	1

$D_0^{(n)}$:

$Q_1^{(n)} Q_0^{(n)}$	F		G	
0 0	0	0	1	1
0 1	0	0	0	0
1 1	0	0	0	0
1 0	0	1	1	0

From these maps we obtain the algebraic expressions easily:

$$D_1^{(n)} = \bar{Q}_1^{(n)} Q_0^{(n)} + Q_1^{(n)} \bar{Q}_0^{(n)}$$

$$D_0^{(n)} = \bar{Q}_1^{(n)} \bar{Q}_0^{(n)} F + Q_1^{(n)} \bar{Q}_0^{(n)} G$$

FOR *R-S* FLIP-FLOPS

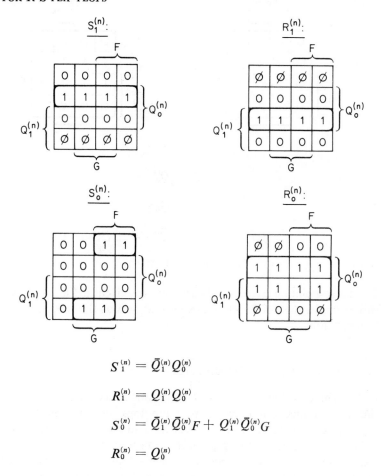

$$S_1^{(n)} = \bar{Q}_1^{(n)} Q_0^{(n)}$$

$$R_1^{(n)} = Q_1^{(n)} Q_0^{(n)}$$

$$S_0^{(n)} = \bar{Q}_1^{(n)} \bar{Q}_0^{(n)} F + Q_1^{(n)} \bar{Q}_0^{(n)} G$$

$$R_0^{(n)} = Q_0^{(n)}$$

FOR *J-K* FLIP-FLOPS

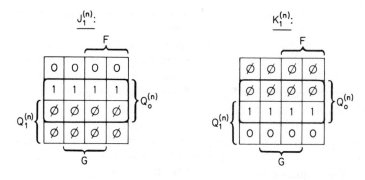

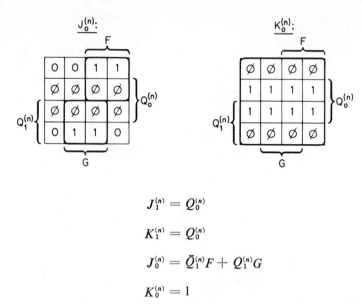

$$J_1^{(n)} = Q_0^{(n)}$$

$$K_1^{(n)} = Q_0^{(n)}$$

$$J_0^{(n)} = \bar{Q}_1^{(n)}F + Q_1^{(n)}G$$

$$K_0^{(n)} = 1$$

The corresponding circuit configurations are given in Fig. 3-32. At least in this example, the *J-K* flip-flop implementation was the simplest. This is true in most cases.

3.7.   Linear Sequential Circuits

There is a special form of synchronous sequential circuits in which the logic connections between most flip-flops are particularly simple: the flip-flops are connected as a shift register, and the new information, which is fed into the first or last stage, is a Boolean function of the present states as well as possibly of some external signals. Figure 3-33 shows the general arrangement with a right-shifting register. Such a system is also called a *linear sequential circuit.*

Linear sequential circuits can be used as special counters. Modified versions with external inputs can be used as error-detection-and-correction devices in digital communication systems, and as code converters.

It can be shown that for any length n of a shift register without external inputs there exists a corresponding combinational circuit such that the register periodically makes all transitions through $2^n - 1$ different states. The starting state can naturally be any of these.

If we regard the contents of the shift register as a binary number and the register is right shifting (i.e., new information enters at the most significant end), the numerical contents C of the register have a rather random-looking sequence (see Table 3-22). Based on this fact, this type of reg-

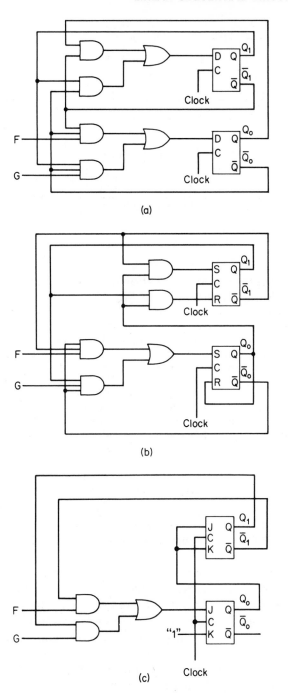

Fig. 3-32. Circuit implementations for Example 3-4.

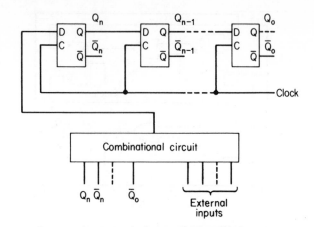

Fig. 3-33. Linear sequential circuit.

Table 3-22. State Sequence of the Four-Bit Linear Sequential Circuit of Fig. 3-28

Shift No.	Q_3	Q_2	Q_1	Q_0	C^*	Shift No.	Q_3	Q_2	Q_1	Q_0	C^*
1	0	0	0	1	1	9	0	1	0	1	5
2	1	0	0	0	8	10	1	0	1	0	10
3	0	1	0	0	4	11	1	1	0	1	13
4	0	0	1	0	2	12	1	1	1	0	14
5	1	0	0	1	9	13	1	1	1	1	15
6	1	1	0	0	12	14	0	1	1	1	7
7	0	1	1	0	6	15	0	0	1	1	3
8	1	0	1	1	11	(16	0	0	0	1	1)
								. . .			

$^*C$, is the decimal equivalent of the binary number $Q_2 Q_2 Q_1 Q_0$.

ister is used extensively as a *pseudo random number generator* for the production of artificial noise. Such circuits are also called *maximum-length sequence generators*.

As a representative of these, we show a 4-bit linear sequential circuit. The necessary Boolean expressions can usually be implemented with EXCLUSIVE OR circuits. Figure 3-34 is the diagram and Table 3-22 the state sequence of this circuit. Each state occurs only once in a period of $2^4 - 1 = 15$ states, and the state 0000 is excluded. The proof is straightforward; the reader may form the value of the input function F for every value combination of the register to obtain the new state of the input flip-flop (Q_3).

Table 3-23 shows feedback connections for some maximum-length

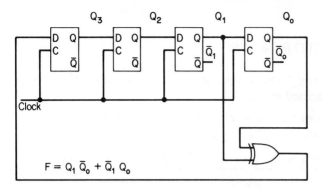

Fig. 3-34. Four-bit maximum-length sequence generator.

sequence generators which require only one EXCLUSIVE OR gate, connected between the outputs of two stages in the same way as in Fig. 3-34. For register lengths $(n + 1 < 31)$ not occurring in this table, a more complicated combinational circuit is needed.[1] Table 3-23 gives possible solutions but not all of them.

Table 3-23. The Two Stages Q_i and Q_j in a Linear Sequential
Circuit between Which an EXCLUSIVE OR Gate
Should Be Connected to Produce the Necessary
Feedback for a Maximum-Length Sequence
(Input Stage Q_n)

$n + 1$*	i	j	$n + 1$*	i	j
2	1	0	17	14	0
3	2	0	18	11	0
4†	3	0	20	17	0
5	3	0	21	19	0
6	5	0	22	21	0
7	6	0	23	18	0
9	5	0	25	22	0
10	7	0	28	25	0
11	9	0	29	27	0
15	14	0	31	28	0

* Number of stages.
† Alternative for Fig. 3-34.

[1] Feedback connections from more than two bits are discussed by Gill [41]. Notice that in most textbooks on linear sequential circuits, stages are numbered from Q_1 to Q_{n+1} and Q_1 is the input stage.

PROBLEMS

3-1 Using truth tables, explain the operation of the toggle flip-flops depicted in Fig. 3-4.

3-2 Draw the waveform for the Boolean function $F = (A \cdot T + F) \cdot \overline{A \cdot T}$ when A and T are as given by Fig. P3-2.

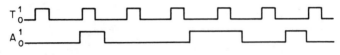

Figure P3-2

3-3 Find Boolean functions describing a circuit with output Q to which one of the inputs A, B, or C is copied and stored when one and only one of the corresponding control signals T_A, T_B, and T_C is 1, respectively. (*Hint*: Use the idea of the bistable latch given in Fig. 3-11.)

3-4 (*a*) Draw the waveforms representing Boolean functions E, F, and G defined as follows when S and R are as given by Fig. P3-4:

$$E = S \cdot \bar{R}$$

$$F = (S + F) \cdot \bar{R}$$

$$G = \bar{R} \cdot G + S$$

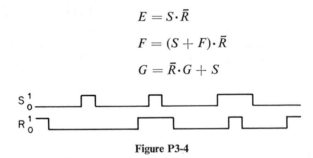

Figure P3-4

(*b*) What are the possible combinations of output variables E, F, and G? How many state variables are there in the system?

(*c*) Design a circuit using NAND gates to implement the system shown.

3-5 Tabulate $X(t + \tau)$ of the circuit of Fig. P3-5 for all values of $X(t)$, $A(t)$, and $B(t)$, where τ is larger than the propagation delay of signals.

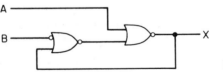

Figure P3-5

3-6   Draw all transitions in the excitation table of Fig. P3-6.

X	Y	Z	A	
0	0	0	011	001
0	0	1	000	011
0	1	1	011	010
0	1	0	011	010
1	1	0	101	000
1	1	1	111	101
1	0	1	101	100
1	0	0	101	111

Figure P3-6

3-7   For what values of C and D are there hazards in the circuit shown in Fig. P3-7 when A or B is changing from one state to another?

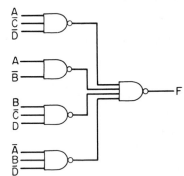

Figure P3-7

3-8   Point out the hazard existing in the asynchronous sequential circuit described by the equation

$$X = \bar{A}B + \bar{B}X$$

and how can it be eliminated.

3-9   Design an asynchronous sequential circuit which gives an indication if one of two signals totally occurs during the other. Discuss a sequence depicted in Fig. P3-9.

Figure P3-9

3-10 (a) Construct the simplest J-K flip-flop of a D flip-flop using a suitable combinational circuit at its input.

(b) Repeat the problem using an R-S flip-flop and a suitable combinational circuit at its inputs.

3-11 The state graph of a synchronous sequential circuit consisting of J-K flip-flops is given in Fig. P3-11. Find

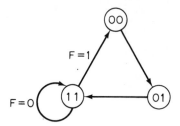

Figure P3-11

(a) the transition table and the Karnaugh maps for J-K flip-flops.
(b) the corresponding Boolean functions.
(c) the circuit implementation.

3-12 Repeat Problem 3-11 for the synchronous sequential circuit described by Fig. P3-12.

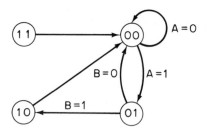

Figure P3-12

3-13 By means of a transition map and a state graph, explain how the circuit shown in Fig. P3-13 (a) works. Also draw the waveforms of X and Y for both values of A when the clocking signal CP is given by Fig. P3-13 (b).

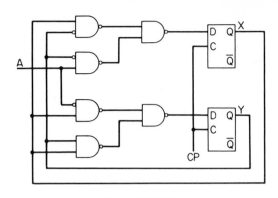

Figure P3-13 (a)

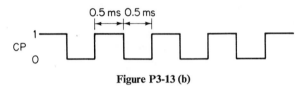

Figure P3-13 (b)

3-14 Write down the successive states of a four-bit maximum-length sequence generator, defined in Table 3-23, starting with the state 0001.

REFERENCES

In addition to the books of Maley and Earle, McCluskey, and Lewin mentioned at the end of Chapter 2, and of Wickes, we should like to mention

    Phister, *Logical Design of Digital Computers* [104],

in which sequential circuits are discussed from an analytical point of view. A more formal approach, representing a theory of mathematical machines can be found in

    Hartmanis and Stearns, *Algebraic Structure Theory of
    Sequential Machines* [56].

The area of sequential circuits has been in a state of intense study, and numerous articles can be found in the most recent issues of *IEEE Transactions on Computers*, as mentioned in the Bibliography.

chapter 4

Computing Circuits

4.1. General

Digital computing involves several rules for the manipulation of symbols. Actually every elementary phase in computing is a kind of table lookup operation in which the operands, or the sets of signals representing symbols, occur as entries and tabulated values constitute the desired output information. For the electronic implementation of tables there are usually various possibilities; logic circuits are one form and digital information stored in addressed memory locations is another. Tables can also be implemented by computing programs (algorithms). There is a certain amount of tradeoff between programmed and wired computing operations. Rules for some operations (e.g., searching for a digit that results from the addition of two digits) are so simple that the output signals are directly formed from the input signals by a combinational logic circuit. Most arithmetic operations, however, are implemented in many steps, in which the output signals of one step are used as the input signals for the next step. This feature was found in synchronous sequential circuits.

Counters are the simplest arithmetic circuits which change the stored information by increments of the least representable unit. For example, the addition of two numbers by a counter would require the accumulation of a corresponding number of counts.

Addition and subtraction are the basic arithmetic operations on which all other computations are based, and in most computers there are circuits that directly form the sum and the difference of two numbers. For *multiplication* and *division*, as well as for such *special functions* as the trigonometric and exponential functions, either wired arithmetic circuits can be designed or these functions can be calculated by stored programs.

129

If the speed of computing is not critical, many special functions can be computed digitally with the aid of special computing systems, called *binary rate multipliers* (BRM's) and *digital differential analyzers* (DDA's). These are versatile units in real-time systems such as on-line coordinate resolvers, and operational units are available as integrated circuits. Such computing is called *incremental*.

The purpose of this chapter is to provide solutions for general digital design problems and to aid in the understanding of digital arithmetic circuits. Special problems, such as the design of optimal arithmetic circuits for general-purpose digital computers, fall outside the scope of this book.

4.2. Representation of Integers in a Modulus

In a base R the largest integer that is representable with n digits is $R^n - 1$, or, using the same symbols for digits as in Chapter 1, the representation reads

$$d_{R-1}d_{R-1}\cdots d_{R-1} \qquad (n \text{ digits}) \qquad [4\text{-}1]$$

For example, 11...1 and 99...9 (n digits) are the largest integers that can be represented with n binary and decimal digits, respectively. If, however, the assignment of representations is continued for numbers larger than $R^n - 1$, there is no place for the most significant digits, and it is said that they *overflow*. The representation left in the register is called the *least positive residue* and it is the same for all numbers which differ from it by an integral multiple of R^n. The number R^n is here called the *modulus* of representation.

An integer A is said to be *congruent* to another integer B modulo m, or $A \equiv B \bmod m$ if $A = B + km$ where k is an integer (positive or negative). For example, all the following congruence relations hold:

$$31 \equiv 21 \bmod 10 \qquad \text{or} \qquad 21 \equiv 31 \bmod 10$$

$$31 \equiv 11 \bmod 10 \qquad \text{or} \qquad 11 \equiv 31 \bmod 10$$

$$31 \equiv 1 \bmod 10 \qquad \text{or} \qquad 1 \equiv 31 \bmod 10$$

In the last congruence, 1 is the least positive residue of 31 modulo 10.

Some properties of congruences are utilized when computing with positive and negative numbers, as well as with numbers that exceed the register capacity. These are based on the following rules, which are mentioned without proof: Given two congruences

$$A_1 \equiv B_1 \bmod m$$

$$A_2 \equiv B_2 \bmod m$$

we have, for example,

$$A_1 + A_2 \equiv (B_1 + B_2) \bmod m$$

$$A_1 - A_2 \equiv (B_1 - B_2) \bmod m$$

$$A_1 \cdot A_2 \equiv B_1 \cdot B_2 \bmod m$$

4.3. Counters

Counters are simply systems which can store and indicate the number of accumulated *counts* or input pulses. Usually we define an input pulse as an excursion $0 \to 1 \to 0$ or $1 \to 0 \to 1$ of a logic signal, whereby the duration of the intermediate value shall not be pertinent. A count is thus a single event. It is customary to discuss the two basic types of counters as separate families: (1) parallel counters and (2) ripple counters.

There are *binary counters* in which the stored numbers of counts are indicated as a binary number, and there are *counters in a base R* in which the digits are represented as groups of bits; the input pulses are led to the least significant digit position, the digit codes are sequenced in an ascending or descending sequence, and a carry or borrow digit is propagated to the adjacent digit position, thus acting as a counting input for this digit, etc.

If the counter has m distinguishable states through which it cyclically makes all transitions, it is called a *modulo m counter*.

4.3.1. Parallel Counters

General Parallel Counters. By a parallel counter we mean a synchronous sequential circuit that does not contain external input signals other than the clocking inputs, and the graph of which forms a ring. The parallel counter thus counts the number of clock pulses.

Generally, if we have a set of m different state symbols (with n flip-flops, $m \le 2^n$) which are unconditionally connected to a ring in an arbitrary order, we have the graph of a general modulo m parallel counter. There is an example of one such counter in Fig. 4-1 — a modulo 3 counter.

Binary number codes can be assigned to the state symbols in ascending order, in which case we have an *up counter* (forward counter). The codes can also be assigned in descending order, in which case we get a *down counter* (backward counter).

Since parallel counters are synchronous sequential circuits, their design, commencing with the completely defined state graph, follows the same lines as the general design procedure of a synchronous sequential circuit. Since there are no external preparatory input signals, the transition map can be represented more compactly in a two-dimensional form in the same way as Karnaugh maps. This is exemplified below.

Example 4-1. Design a synchronous modulo 3 parallel counter implementing the state sequence given in Fig. 4-1. The two memory cells X_1 and X_0 consist of J-K flip-flops. First we draw the transition map, which for this simple example is shown in the conventional form (a) and in the two-dimensional form (b) in Table 4-1.

State symbol: X_1X_0

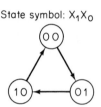

Fig. 4-1. State graph of a modulo 3 counter.

Table 4-1. Transition Maps for a Modulo 3 Counter

$X_1\ X_0$		
0 0	0	1
0 1	1	0
1 1	$\varnothing$	$\varnothing$
1 0	0	0

(a)

	X_0	
X_1	0	1
0	0 1	1 0
1	0 0	$\varnothing$ $\varnothing$

(b)

Next we have the control maps for J_1, K_1, J_0, and K_0 (Table 4-2), in which the unused state of Table 4-1 has been denoted by x for easy reviewing. The circuit is implemented in Fig. 4-2.

Table 4-2

J_1:	X_0		K_1:	X_0		J_0:	X_0		K_0:	X_0	
	0	1		$\varnothing$	$\varnothing$		1	$\varnothing$		$\varnothing$	1
X_1	$\varnothing$	x	X_1	1	x	X_1	0	x	X_1	$\varnothing$	x

$$J_1 = X_0$$
$$J_0 = \overline{X}_1$$
$$K_1 = K_0 = 1$$

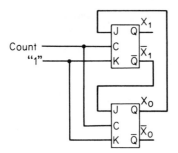

Fig. 4-2. Modulo 3 parallel counter.

In this case, as in every design, we must check what happens if the counter for some reason gets to a forbidden state (i.e., 11 in this example). This might be the case after switching the supply voltages on, or the circuit may get there as the result of a disturbance. In order that the counter should not be trapped in the forbidden state, there must be an unconditional transition to some of the allowed states at the next count. If $X_0 = X_1 = 1$, we have $J_0 = 0$, $K_0 = 1$, $J_1 = K_1 = 1$, and the successor state will be 00.

The counter should be reset before use. The resetting can be done with the aid of external control signals such as those shown in the next section. Some flip-flops have separate resetting inputs.

Up/Down Counters. Since the number-code sequence in parallel counters can be made arbitrary, counting up (forward) and down (backward) has no difference in principle; for example, in Fig. 4-1 we could have reversed the arrows and performed another design procedure for this case. However, there are applications in which the same counter is made to count up or down depending on particular logic conditions. Therefore, two-way arrows with different logic conditions in the two directions must be assigned to the graph. If the counter must halt with a third logic condition, return arrows for each state must be added.

Example 4-2. Design a modulo 4 counter $(X_1 X_0)$ which counts up when $U = 1$, $D = 0$ and down when $U = 0$, $D = 1$. With $U = 0$, $D = 0$ the counter shall halt, and the combination $U = 1$, $D = 1$ is forbidden. First we draw the state graph (Fig. 4-3), and the implementation (Tables 4-3 and 4-4) is straightforward:

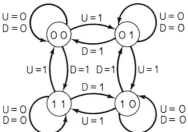

Figure 4-3

Table 4-3. Transition map for Example 4-2

X_1 X_0				U			
0 0	0 0	1 1	$\varnothing$	$\varnothing$	0 1
0 1	0 1	0 0	$\varnothing$	$\varnothing$	1 0
1 1	1 1	1 0	$\varnothing$	$\varnothing$	0 0
1 0	1 0	0 1	$\varnothing$	$\varnothing$	1 1

D

Table 4-4

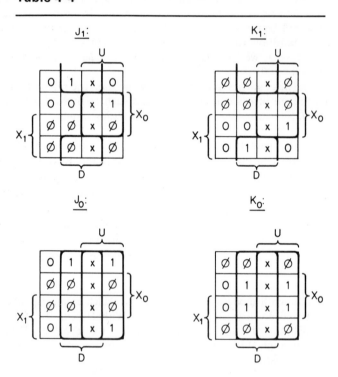

$$J_1 = K_1 = UX_0 + D\bar{X}_0$$

$$J_0 = K_0 = U + D$$

Ring Counters and Johnson Counters. An end-around shift register with n bits can act as a modulo n counter, called a *ring counter*, if it advances through n distinguishable states. The shift register is a synchronous sequential circuit, and so the ring counter is a parallel counter.

Every linear sequential circuit without external inputs is a parallel counter, too. However, the *Johnson counter*, often called the *twisted ring counter* or the *switch-tail ring counter*, also belongs to the class of linear sequential circuits. It is made from an end-around shift register by copying the negation of the last stage to the first stage, whereby a sequence of $2n$ different states can be obtained. A modulo 10 Johnson counter is shown in Fig. 4-4, and its counting sequence is shown in Table 4-5.

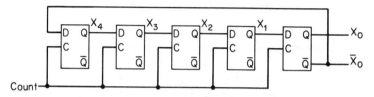

Fig. 4-4. Modulo 10 Johnson counter.

Table 4-5. State Sequence of a Modulo 10 Johnson Counter

Counts	X_4	X_3	X_2	X_1	X_0
—	0	0	0	0	0
1	1	0	0	0	0
2	1	1	0	0	0
3	1	1	1	0	0
4	1	1	1	1	0
5	1	1	1	1	1
6	0	1	1	1	1
7	0	0	1	1	1
8	0	0	0	1	1
9	0	0	0	0	1
10	0	0	0	0	0

Comment. The state decoders of a Johnson counter are two-input AND gates, because any legitimate state can be recognized by two particular bits selected properly for each decoder. For example, after seven counts we have $X_3 = 0$, $X_2 = 1$, and the decoder of state 7 is $P_7 = \bar{X}_3 X_2$.

Modified Johnson Counters. In order to implement a modulo $2n - 1$ counter with the Johnson counter principle, it is necessary to bypass one state of the modulo $2n$ sequence. This state is, for example, the one full of 1's which can be anticipated by decoding the previous state, $111 \cdots 110$, and by controlling the input of the shift register so that the next state will be $011 \cdots 111$. The auxiliary feedback circuit is shown in Fig. 4-5.

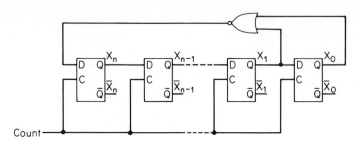

Fig. 4-5. Modified Johnson counter.

Self-Correcting Johnson Counters. A drawback of many modulo m parallel counters ($m < 2^n$) is that if they are set to nonlegitimate states due to noise or during the time the power supply is being turned on, the system may be caught in these states or oscillate between them. In other words, the nonlegitimate states may form an isolated graph. The state graphs of general synchronous counters can always be defined in such a way that all nonlegitimate states have an unconditional transition to a legitimate state. This is called *self-correcting*. Special methods will be used to introduce the self-correcting feature.

We give, without proof, a few specific examples of self-correcting Johnson counters.

For modulo 2n Johnson counters the J input of the first stage is connected to the $\bar{Q}$ output of the last stage, whereas the K input of the first stage is obtained from an AND circuit. There are p inputs to this circuit, where p is the next larger integer to $n/3$. The inputs are taken from the p last Q outputs of the counter. A self-correcting modulo 10 counter is shown in Fig. 4-6. Note that the gate does not change a correct counting sequence,

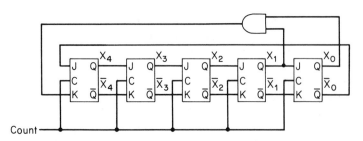

Fig. 4-6. Self-correcting modulo 10 Johnson counter.

but if there is a nonlegitimate state in the counter, it will gradually change to a legitimate state in a few steps. (See [181].)

Some *self-correcting modulo 2n — 1 Johnson counters* can be implemented by J-K flip-flops without extra gates. Examples are the modulo

5, modulo 7, and modulo 9 counters shown in Fig. 4-7 with their respective state sequences. (See the References for this chapter.)

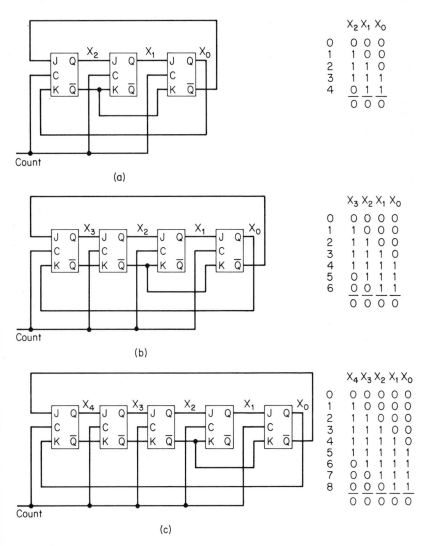

Fig. 4-7. Some modulo $2n - 1$ self-correcting Johnson counters: (a) modulo 5; (b) modulo 7; (c) modulo 9.

4.3.2. Ripple Counters

With parallel counters, any counting sequence can be implemented. There must be special reasons to make use of other methods. One reason might be the simplicity of circuits, another the avoiding of many parallel clocking inputs, which would cause a heavy loading on the counting pulses. *Ripple counters* usually have advantages in both these respects. Ripple counters, however, have also drawbacks: one of them is that the flip-flops do not change their states simultaneously. This can lead to hazards as discussed later in this section.

In a ripple counter the counting pulses are usually brought to the clocking input of the least significant bit position, the output of the corresponding flip-flop is used as a further clocking signal for the next position, and so on. The clocking signals of other stages may also be obtained as Boolean functions of the outputs, but a common feature of all ripple counters as compared to parallel counters is that the flip-flops change their states one at a time in a sequence ("ripple down"). In a long register there might be a substantial signal propagation delay through all stages, and false total states of the register may be recognized during the propagation of information. This is why ripple counters are seldom utilized in control circuits. Because of their simplicity, they may be used in frequency counters, for example. The simplest ripple counter is the binary counter discussed in Chapter 3 (see Fig. 3-15).

Design Procedure for True Ripple Counters. True ripple counters are made of toggle flip-flops without preparatory inputs. Let us assume that all flip-flops change their states with the transition $1 \to 0$ of the clocking signal, and it is not pertinent at which time this signal is returned back to the value 1. Here we discuss the design procedure within the frame of a particular example.

Example 4-3. We have to design a 8421-coded modulo 10 counter (up decade counter) using only clocking inputs to toggle the flip-flops. The most important task is obviously to determine at which time each of the flip-flops has to halt and when to toggle. This is seen either from a timing

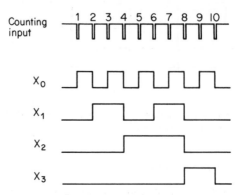

Fig. 4-8. Timing diagram of a 8421 up decade counter.

diagram (Fig. 4-8) or from a state sequence. The input T_i of a flip-flop X_i before any change in X_i must be 1 and after the change it must be 0. In the meantime, before the next change in X_i, the value of T_i is returned to 1, which can be done at any time. We start by writing the obligatory values 1 and 0 for T_i's to the sequence table of the X_i's (Table 4-6), which at the same time serves as an incompletely defined truth table. All vacant places are filled by a kind of "don't-care" symbol y_i ($y = a, b, c, \ldots$ and $i = 1,$ $2, 3, \ldots$). A different letter is used in each succession of symbols.

Table 4-6. Sequence Table

Count	X_3	X_2	X_1	X_0	T_3	T_2	T_1
0	0	0	0	0	0	b_2	d_2
1	0	0	0	1	a_1	b_3	1
2	0	0	1	0	a_2	b_4	0
3	0	0	1	1	a_3	1	1
4	0	1	0	0	a_4	0	0
5	0	1	0	1	a_5	c_1	1
6	0	1	1	0	a_6	c_2	0
7	0	1	1	1	1	1	1
8	1	0	0	0	0	0	0
9	1	0	0	1	1	b_1	d_1
10	0	0	0	0			
			. . .				

The "don't cares," however, are not independent of each other. In a succession of one kind of letter y_i, all of them may be 0's or 1's, or the first ones must be 0's and the last ones 1's. For example, if a_1 to a_i are 0, then a_{i+1} to a_6 must be 1, for arbitrary $i = 1$ to 5.

The simplest choice for the "don't cares" is determined by the method of incompletely defined Karnaugh maps of T_3, T_2, and T_1 (Table 4-7) (T_0 is the counting input). The interdependency of "don't cares" stated before must be taken into account when terms are combined. Notice that there are also forbidden states, denoted by $\varnothing$, in the maps.

Thus three gates are needed (Fig. 4-9). It should be mentioned that a parallel up decade counter also needs three gates if J-K flip-flops are used. It would seem that $T_3 = X_2 + X_1 + X_0$ is an even simpler choice. A check at the fourth count shows, however, that X_2 is changing from 0 to 1, whereas X_1 and X_0 are changing from 1 to 0. Since X_2 is changing later than X_1 and X_0, there would be a hazard in the sense that for a very short time T_3 makes an excursion $1 \rightarrow 0 \rightarrow 1$, possibly causing a false triggering of X_3. The hazard should be avoided by an additional delay of, say, X_2, which could be accomplished by taking $\bar{X}_2$ through an inverter. Then, again, we would have as many gates as in the choice given above, which has no hazards.

Table 4-7

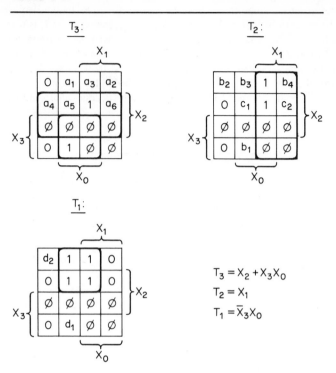

$$T_3 = X_2 + X_3 X_0$$
$$T_2 = X_1$$
$$T_1 = \overline{X}_3 X_0$$

Because of the danger of hazards, the design of ripple counters needs special care, and it is advisable to use them mainly if optimization of hardware in integrated assemblies is pertinent.

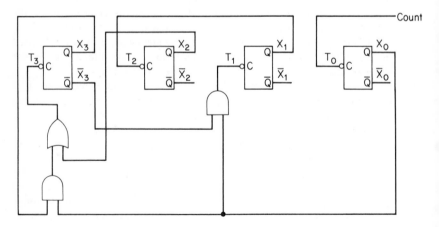

Fig. 4-9. True 8421 ripple up decade counter.

Ripple Down and Ripple Up/Down Counters. The previous procedure can be easily extended for the design of down and up/down counters. In a down counting operation, the state sequence table is read in the reverse order; therefore, the 1's and 0's in the T_i's must be interchanged, as compared to corresponding up counting. The "don't-care" fillers x_i are also numbered in reverse order. An up/down counter has a truth table for the T_i's which is a combination of truth tables of the up and down counters, conditioned by up/down control signals.

Mixed Counters. Very often neither parallel nor serial counters are the simplest ones. As an example, let us take a 8421-coded up decade counter that can be implemented with four J-K flip-flops and only one additional gate. The proof of the correct operation can be left as an exercise; let it suffice to give the circuit diagram in Fig. 4-10, from which we can conclude

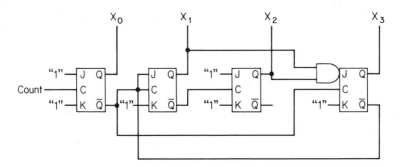

Fig. 4-10. Mixed (nonsynchronous) 8421 up decade counter.

that in some state transitions the flip-flops are directly toggled by a ripple counter method, whereas between some states the X_1 and X_3 flip-flops are also controlled by J and K inputs.

4.4. Digital Arithmetic

4.4.1. Representation of Signed Numbers

Representation of Negative Integers in Counting Registers. Let us consider a four-digit decimal register used to accumulate single counts. When a 1 is added to 9999 we obtain 0000 as the least positive residue of 10,000 modulo 10,000. If we *subtract* a 1 from 0000 and if there is no way to show the sign, the most natural selection for the assignment of a representation for -1 would be 9999. Thus we can count up and down starting from 0000. To keep the positive and negative numbers separated, we require that the largest representable positive integer is, for example, 4999 and the representation 5000 thus belongs to the number -5000.

Representation of Signed Binary Numbers in Computers.[1] Scientific and engineering computing systems are usually based on binary numbers, because in a binary number system, the arithmetic rules are simplest to implement with logic circuits. It is a natural requirement that positive as well as negative numbers must be operated by the same circuits. In the following, the sign of the number is always represented by a *sign bit* called x_0 (0 for $+$, 1 for $-$). The leftmost storage cell in a register is reserved for the sign bit. The rest of the register gives the *magnitude* (absolute value) of the number, but there are various possibilities: The representation of magnitudes may be the same for positive as well as negative numbers, in which case we have the *signed magnitude representation*, or the magnitude of negative numbers may be given as some kind of complement of the corresponding positive value. For binary numbers there are two basic types of complement representations: the *1's-complement representation* and the *2's-complement representation*. Each has advantages and drawbacks, and all of them are found in contemporary computers.

The radix point (binary point) is usually not shown in the machine; instead, a convention is made about its position. The computing is usually done with either *integers* or *fractions*. In the first case the radix point must be imagined after the last bit; in the second case, between the sign bit and the magnitude part. Since arithmetic rules are identical regardless of whether the words are interpreted as integers or fractions (in other words, the same hardware is utilized), we use only fractional numbers in the following discussion.

Let us denote the binary contents of a register

$$x_0.x_1x_2 \cdots x_N \qquad [4\text{-}2]$$

The leftmost bit (x_0) is the sign bit. This word will now be interpreted in different ways in different representations. In the above representation the x_i's $(i = 1$ to $N)$ are to be regarded as Boolean variables. If x_0 is regarded as a *digit* with a value 2^0, the *apparent numerical value* of the binary word would be

$$V = \sum_{i=0}^{N} x_i 2^{-i} \qquad [4\text{-}3]$$

---

[1] All representations of numbers discussed in this book are called *fixed-point representations*. Numbers in scientific computations are usually represented as *floating-point numbers* (i.e., in the form $X = a \cdot 2^b$ where a is the *fractional part* ($\frac{1}{2} \le a < 1$) and b the *exponent*). See, for example, the books by Flores [35] and Chu [18].

The *true numerical value* X of a number represented by V is *in the signed magnitude representation,*

$$X = (-1)^{x_0} \sum_{i=1}^{N} x_i 2^{-i} \qquad [4\text{-}4]$$

One's-Complement Representation. In the 1's-complement representation, a positive number is identical with the signed magnitude representation, whereas the representation for negative numbers is obtained by taking 1 for x_0 and *inverting all bits of the magnitude part.* A reason for using any complement representation is that the sign bit x_0 is operated in arithmetic circuits in the same way as the number bits. The inverse operation is to find the numerical value X of a representation; this is

$$X = \sum_{i=1}^{N} x_i 2^{-i} \qquad \text{for positive numbers } (x_0 = 0)$$

$$\qquad [4\text{-}5]$$

$$X = -x_0 + \sum_{i=1}^{N} x_i 2^{-i} + 2^{-N} \qquad \text{for negative numbers } (x_0 = 1)$$

Example 4-4. The 1's-complement representation of $X = -0.0001$ is, by the previous rule, 1.1110, and the numerical value X is obtained from this expression as

$$X = -1 + 0.1110 + 0.0001 = -0.0001$$

Two's-Complement Representation. In the 2's complement representation, a positive number is also identical with the representation in signed magnitude, whereas the representation of negative numbers is obtained by taking 1 for x_0 and $1 - X$ for the numerical part. Another rule is to form first the 1's complement of X and then increment the apparent numerical value by 2^{-N}, i.e., by one unit of the least significant bit. From the definition it follows that the numerical value X is obtained from its representation as

$$X = -x_0 + \sum_{i=1}^{N} x_i 2^{-i} \qquad [4\text{-}6]$$

which is true for positive as well as for negative numbers.

Example 4-5. The 2's-complement representation of $X = -0.0001$ by the application of the previous rule is equal to 1's complement $(= 1.1110)$ incremented by 0.0001, which makes 1.1111; the inverse operation, finding the true numerical value X, is

$$X = -1 + 0.1111 = -0.0001$$

Comparison of the Three Signed Representations. All numbers X which can be represented by a register are shown on the circumference of a circle in Fig. 4-11. In the three signed representations (signed magnitude, 1's complement, and 2's complement) the numbers are mapped in different ways on this circle. Positive numbers are identical in all three representations.

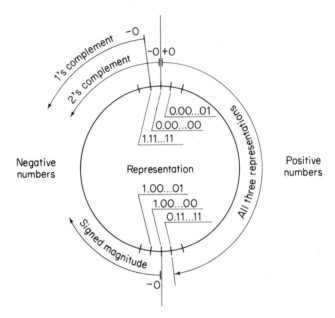

Fig. 4-11. Representations of signed numbers.

For negative numbers the zero points and the directions of growth are different. It should be noted that for the signed magnitude representation we have two zero points, "plus zero" $= 0.000 \ldots 0$ and "minus zero" $= 1.000 \ldots 0$; also in 1's-complement representation we have a "plus zero" $= 0.000 \ldots 0$ and a "minus zero" $= 1.111 \ldots 1$, whereas in 2's-complement representation, the only zero is $0.000 \ldots 0$.

4.4.2. Arithmetic Algorithms for Signed Binary Numbers

The purpose of this section is to describe some representative algorithms for arithmetic operations on signed binary numbers. If two numbers are expressed in a particular representation, it would be natural to require that the arithmetic algorithm should yield the result in the same representation. Some algorithms indeed satisfy this requirement, whereas simple corrections are allowed in others. In the choice of an algorithm, a compromise between the simplicity of representations and the needed correction must be made.

Addition and Subtraction Tables for Binary Digits. Addition and subtraction in an arbitrary number system with a base R can be performed by the "pencil-and-paper method" in a fashion similar to the familiar decimal operations. In addition, the augend, the addend, and the carry digits of the operands are handled in the normal way. A missing carry is indicated with a digit 0. In subtraction, the minuend, the subtrahend, and the borrow are indicated by their respective digits. In the following we are dealing with positive fractions, and we assume, for the moment, that $X \geq Y$. Assume also that there is no overflow in the addition.

Addition:	Augend	$X = .x_1 x_2 \ldots x_N$
	Addend	$Y = .y_1 y_2 \ldots y_N$
	Carry	$C = .c_1 c_2 \ldots c_N$
	Sum	$S = .s_1 s_2 \ldots s_N$

Subtraction:	Minuend	$X = .x_1 x_2 \ldots x_N$
$(X \geq Y)$	Subtrahend	$Y = .y_1 y_2 \ldots y_N$
	Borrow	$B = .b_1 b_2 \ldots b_N$
	Difference	$D = .d_1 d_2 \ldots d_N$

The following relations define the addition and subtraction of digits:

$$s_i = s_i(x_i, y_i, c_i) \qquad i = 1 \text{ to } N, \ c_N = b_N = 0$$

$$c_{i-1} = c_{i-1}(x_i, y_i, c_i)$$

$$d_i = d_i(x_i, y_i, b_i) \tag{4-7}$$

$$b_{i-1} = b_{i-1}(x_i, y_i, b_i)$$

These relations are usually defined in terms of *addition and subtraction tables* (Table 4-8). In the *binary number system* these tables are particularly simple and can be interpreted as truth tables (or Karnaugh maps) for the Boolean variables s_i, c_{i-1}, d_i, and b_{i-1}.

A couple of remarks should be made here. The tables for the sum and the difference are identical with $c_i = b_i$, and they define the *modulo 2 sum* of x_i, y_i, and c_i (i.e., s_i and d_i are the least positive residues mod 2 of the true sum of respective digits). Also, it may be noted that the table of b_{i-1} is obtained from the table of c_{i-1} by using $\bar{x}_i$ instead of x_i in the entry, with $b_i = c_i$.

Table 4-8

Addition Tables

Sum digit (s_i)			Carry digit (c_{i-1})		
	c_i			c_i	
x_i y_i	0	1	x_i y_i	0	1
0 0	0	1	0 0	0	0
0 1	1	0	0 1	0	1
1 1	0	1	1 1	1	1
1 0	1	0	1 0	0	1

Subtraction Tables

Difference digit (d_i)			Borrow digit (b_{i-1})		
	b_i			b_i	
x_i y_i	0	1	x_i y_i	0	1
0 0	0	1	0 0	0	1
0 1	1	0	0 1	1	1
1 1	0	1	1 1	0	1
1 0	1	0	1 0	0	0

We shall see also that in the binary number system Eqs. [4-7] can be stated as Boolean expressions.

Addition and Subtraction in Two's-Complement Representation. As the first example of arithmetic algorithms for signed numbers we discuss the 2's-complement representation, in which additions and subtractions with arbitrary operands are particularly simple:

> *The representations, including the sign bits, are directly added or subtracted, whereby the result is correctly obtained in 2's-complement representation.*

A simple proof for this algorithm can be given without algebraic considerations if we imagine that the original contents of registers are obtained by counting up (for positive numbers) or down (for negative numbers) by increments of 2^{-N} (the least significant position), starting from zero. (The true electronic operation is done by other methods.) For example, one count down from 0.00 ... 00 yields 1.11 ... 11, which is the correct representation of $-0.00 ... 01$. The addition of a positive number and subtraction of a negative number are equivalent to counting up from these values by an amount given, respectively, by the magnitude of the addend or the subtrahend. Similarly, subtraction of a positive number and addition of a negative number are equivalent to counting down by the corresponding amount. As with counters, no correction is needed when passing the value zero during counting.

A more formal proof may be given by simultaneously considering operations on true numbers and on representations. Let us denote the numerical part of the representation of X by

$$X^* = \sum_{i=1}^{N} x_i 2^{-i} \qquad\qquad [4\text{-}8]$$

with similar starred notations for other variables. For true values

$$X = -x_0 + X^*$$

$$Y = -y_0 + Y^*$$

$$S = X + Y = -(x_0 + y_0) + (X^* + Y^*)$$
$$= -s_0 + S^*$$

$$D = X - Y = -(x_0 - y_0) + (X^* - Y^*)$$
$$= -d_0 + D^*$$

always hold, whereas for representations we have

$$(x_0 + X^*) + (y_0 + Y^*) = (x_0 + y_0) + (X^* + Y^*)$$
$$= s_0' + S^*$$

$$(x_0 + X^*) - (y_0 + Y^*) = (x_0 - y_0) + (X^* - Y^*)$$
$$= d_0' + D^*$$

Thus we see that S^* and D^* are always given correctly in 2's-complement representation, whereas a separate discussion is needed for the sign bits. In order to avoid overflow, we usually require $|X|, |Y| < 0.1 (= 2^{-1})$. In the addition of X^* and Y^*, a carry may be propagated to s_0'. Similarly, d_0' may be changed by a borrow coming from the number bits. If X and Y now have the same signs and $|X| + |Y| < 1$, there will be *no* carry in the addition of positive numbers but *always* a carry for negative numbers. Thus s_0', being the modulo 2 sum of x_0, y_0, and the carry c_0, has the correct value s_0. If X and Y have different signs, there will be a carry only if the negative operand has a smaller magnitude. For example, for $X > 0$, $Y < 0$, we have $Y^* =$

$1 - |Y|$, $X^* = |X|$. If now $|X| > |Y|$, then $X^* < 1 - Y^*$, and $X^* + Y^* > 1$. Thus s_0' becomes 0, as expected. With no carry $s_0' = 1$ and thus $s_0' = s_0$.

Similar considerations for the difference show that $d_0' = d_0$. Note that in modulo 2 subtraction, $0 - 1 \equiv 1 \bmod 2$.

Addition in One's-Complement Representation. The sum of two positive numbers in 1's-complement representation is obviously represented correctly as long as the sum does not exceed the number capacity. The same is not generally true if negative numbers are included.

For the addition of two numbers, the following simple rule can be established:

> *The representations of the augend and the addend, including the sign bits, are added directly as positive numbers. Whenever a carry is obtained after the addition of sign bits, it is transferred as an end-around carry to the least significant position. (In an extreme case, the carry may once more be propagated through the partial sum.)*

Example 4-6. The following addition is done by the given algorithm:

Augend	$X =$	0.1101
Addend	$Y =$	1.0110
Carry	$C =$	(1)1.1000
Partial sum		0.0011
End-around carry		1
Sum	$S =$	0.0100

We omit the proof of this and the next algorithm by stating that the end-around correction is needed because the zero point of negative numbers is shifted by 2^{-N} with respect to 2's-complement representation, and this difference must be taken into account when passing zero.

Subtraction in One's-Complement Representation. For arbitrary operands the subtraction algorithm in 1's-complement representation reads as follows:

> *The representation of the subtrahend is subtracted from the representation of the minuend, whereby sign bits are treated as usual digits. Whenever a borrow is obtained after the subtraction of the sign bits, an end-around borrow is transferred to the least significant position. (In an extreme case, the borrow may once more be propagated through the digits of the partial difference.)*

Example 4-7. The following subtraction is performed in 1's-complement representation:

Minuend	$X =$	1.1001
Subtrahend	$Y =$	1.1110
Borrow	$B =$	(1)1.1100

Partial difference		1.1011
End-around borrow		1

Difference	$D =$	1.1010

Addition in Signed Magnitude Representation. If X and Y have the
same sign, the digits of the magnitude are added, and as far as the capacity
of the numerical part is not exceeded, the result is correct if provided with the
same sign as x_0. If X and Y have different signs, the representation of $|Y|$
is subtracted from the representation of $|X|$ without prior examination of
the signs. If a borrow $b_0 = 0$ is obtained from the most significant number
bits, we know that $|Y| < |X|$ and the result has the same sign as X. If $b_0 =$
1, we find that $|Y| > |X|$, and the result has the opposite sign as X. In the
former case the number digits are obtained correctly, whereas in the latter
case, the numerical part of the result is the 2's complement of the correct
value. The result must be complemented in a separate operation to produce
a correct result.

We see that the addition algorithm for signed magnitude representation is rather complicated; it requires examination of signs and borrow digits
and may imply complementation of the result.

Example 4-8. The following operations are performed following previously discussed rules:

(a) $X = 1.1100$ $X < 0$ $|X| > |Y|$

        $Y = 0.1010$ $Y > 0$

.1100 − .1010 = .0010 and $b_0 = 0$. The result is 1.0010.

(b) $X = 0.1100$ $X > 0$ $|X| < |Y|$

        $Y = 1.1110$ $Y < 0$

.1100 − .1110 = .1110, where a borrow, $b_0 = 1$, is needed. The 2's complement of number digits is .0010. The result is 1.0010.

Subtraction in Signed Magnitude Representation. Since the subtraction of a number is equivalent to addition of the negative of this number, it is simplest in signed magnitude representation to change the sign of the subtrahend and perform an addition according to previous rules.

Multiplication in Signed Magnitude Representation. It is obvious that the multiplication of signed numbers is simplest in signed magnitude representation:

$$X = (-1)^{x_0} X^*$$

$$Y = (-1)^{y_0} Y^* \qquad [4\text{-}9]$$

$$X \cdot Y = (-1)^{x_0+y_0} X^* \cdot Y^*$$

The product is thus positive if the operands have the same signs, whereas the product is negative if the signs are different. In the multiplication of X^* and Y^* we obtain $2N$ number digits, and therefore a double-length register for the accumulation of products is usually necessary.

The organization of an arithmetic unit will be discussed later. Let it be mentioned here that in multiplication, as well as in division, three registers are needed; the double-length register is formed of two parts according to Fig. 4-12. The left portion is called the *accumulator;* numbers can be added

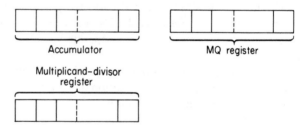

Accumulator   MQ register

Multiplicand–divisor
register

Figure 4-12

to it and subtracted from it. The right portion is a register which can be shifted right or left. This register is called the *MQ (multiplier–quotient) register.* In multiplication, the accumulator and the MQ register are used as a single *double-length product register* which during right-shifting operations behaves like a single register. The multiplier may be initially stored in the MQ register, which we shall discuss later. A third register, called the *multiplicand–divisor register,* is also needed.

The simplest multiplication algorithm for $X^* \cdot Y^*$ is long multiplication, which is, in fact, similar to the familiar pencil-and-paper method. Since

$$X^* \cdot Y^* = 2^{-1}(x_1 Y^*) + 2^{-2}(x_2 Y^*) + \cdots + 2^{-N}(x_N Y^*)$$
$$= 2^{-1}(x_1 Y^* + 2^{-1}(x_2 Y^* + \cdots + 2^{-1}(x_N Y^*) \cdots)) \qquad [4\text{-}10]$$

we can formulate this algorithm in a suitable form for automatic computations: Start with the least significant bit of the multiplier. Add $x_N Y^* = Y^*$ or 0 to the accumulator and shift this number once to the right, whereby $2^{-1}(x_N Y^*)$ is formed. If $x_N = 0$, no addition would be needed, but a shift is implemented anyway, in order that this step of computation is formally similar to the ones that follow. Continue with the next bit by adding $x_{N-1} Y^*$ to the previous contents of the accumulator and shift this number once to the right. Now $2^{-1}(x_{N-1} Y^* + 2^{-1}(x_N Y^*))$ is formed. When all bits of the multiplier have been used in this way, we have $X \cdot Y^*$ in the accumulator and the MQ register.

Example 4-9. The following numbers are to be multiplied:

$$Y = 0.0110 \qquad \text{multiplicand}$$

$$X = 0.1011 \qquad \text{multiplier}$$

The phases of this computation are explained below; the accumulator and the MQ register are shown together as a product register. The multiplier bits, initially stored in the MQ register, are underlined.

i	x_i	Accumulation of the partial products	
4	1	0. 0 0 0 0 0 <u>1 0 1 1</u>	Clear product register
		+0. 0 1 1 0	Add　$x_4 Y^*$
		0. 0 1 1 0 0 <u>1 0 1 1</u>	
			Shift
		0. 0 0 1 1 0 0 <u>1 0 1</u>	
3	1	+0. 0 1 1 0	Add　$x_3 Y^*$
		0. 1 0 0 1 0 0 <u>1 0 1</u>	
			Shift
		0. 0 1 0 0 0 1 0 <u>0 1 0</u>	
2	0	+0. 0 0 0	Add　$x_2 Y^*$
		0. 0 1 0 0 0 1 <u>0 1 0</u>	
			Shift
		0. 0 0 1 0 0 0 1 <u>0 1</u>	
1	1	+0. 0 1 1 0	Add　$x_1 Y^*$
		0. 1 0 0 0 0 0 1 <u>0 1</u>	
			Shift
		0. 0 1 0 0 0 0 0 1 <u>0</u>	
		$= X \cdot Y^*$	

Multiplication in Two's-Complement Representation (Booth's Algorithm). The following method is not essentially shorter than the previous one but it uses operands which have been expressed in 2's complement representation. The sign bits are operated as number digits without prior examination. This method, called *Booth's algorithm*, can be derived in the following way: Using the identity $2^{-i+1} - 2^{-i} = 2^{-i}$, we have

$$X \cdot Y = (-x_0 + \sum_{i=1}^{N} x_i 2^{-i})Y$$

$$= [-x_0 2^0 + x_1(2^0 - 2^{-1}) + \cdots + x_N(2^{-N+1} - 2^{-N})]Y$$

$$= [(x_1 - x_0)2^0 + (x_2 - x_1)2^{-1} + \cdots + (x_{N+1} - x_N)2^{-N}]Y$$

$$= [(x_1 - x_0) + 2^{-1}((x_2 - x_1) + 2^{-1}((x_3 - x_2) + \cdots$$

$$+ 2^{-1}(x_{N+1} - x_N) \cdots))]Y \qquad \text{where } x_{N+1} = 0 \qquad [4\text{-}11]$$

In Eq. [4-11] we have to multiply 2's-complement numbers by 2^{-1}. For positive numbers there is no difference between the three signed representations, and so multiplication by 2^{-1} is equivalent to shifting right once (including the sign bit since it is 0). An extra difficulty arises when multiplying negative numbers by 2^{-1} in the 2's-complement representation; after the operation, the sign bit will be 1, and the number bits are found in the following way:

Denote $Z = -1 + \sum_{i=1}^{N} z_i 2^{-i}$, which is a negative number. Now

$$2^{-1}Z = -2^{-1} + \sum_{i=1}^{N} z_i 2^{-(i+1)} = -1 + \sum_{i=1}^{N} z_i' 2^{-i} \qquad [4\text{-}12]$$

where

$$z_1' = 1$$

$$z_{i+1}' = z_i \qquad \text{for } i = 1 \text{ to } N - 1 \qquad [4\text{-}13]$$

According to Eq. [4-12], the z_j' are number bits of $Z' = 2^{-1}Z$ in 2's-complement representation, and Eqs. [4-13] define the new bits. Thus we can express the *arithmetic right shifting algorithm in 2's-complement representation:*

The word, including the sign bit, is shifted right in the normal way, but the new sign bit is the same as the old one.

The accumulation of partial products resembles the corresponding operation in signed magnitude representation but differs from it in two respects:

1. Since Y is multiplied by $(x_{i+1} - x_i)$ instead of x_i, a comparison of two adjacent bits of the multiplier is needed at the ith step. If $x_i = 0$, $x_{i+1} = 1$, Y is *added* to the previous contents of the accumulator. If $x_i = 1$, $x_{i+1} = 0$, Y is *subtracted* from the previous contents of the accumulator. If $x_{i+1} = x_i$, *no operation* is performed.

2. After the previous step, the contents of the product register, including the sign bit, are shifted once to the right using the arithmetic shifting algorithm given before; after the last arithmetic operation no shifting is done.

Comment. It is not generally known that Booth's algorithm yields a false result if $Y = 1.00 \ldots 00 \, (= -1)$. Thus this value of Y must be excluded.

Example 4-10. As an example of the application of Booth's algorithm, consider the multiplication of X and Y,

$$Y = 1.0110 \qquad \text{multiplicand}$$

$$X = 0.1101(0) \qquad \text{multiplier}$$

Accumulation of the
partial products

i	x_i	x_{i+1}		
4	1	0	0. 0 0 0 0 0 1 1 0 1 0	
			− 1. 0 1 1 0	Subtract Y
			0. 1 0 1 0 0 1 1 0 1 0	
				Shift
3	0	1	0. 0 1 0 1 0 0 1 1 0 1	
			+ 1. 0 1 1 0	Add Y
			1. 1 0 1 1 0 0 1 1 0 1	
				Shift
2	1	0	1. 1 1 0 1 1 0 0 1 1 0	
			− 1. 0 1 1 0	Subtract Y
			0. 0 1 1 1 1 0 0 1 1 0	
				Shift
1	1	1	0. 0 0 1 1 1 1 0 0 1 1	
			0. 0 0 0 0	No operation
			0. 0 0 1 1 1 1 0 0 1 1	
				Shift
0	0	1	0. 0 0 0 1 1 1 1 0 0 1	
			+ 1. 0 1 1 0	Add Y
			1. 0 1 1 1 1 1 1 0 0 1	
			$= X \cdot Y$	

Division in Signed Magnitude Representation. Sign rules in division are identical with corresponding rules in multiplication. Division is simplest in the signed magnitude representation because the main procedure is always done with magnitudes that are positive numbers; only this case will be discussed here. For simplicity, the operands are scaled so that the divisor is larger than the dividend. The main phases of this procedure are the following:

1. Transfer the dividend into the accumulator and the MQ register. Shift once to the left.
2. Subtract the divisor from the accumulator. *If the partial remainder is negative*, restore the original contents of the accumulator by adding the divisor. The most significant bit of the quotient (q_1) is in this case 0. *If the partial remainder is positive or zero*, no restoring is performed, and the most significant bit of the quotient (q_1) is 1.
3. Shift the contents of the accumulator and the MQ register once to the left, and record the most significant bit of the quotient (q_1) to the right end of the MQ register.
4. Continue by subtracting the divisor from the accumulator; etc.

Although this algorithm is easily recognized to be equivalent to the familiar pencil-and-paper method, a formal proof will be given in the following. Denote

$$X = \text{dividend}$$
$$Y = \text{divisor} \ (Y \text{ is taken} > X)$$
$$Q = \text{quotient}$$
$$r_i = i\text{th partial remainder} \ (r_0 = X \text{ by definition})$$
$$R = \text{remainder of the division}$$

Denote the ith bit of the quotient by q_i. The division is based on the familiar cut-and-try comparison of the partial remainder r_i and the product $q_{i+1}Y$. According to the previous discussion, the new partial remainder r_{i+1} is obtained from the recursive formula

$$r_{i+1} = 2r_i - q_{i+1}Y \qquad [4\text{-}14]$$

where $q_{i+1} = 0$ if $2r_i < Y$, and $q_{i+1} = 1$ if $2r_i \geq Y$. The algorithm is initiated with $i = 0$, whereby $r_0 = X$. Multiplying both sides by $2^{-(i+1)}$ we get

$$r_{i+1}2^{-(i+1)} = r_i 2^{-i} - q_{i+1}2^{-(i+1)}Y \qquad [4\text{-}15]$$

By a repeated application of this formula from $i = 0$ to $N - 1$ we obtain

$$r_N 2^{-N} = r_0 2^0 - \sum_{i=0}^{N-1} q_{i+1}2^{-(i+1)}Y$$

where $r_0 2^0 = X$, and hence (with $i + 1 = j$)

$$\frac{X}{Y} = \sum_{j=1}^{N} q_j 2^{-j} + \frac{2^{-N}r_N}{Y}$$

$$= Q + \frac{R}{Y} \qquad [4\text{-}16]$$

by which R is defined as

$$R = 2^{-N} r_N \qquad\qquad [4\text{-}17]$$

This completes the proof.

Example 4-11. Division of

$$X = 0.10100011 \qquad \text{dividend}$$

$$Y = 0.1011 \qquad \text{divisor}$$

is shown below. The digits in parentheses have the weight $2^0 = 1$, and they are temporarily buffered in the sign position of the accumulator. The bits shifting into the MQ register from the right represent the quotient, and they are underlined.

	Accumulator	MQ register	
	. 1 0 1 0	0 0 1 1	$r_0 = X$ in accumulator
	(1) . 0 1 0 0	0 1 1 0	Shift left (multiply X by 2)
$-$	. 1 0 1 1		Subtract Y
	(0) . 1 0 0 1	0 1 1 0	$Y < 2r_0,\;\; q_1 = 1$
	(1) . 0 0 1 0	1 1 0 <u>1</u>	Shift left, q_1 enters MQ register
$-$	. 1 0 1 1		Subtract Y
	(0) . 0 1 1 1	1 1 0 <u>1</u>	$Y < 2r_1,\;\; q_2 = 1$
	(0) . 1 1 1 1	1 0 <u>1 1</u>	Shift left, q_2 enters MQ register
$-$	. 1 0 1 1		Subtract Y
	(0) . 0 1 0 0	1 0 <u>1 1</u>	$Y < 2r_2,\;\; q_3 = 1$
	(0) . 1 0 0 1	0 <u>1 1 1</u>	Shift left, q_3 enters MQ register
$-$	. 1 0 1 1		Subtract Y
Overflow due	(1) . 0 1 1 0	0 <u>1 1 1</u>	$Y > 2r_3,\;\; q_4 = 0$
to borrow $+$	. 1 0 1 1		Restore r_4
	(0) . 1 0 0 1	0 <u>1 1 1</u>	r_4 in accumulator
	(1) . 0 0 1 0	<u>1 1 1 0</u>	Shift left, q_4 enters MQ register, end of computation; remainder is left in accumulator

$$Q = 0.1110$$

$$R = 2^{-4} r_4 = 0.00010010$$

4.4.3. Addition and Subtraction Circuits

Most arithmetic operations are based on the addition and subtraction of individual digits, usually starting with the least significant digit position, whereby carry or borrow digits are propagated to next positions. In this book we shall discuss two types of addition and subtraction, called *serial* and *parallel operation*, respectively. In a serial addition or subtraction there is a combinational circuit which operates on one digit position at a time. The corresponding digits of the two operands are shifted (e.g., using shift registers) to the inputs of this circuit in succession, whereby the carry or borrow digit formed of this position is temporarily stored and utilized when the next digit position is due. In parallel operations there are separate addition and subtraction circuits for each digit position, and the operands as well as the result are represented by parallel logic signals. Each circuit accepts the carry or borrow signal from the adjacent digit. However, as in the series circuit, an output digit in the parallel circuit cannot attain its final value until the carry or borrow information from all lower digit positions has been propagated up to this position. A parallel arithmetic operation is faster than a corresponding operation performed in series. Because of the necessary propagation of carry and borrow in both cases, however, the difference in speed is not large; in many special computing systems, material savings obtained by serial operations may be decisive, whereas in general-purpose computers, in which high speed is most important, parallel circuits are preferred. In the largest computers there are often arithmetic circuits which operate on several digits at a time. The carry and borrow conditions are studied simultaneously over the digit groups, and the delays then are significantly reduced.

Full Adder and Half Adder. In binary arithmetic the first task is to implement the addition and subtraction tables. Being Boolean functions of input variables, the sum and carry outputs are defined, for example, by the following expressions:

$$s_i = x_i \bar{y}_i \bar{c}_i + \bar{x}_i y_i \bar{c}_i + \bar{x}_i \bar{y}_i c_i + x_i y_i c_i$$

$$\bar{c}_{i-1} = x_i \bar{y}_i \bar{c}_i + \bar{x}_i y_i \bar{c}_i + \bar{x}_i \bar{y}_i$$

[4-18]

This algebraic form has the advantage that there are two common terms in s_i and $\bar{c}_{i-1}$.

A circuit implementing Eqs. [4-18] or their equivalent is called a *full adder* (FA). More commonly a full adder is made of operational blocks called a *half adder* (HA). The half adder is a circuit that has only two inputs, x and y, and produces two output signals, s and c. This circuit is a combinational circuit, defined by Table 4-9.

Table 4-9. Half Adder

x	y	s	c
0	0	0	0
0	1	1	0
1	0	1	0
1	1	0	1

$$s = \bar{x}y + x\bar{y} = \overline{xy + \bar{x}\bar{y}}$$

$$c = xy$$

[4-19]

Using NAND gates and two inverters, the half adder may be implemented as in Fig. 4-13. If there is no carry in the addition, a half adder can be used; such cases occur, for example, in the addition of the least significant digits, and also in all positions if the addend is $.00\ldots01$ (up counting).

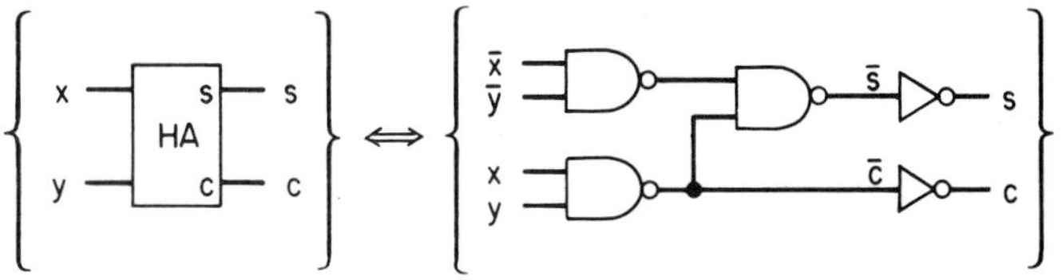

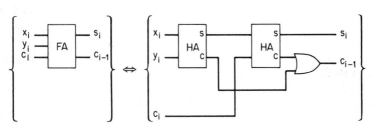

Fig. 4-13. Half adder.

A full adder is built of two half adders and one OR gate, as in Fig. 4-14.

Fig. 4-14. Full adder.

Implementation of the Subtraction Table. We mentioned previously that the addition table can be converted into a subtraction table using $\bar{x}_i$ instead of x_i and interpreting c_i as b_i. Since the interconnections for carry and borrow obviously are the same, a full adder can be converted into a *full subtractor* (FS) using a selector circuit, depicted in Fig. 4-15 in front of the x_i input; usually x_i and $\bar{x}_i$ are both available as register outputs, and one of the logic signals ADD (= addition) or SUB (= subtraction) is true, so x_i' will be equal to x_i or $\bar{x}_i$, respectively.

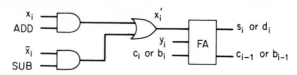

Fig. 4-15. Full adder/subtractor combination.

Addition in Series and in Parallel. In *serial addition* we assume that the augend X and the addend Y are available as binary series signals, their bits being delimited by clocking intervals. A pair of values (x_i, y_i) enters the addition circuit at a clocking signal CP_i. The same signal CP_i is used to control a D flip-flop which acts as a carry storage. The serial-addition circuit is depicted in Fig. 4-16. FA is a combinational circuit, so it will produce the signals s_i and c_{i-1} simultaneously with the input values x_i, y_i, and c_i (not

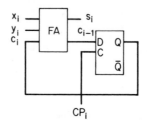

Fig. 4-16. Serial-addition circuit.

regarding a minor propagation delay). On the other hand, c_{i-1} is a preparatory input to the D flip-flop, so its value is not transferred to the input until the next clocking signal, CP_{i+1}, is due. In this way the series addition circuit is a kind of synchronous sequential circuit and will produce the sum digits in the form of binary serial signal.

In a *parallel addition* the operands enter operational circuits in parallel form as static binary signals, and there are separate logic circuits for each (x_i, y_i). There is no logical feedback in the sense of sequential circuits in this implementation. Figure 4-17 shows the interconnections.

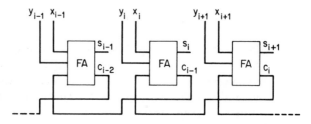

Fig. 4-17. Parallel addition circuit.

Accumulators. In the previous addition circuits, the operands could have been taken from operand registers and the result directed to a third register. It is common in arithmetic operations that the result is written back to one of the operand registers, which is then called the *accumulator register.* This name comes from desk calculators, where subsequent additions or subtractions are done on the contents of the main register bearing this name. In electronic computers, the accumulator register, together with the addition circuits, should be called the *accumulator*, although this name is often used to mean the accumulator register only.

An accumulator is most simply implemented in serial form (see Fig. 4-18). When clocking pulses are given to the shift register, comprising

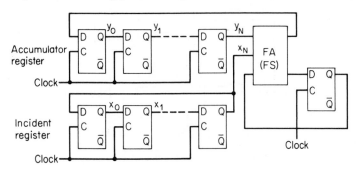

Fig. 4-18. Serial accumulator.

the accumulator register, and the addend (subtrahend) register, here called the *incident register,* their contents are shifted to the right and subsequent bits enter the FA. At the same time carry digits are delayed by the D flip-flop. Since the operand in the accumulator register is not restored, vacant positions in the most significant end become available for the result at each shift pulse. After $N + 1$ shifts the result is in the accumulator, and the other operand is also restored in the end-around shifting incident register.

Clocked flip-flops make the implementation of the *parallel accumulator* straightforward (Fig. 4-19).

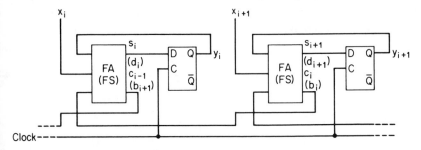

Fig. 4-19. Parallel accumulator using clocked flip-flops.

Without using full adders, an (adding) parallel accumulator can be implemented with a smaller consumption of material as shown in Fig. 4-20, in which one bit position of the accumulator is shown. This circuit, sometimes also called the *gated carry accumulator*, uses *two* clock pulses *CP* for one addition operation and is correspondingly slower. Assume for the present that the signal *COMPLEMENT* is 0 and need not be considered in the addition.

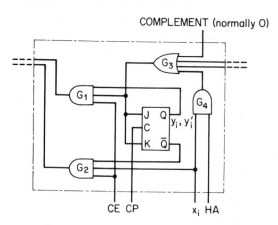

Fig. 4-20. Two-phase parallel accumulator.

At the first *CP* pulse the previous contents of the *J-K* flip-flop (accumulator bit y_i) are added with the corresponding bit x_i of the addend *without taking the carry into account*. This phase is called *half addition*, and during it the control signal *HA* is 1, whereas *CE* (*carry enable*) is 0. At the second *CP* pulse all resulting carries are added, whereby $HA = 0$ and $CE = 1$. The analysis of this circuit is based on the transitive properties of modulo 2 addition, and its proof will be given in the following discussion.

The operation of this circuit can be explained as follows: When $CE = 0$, $HA = 0$, the signals coming from the adjacent position on the right (horizontal lines) are 0. The *J* and *K* inputs now have the value x_i defined by the gates G_3 and G_4. Thus if $x_i = 0$, the original state y_i in the flip-flop is retained at the *CP* signal, whereas for $x_i = 1$ the flip-flop will toggle. Calling the new state in the flip-flop y_i', we have its truth table in Table 4-10.

Table 4-10

x_i	y_i	y_i'
0	0	0
0	1	1
1	0	1
1	1	0

Thus, indeed, the bits x_i and y_i are added modulo 2. On the next phase $HA = 0, CE = 1$, and the output of G_4 is 0. The J and K inputs are now defined by the signals coming from the G_1 and G_2 gates of the next adjacent position on the right. We shall now give an inductive proof for the addition of carries: We shall show that the logical sum of the outputs of G_1 and G_2 is the correct carry which, when added to y'_i, will transmit a new correct carry to the left.

If $y'_i = 0$ ($\bar{y}'_i = 1$), we consider separately the cases $x_i = 1$ and $x_i = 0$. When $x_i = 1$, y_i was 1, and the carry must be 1. Indeed it is formed through G_2. When $x_i = 0$, y_i was 0, and the carry must be 0. But G_1 and G_2 then give a 0 which is correct, too. If now $y'_i = 1$ ($\bar{y}'_i = 0$) which means that x_i is 1 and y_i was 0, or x_i is 0 and y_i was 1, gate G_2 is disabled and G_1 gives an output that is the same as the incoming carry. This is obviously correct. To complete the inductive proof, it is only necessary to show that the formation of a carry starts correctly in the least significant position. This is obviously true because $c_N = 0$ ($N =$ last position).

It remains to show that the new y_i is obtained correctly in the carry addition. If the carry is 1, then $J = K = 1$ and the flip-flop shall toggle, which is equivalent to a modulo 2 addition. If the carry is 0, y'_i is not changed since $J = K = 0$. This, also, is obviously right. The various cases are collected in Table 4-11, which is easily recognized as the addition table.

Table 4-11

c_i	x_i	y_i	y'_i	s_i		c_{i-1}
0	0	0	0	0 ⎫		0
0	0	1	1	1 ⎪		0
0	1	0	1	1 ⎬ y'_i		0
0	1	1	0	0 ⎭		1
1	0	0	0	1 ⎫		0
1	0	1	1	0 ⎪ y'_i toggled		1
1	1	0	1	0 ⎬ with $c_i = 1$		1
1	1	1	0	1 ⎭		1

Let us now consider the signal *COMPLEMENT*, which was 0 before. If it is made 1 and we have $HA = CE = 0$, the only gate which is enabled is G_3, and it will produce 1 to the J and K inputs. Since no carry will be propagated, all flip-flops (y_i) are toggled at the next clock pulse. Thus we get the 1's complement of the previous contents to the accumulator. If the 2's complement were needed, the number would be incremented by 2^{-N}. The incrementing may be effected, for example, by making $c_N = 1$, $x_i = 0$ and having $CE = 1$ at the next clock pulse.

The provision for complementing may be utilized in subtraction, which is done by adding the 2's complement of the subtrahend to the minu-end. We shall later exemplify the steps for the subtraction of a number from the accumulator.

Organization of an Arithmetic Unit. The arithmetic unit is that part of a computer in which the processing of digital information is done, so much care is usually devoted to the definition of its main features. Even in the simplest systems one can usually afford reasonable consumption of material by arithmetic circuits. It is a rather common requirement in general-purpose as well as specialized computing that at least the four basic arithmetic operations—addition, subtraction, multiplication, and division—should be implemented automatically without memory reference instructions during the computing operation. In contemporary computers this requirement is more enhanced than it was before, because elementary electronic addition and subtraction operations can be easily executed in a time that is roughly one tenth of one read/write memory cycle of ferrite-core memories.

The simplest, yet effective, set of operational registers in an arithmetic unit includes the following elements:

1. *Accumulator* (A), length $N + 1$ bits, which are denoted A_0 to A_N. The accumulator has provisions for addition and subtraction, as well as right and left shifting. (In most computers there are several accumulators, however.)

2. *MQ register* (Q), length $N + 2$ bits. The number bits are denoted by Q_0 to Q_{N+1}. The set of flip-flops $\{A_0, \ldots, A_N, Q_0, \ldots, Q_{N+1}\}$ forms a right- and left-shifting register.

3. *Operand register* (R), length $N + 1$ bits, which are denoted by R_0 to R_N. This register stores one of the operands (addend, subtrahend, multiplicand, or divisor), but it does not perform any register operations.

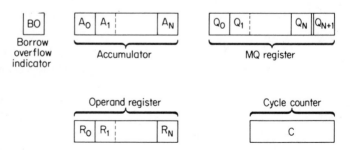

Fig. 4-21. Organization of registers in an arithmetic unit.

4. *Cycle counter* (*C*), which counts up or down but which after *N* counts always gives an indication in the form of a logic signal.

5. *Borrow overflow* (*BO*) flip-flop of the accumulator.

This set of registers is shown in Fig. 4-21. The associated bit Q_{N+1} of the MQ register is distinguished by double lines but is, in fact, an integral part of the shift register.

4.5.   Case Study of Arithmetic Algorithms as Register Operations

It is not possible for us in this book to review all the various arithmetic operations occurring in general-purpose computers. Instead, to give at least an idea of the operation of complete digital systems, we shall discuss the implementation of the usual arithmetic operations in an arithmetic unit that is reasonable effective (fast) and yet moderately simple. Assume that the set of registers described in Section 4.4 is used and that the accumulator is the two-phase type. We shall assume that initially all numbers are expressed in 2's-complement representation, and the subtraction is done by the addition of the negative of the subtrahend, which is found by complementing. The operands are assumed in the beginning to exist in *A* and *R*, respectively.

Addition and Subtraction. A few special notations are introduced in the following to facilitate a compact description. In the two-phase addition, the half-addition phase of the operand *R* (corresponding to the bits x_i in the previous section) and *A* (corresponding to the bits y_i before) is, in fact, a modulo 2 addition of all respective bits and is denoted here as

$$A \Leftarrow (A) \oplus (R) \qquad [4\text{-}20]$$

In the next phase the addition of all resulting carries is executed, and we shall denote this by

$$A \Leftarrow (A) + C_R \qquad [4\text{-}21]$$

where C_R denotes a number formed of the carries in the half-addition of *R*.

The negative of *A* in 2's-complement representation is found by complementing all bits and then adding 2^{-N} to the result: The notation

$$A \Leftarrow (\bar{A}) \qquad [4\text{-}22]$$

means complementing of all bits of *A*, and the operation

$$A \Leftarrow (A) + 2^{-N} \qquad [4\text{-}23]$$

is executed by having zero in R, holding the carry bit c_N of the least significant position of A at the value 1, and executing one addition of the carries ($CE = 1$). It is now possible to describe the subtraction of A from R in the following way:

$$1. \quad A \Leftarrow (\bar{A})$$

$$2. \quad A \Leftarrow (A) \oplus (R) \qquad\qquad [4\text{-}24]$$

$$3. \quad A \Leftarrow (A) + C_R + 2^{-N}$$

In the following we shall denote successive clock phases by P_i, S_i, or T_i ($i = 0, 1, 2, \ldots$). We shall return to the latter problem in Chapter 5.

REGISTER OPERATIONS FOR ADDITION $[A \Leftarrow (A) + (R)]$

Control signal conditions*

P_0: $\quad A \Leftarrow (A) \oplus (R) \qquad HA = 1$

P_1: $\quad A \Leftarrow (A) + C_R \qquad CE = 1$

REGISTER OPERATIONS FOR SUBTRACTION $[A \Leftarrow (A) - (R)]$

P_0: $\quad A \Leftarrow (\bar{A}) \qquad\qquad\qquad COMPLEMENT = 1$

P_1: $\quad A \Leftarrow (A) \oplus (R) \qquad\qquad HA = 1$

P_2: $\quad A \Leftarrow (A) + C_R + 2^{-N} \qquad CE = 1, c_N = 1$

P_3: $\quad A \Leftarrow (\bar{A}), R \Leftarrow 0 \qquad\quad COMPLEMENT = 1$

P_4: $\quad A \Leftarrow (A) + 2^{-N} \qquad\quad CE = 1, c_N = 1$

Multiplication. Multiplication by Booth's algorithm is executed in the following steps: First the multiplier is transferred to the MQ register (Q) and $Q_{N+1} = 0$. The bits Q_N and Q_{N+1} are decoded to indicate whether this pair (Q_N, Q_{N+1}) is (0, 1), (1, 0), (0, 0), or (1, 1). Corresponding to the pairs of values (x_i, x_{i+1}) occurring in the discussion of Booth's algorithm in Section 4.4, the first condition is now used to define an addition, the second a subtraction, and the two last pairs no operation, respectively. The multiplier is shifted to the right at every accumulation of partial products and so new bits always enter the Q_N, Q_{N+1} positions.

The multiplicand is in the R register, and the partial products will be accumulated in A. While the length of the product grows and its least significant bits are intruding further into the Q register, vacant places for the product are yielded by the recessing, right-shifting multiplier.

*All control signals not shown attain the value 0.

In terms of addition and subtraction routines described before, the multiplication algorithm is executed as the following register operations, and the conditional clock phases are expressed as corresponding Boolean functions standing on the left:

REGISTER OPERATIONS FOR MULTIPLICATION $[A \Leftarrow (Q)\cdot(R)]$

Assume that the multiplier is in A and has first to be transferred into Q. A stop signal $STM = 1$ is given when the contents of the cycle counter become $(C) = N$.

$$S_1: \quad Q \Leftarrow (A), Q_{N+1} \Leftarrow 0, C \Leftarrow 0, A \Leftarrow 0$$

$$\bar{Q}_N Q_{N+1} S_2: \quad A \Leftarrow (A) + (R) \Big\}$$
$$Q_N \bar{Q}_{N+1} S_2: \quad A \Leftarrow (A) - (R) \Big\} \text{ (no operation if } Q_N = Q_{N+1})$$

$$STM\cdot S_3: \quad \text{computation stops}$$

$$\overline{STM}\cdot S_3: \quad \text{arithmetic shift of } A \text{ and } Q \text{ to the right, including}$$

$$\begin{cases} A_0 \Leftarrow (A_0), A_{i+1} \Leftarrow (A_i) & \text{for } i = 0 \text{ to } N - 1 \\ Q_0 \Leftarrow (A_N), Q_{i+1} \Leftarrow (Q_i) & \text{for } i = 0 \text{ to } N \end{cases}$$
$$C \Leftarrow (C) + 1$$

This cycle is repeated from clock phase S_2 until computation stops.

Division. Initially, the dividend is in A and the MQ register and the divisor in R. For the division in signed magnitude representation, operands that are expressed as 2's complements must first be transformed to signed magnitude representation. Assume that the signs are stored somewhere else and that the sign bits of A and R are 0.

When the sign bits are operated as the usual number bits, a borrow resulting from them in subtraction is indicated as borrow overflow and denoted by BO in the following. A special BO flip-flop is reserved for this purpose, and its output is used as a logic condition for a restoring operation. In short, the division algorithm is executed as follows:

REGISTER OPERATIONS FOR DIVISION

$$T_0: \quad C \Leftarrow 0, \text{ shift left, including}$$

$$\begin{cases} A_i \Leftarrow (A_{i+1}) & \text{for } i = 0 \text{ to } N - 1, A_N \Leftarrow (Q_0) \\ Q_i \Leftarrow (Q_{i+1}) & \text{for } i = 1 \text{ to } N \quad (Q_{N+1} \Leftarrow 0) \end{cases}$$

$$T_1: \quad A \Leftarrow (A) - (R), Q_{N+1} \Leftarrow 1$$

$$BO\cdot T_2: \quad A \Leftarrow (A) + (R), Q_{N+1} \Leftarrow 0 \text{ (no operation if } BO = 0)$$

$$T_3: \quad \text{shift left, } C \Leftarrow (C) + 1, BO \Leftarrow 0$$

This cycle is repeated from clock phase T_1 until $(C) = N$, after which computation stops.

4.6. Incremental Computing

Especially in real-time computing systems in which variables are continuous, slowly varying functions of time and where the speed of computing is not critical, arithmetic operations and generation of special functions can be made by special means, called *incremental computing*. The idea underlying incremental computing is the recursiveness of expressions: The new values of variables are obtained from the old ones by adding or subtracting small corrections to them. The examples discussed in this section clearly show that the differential corrections are usually related in a very simple manner to the previous values of machine variables and can be implemented using the simplest arithmetic operations. The corrections in incremental computers are made in small steps on all occurring machine variables, in successive *iterations*.

In this section we shall first discuss a special computing circuit called *the binary rate multiplier*. Then a more versatile device, called *the digital differential analyzer*, is described.

4.6.1. Binary Rate Multiplier

The product of two numbers can be formed in many ways. In a *binary rate multiplier* (BRM) the *multiplicand is represented by the number of pulses in a pulse train; the multiplier controls a gating circuit which picks up a corresponding fraction of these pulses to counting circuits.*

An arbitrary fraction of a pulse frequency can be selected by expanding the original pulse train in *binary frequency components*. This idea is illustrated in Fig. 4-22, in which CP represents a train of numbered pulses. The "clock phases" $0, 1, 2, \ldots$ of C_0 are divided between different lines in such a way that the pulse frequencies on the lines are related as $2^{-1}, 2^{-2}, 2^{-3}, \ldots$ and each clock phase occurs on one line only. The pulses on different lines are *noncoincident* and thus *additive*.

Notice that the primary clock pulse train CP need not be evenly spaced in time: the binary frequency components can be obtained from any CP by a counter–decoder method. For this purpose, let us take an *input counter*, often also called *serializing* or *classifying counter*, with CP as its input. Being a binary counter, the input counter forms output signals B_1, $B_2, \ldots$ as depicted in Fig. 4-23, where CP has been chosen periodic for simplicity and clarity. The negative-going edges, denoted by arrows, may be used for ripple counting. The positive-going edges, shown by the darker lines, have a similar spacing in time as pulses in the binary frequency com-

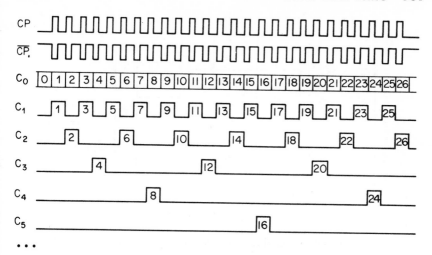

Fig. 4-22. Binary frequency components.

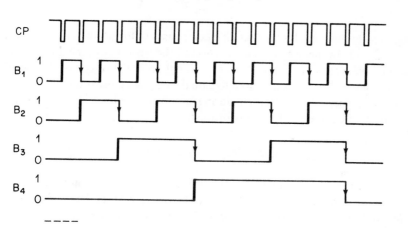

Fig. 4-23. Output signals from a binary counter used in a BRM.

ponents, and never coincide. It is now an easy problem to produce short pulses at the positive-going edges, using, for example, differentiating circuits (as will be discussed in Chapter 8), edge-triggered pulse generators, or perhaps special combinational circuits.

Figure 4-24 shows how the "differentiated" outputs C_1 to C_N from the input counter (i.e., the *binary frequency components*) are gated by the bits of the multiplier. Denote the multiplier by a binary fraction

$$M = .m_1 m_2 \ldots m_N \qquad [4\text{-}25]$$

The bits m_1 to m_N are regarded as Boolean variables, as are the frequency-component signals C_1 to C_N.

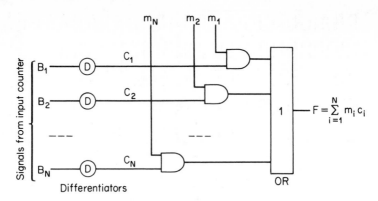

Fig. 4-24. Gating of binary frequency components.

Let us now consider the output logic signal F defined by

$$F = \sum_{i=1}^{N} m_i C_i \qquad\qquad [4\text{-}26]$$

Obviously this signal is "on" at all clock phases C_i for which the corresponding $m_i = 1$. In this way the *duty cycle* (i.e., the fraction of time during which F has the value 1) is directly proportional to the product of M and the frequency of CP. When F is used to gate the original pulse train CP, in which, say, n_0 pulses are transmitted during a sufficiently long interval, then the number of counted pulses n is obviously

$$n = n_0 \sum_{i=1}^{N} m_i 2^{-i} = M n_0 \qquad 0 \le M \le 1 \qquad [4\text{-}27]$$

The complete binary rate multiplier is thus shown in Fig. 4-25, and its symbolic notation is given in Fig. 4-26. Notice that m_1 is the *most significant bit* while B_1 is the *least significant position* of the input counter.

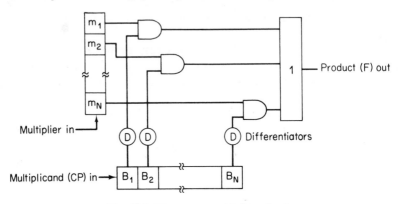

Fig. 4-25. Binary rate multiplier: circuit.

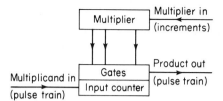

Fig. 4-26. Binary rate multiplier: symbolic notation.

Generation of Elementary Functions by the BRM. As an introduction to incremental computing, let us discuss the generation of some elementary functions by the BRM. To transform the output pulse train into a machine variable, the pulses (increments) must be summed up by a separate register. This can be the input counter of another unit.

Example 4-12. Design a BRM system that computes the function $Y = e^X$. This function is a solution of the differential equation

$$dY = Y \, dX \qquad [4\text{-}28]$$

The multiplier register is used to count the increments dY in the way depicted in Fig. 4-27. The registers are labeled by corresponding machine variables. Up counting may be denoted by a $+$ sign at the input.

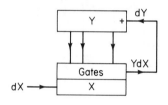

Figure 4-27

Example 4-13. Design a BRM system that computes the functions $X = \cos\theta$ and $Y = \sin\theta$.

Both X and Y are obtained simultaneously from the differential equations

$$dY = X \, d\theta \qquad dX = -Y \, d\theta \qquad [4\text{-}29]$$

provided that initial conditions are given correctly. In the circuit implementation of Eq. [4-29], a common input register is used for two sets of gates (Fig. 4-28). Increments that are to be subtracted are denoted by a $-$ sign at the counter input. A special problem arises when the sign of X or Y is changing. This is most easily handled by external logic circuits and using bidirectional counters for X and Y; an up counter is turned into a down counter and vice versa by logic control.

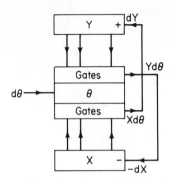

Figure 4-28

Example 4-14. Design a BRM system that computes the reciprocal function $Y = 1/X$.

Differentiation yields

$$dY = -Y^2 dX = -Y \cdot Y \, dX$$

The direct feedback system depicted in Fig. 4-29 will take care of this equation.

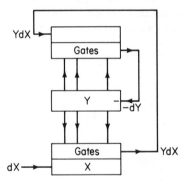

Figure 4-29

Errors in the BRM. If the multiplier is changing, errors may be caused, because previously it was assumed that the multiplier was steady. To minimize this effect, the multiplier register is made longer and only the most significant bits, which are changed more seldom, are used for gating purposes.

4.6.2. Digital Differential Analyzer

Frequently, computations in systems engineering consist of solutions to differential equations. On the other hand, differential equations and differential expressions can be used to define special functions (e.g., polynomial, exponential, logarithmic, and trigonometric functions). *Analog computers* are widely used for such purposes. However, the accuracy of analog computers is limited to a few digits, and *digital differential analyzers* (DDA) have been used to circumvent this difficulty.

The basic component of analog and digital differential analyzers is the *integrator,* which forms an expression

$$Y(t) = Y(0) + \int_0^t f(\tau)\, d\tau \qquad [4\text{-}30]$$

with $f(t)$ as its input. In the DDA, the integrator consists of the system depicted in Fig. 4-30(a). Numbers in it are usually represented as binary fractions. Here Y is a register that is an up/down counter. The increment $(\Delta Y)_i$ is $+2^{-N}$, 0, or -2^{-N}. The accumulator A is a usual arithmetic double-length

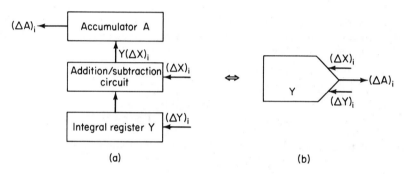

Fig. 4-30. Digital integrator: (a) register organization; (b) symbolic representation.

accumulator; however, in simpler systems several accumulator registers may share common addition and subtraction circuits. The purpose of the addition/subtraction circuit is to form the product $Y(\Delta X)_i$ where $(\Delta X)_i$ is $+2^{-N}$, 0, or -2^{-N}, and to accumulate it to the previous contents of A. Note that no multiplication operation is needed, because Y is only shifted to the right by N positions and conditionally added with or subtracted from A. The carry or borrow overflow of A is detected. A carry overflow is regarded as an output $+2^{-N}$, whereas a borrow overflow is interpreted as an output -2^{-N}. The signal denoted by $(\Delta A)_i$ is the overflow at the cycle i. *The sum of the $(\Delta A)_i$ over a long period is obviously the same as the sum of $Y(\Delta X)_i$ during the same time.* Thus we can deduce that the average value ($\langle \cdot \rangle$) of the increments $(\Delta A)_i$ is

$$\langle (\Delta A)_i \rangle = \langle Y(\Delta X)_i \rangle \qquad [4\text{-}31]$$

If there is no danger of confusion, Eq. [4-31] is expressed as

$$dA = Y\, dX \qquad [4\text{-}32]$$

The increments dA are added (integrated) by the integral register of another unit. The interconnection of digital integrators is made according to principles similar to those of an analog computer. However, the independent value in analog computers is time t, which is a monotonically increasing variable, whereas the independent variable X in DDA's may be incremented up or down or even halted if necessary. The most important aspect of DDA's discussed in this book is their use for the computing of special functions, and this use will be illustrated by a few examples.

Example 4-15. Design a DDA that computes $Y = f(X) = e^X$. First we differentiate Y:

$$dY = e^X \, dX = Y \, dX \qquad [4\text{-}33]$$

As long as the number X can be given as a sum of increments dX to the corresponding input of the digital integrator, the connection in Fig. 4-31 will permit us to solve for Y.

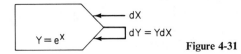

Figure 4-31

Example 4-16. Design a DDA that performs a multiplication by a constant:

$$Y = KX \qquad [4\text{-}34]$$

$$dY = K \, dX \qquad [4\text{-}35]$$

Figure 4-32 shows the implementation, in which the contents of the integral register K are constant.

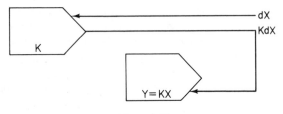

Figure 4-32

Example 4-18. Design a DDA that forms the product of two numbers X and Y given as separate increments dX and dY. Denote

$$Z = XY \qquad [4\text{-}36]$$

$$dZ = X \, dY + Y \, dX \qquad [4\text{-}37]$$

In the implementation of Eq. [4-37] an increment adder is needed. The circular symbol in Fig. 4-33 stands for any practical arrangement for the addition of increments.

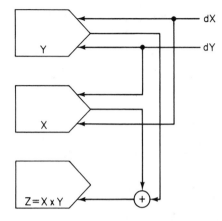

Figure 4-33

Example 4-18. Design a DDA (Fig. 4-34) that forms the function $Y = 1/X$. With this provision, a division can be easily implemented

$$dY = -\frac{1}{X^2} dX = -Y(Y \, dX) = -Y \, dZ \qquad \text{where } dZ = Y \, dX \quad [4\text{-}38]$$

Another special feature occurs with this circuit. A negative of an increment is to be formed, which is a relatively easy operation, because the increment may take only three values: $+1, 0,$ and -1. The (arithmetic) negation is shown with a minus sign written at the output of the integrator.

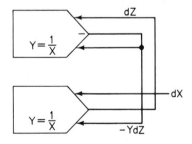

Figure 4-34

Example 4-19. Design a DDA (Fig. 4-35) that computes the trigonometric functions $\sin X$ and $\cos X$, where X is expressed in radians and given by increments dX:

$$Y = \sin X$$

$$Z = \cos X$$

$$dY = \cos X \, dX = Z \, dX$$

$$dZ = -\sin X \, dX = -Y \, dX$$

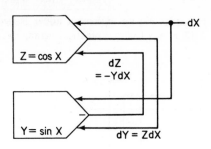

Figure 4-35

PROBLEMS

4-1 Design a parallel counter having the output sequence 000, 100, 111, 011, 001, 000, . . .
(a) using *D* flip-flops.
(b) using *J-K* flip-flops.

4-2 Design all the possible modulo 3 parallel counters having different state graphs and the forbidden state 00, using two *J-K* flip-flops.

4-3 (a) Design a modulo 6 Johnson counter using *J-K* flip-flops only.
(b) Does the counter have forbidden states and, if so, what are they?

4-4 (a) Using additional gates, modify the counter of Problem 4-3 so that it becomes self-correcting.
(b) What is the maximum number of counts after which any error is corrected? Point out the corresponding sequence.

4-5 Design a three-bit binary ripple counter that counts up for a command signal $F = 1$ and down for $F = 0$. Transitions have to occur at the trailing edges of clock signals.

4-6 Perform the additions
(a) $29 + 17$
(b) $29 + (-17)$
(c) $11 + (-29)$
(d) $(-11) + (-29)$
(1) in 2's-complement representation, (2) in 1's-complement representation, and (3) in signed magnitude representation. The operations ought to be carried out on binary fractions in a suitable scale in which all numbers are multiplied by a scaling factor 2^{-6}.

4-7 Perform the subtractions
(a) $29 - (-17)$
(b) $29 - 17$
(c) $11 - 29$
(d) $(-11) - 29$

(1) in 2's-complement representation, (2) in 1's-complement representation, and (3) in signed magnitude representation. Use the same scaling as in Problem 4-6.

4-8   Perform the multiplications
(a) $29 \cdot 17$
(b) $17 \cdot (-11)$
(c) $(-17) \cdot (-11)$
(d) $(-11) \cdot 29$
(1) in 2's-complement representation and (2) in signed magnitude representation. Use the scaling factor 2^{-5}.

4-9   Perform the following divisions in signed magnitude representation:
(a) $\dfrac{17}{29}$

(b) $\dfrac{-11}{29}$

Use the same scaling as in Problem 4-8.

4-10 A full adder represented by Fig. 4-14 is tested and the following truth table is obtained:

x_i	y_i	c_i	s_i	c_{i-1}
0	0	0	0	0
0	0	1	1	0
0	1	0	1	0
0	1	1	0	1
1	0	0	1	0
1	0	1	0	1
1	1	0	0	0
1	1	1	1	0

If one of the half adders is faulty, which one is it?

4-11 Design a BRM that computes the function $Y = \frac{1}{2}X^2$.

4-12 Design a DDA that computes the function $Y = \frac{1}{2}X^2$.

REFERENCES

The basic facts of counters can be found in the application reports of most manufacturers of integrated circuits, and these are the primary source of information of special digital circuit configurations. For example, the circuits of Fig. 4-7 can be found in *Application Memos* by Signetics Corporation, Sunnyvale, Calif.: Signetics Corp., 1968, from p. 4-4 on. Most textbooks on computer design also contain basic principles of counting and arithmetics. The books

Chu, *Digital Computer Design Fundamentals* [17],
Chu, *Introduction to Computer Organization* [18],
Flores, *The Logic of Computer Arithmetic* [35], and
Flores, *Computer Organization* [36]

are good sources of more detailed information on computer arithmetic. Also, the first two volumes of the seven-volume series of

Knuth, *The Art of Computer Programming* [76]

contain much basic information about arithmetics. The fundamentals of the BRM have been explained in a series of publications mentioned in the Bibliography. The DDA, which is more common than BRM, has been extensively described in

Sizer (ed.), *The Digital Differential Analyzer* [119].

(See also the Bibliography.)

chapter 5

Control Operations in Digital Systems

5.1. Micro-operations

Digital operations are composed of elementary tasks such as setting or clearing flip-flops, delivering pulse signals to proper inputs of gates, giving a predetermined number of shift pulses to shift registers, etc. There is usually a rather small choice of elementary operations, but the complexity of digital systems comes from the fact that a large number of sequential operations must be performed in a definite order and extension in time, taking branching conditions into account.

A general *micro-operation*, or an *assignment*, is denoted as

$$f_i: \qquad A_j \Leftarrow g_k(A_l, A_m, A_n, \ldots) \qquad\qquad [5\text{-}1]$$

where f_i is a Boolean function over some set of binary signals in the system and A_j is a bit storage. The signals denoted by $A_l, A_m, A_n, \ldots$ are outputs of other bit storages or external control signals, and g_k is a Boolean function of these values. The symbol $f_i:$ means that the micro-operation defined by Eq. [5-1] is executed when f_i becomes 1. In synchronous digital systems f_i is usually of the form:

$$f_i = F_i T_i \qquad\qquad [5\text{-}2]$$

where the T_i's are derived from a timing signal generator and do not overlap. In this way the micro-operations can be ordered in time.

The assignment may be implemented by direct *R-S* flip-flops provided that the values $A_l, A_m, A_n, \ldots$ are not affected by A_j. This is shown in Fig. 5-1. The *bistable latch* (Fig. 3-12) is a yet simpler implementation.

177

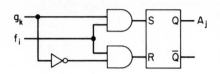

Fig. 5-1. Transfer of information to a direct R-S flip-flop.

Another method using a clocked (D) flip-flop is shown in Fig. 5-2. This method is used if $A_l, A_m, A_n, \ldots$ would be affected by the change in A_j. Two-phase master-slave flip-flops (bistable latches) can also be used.

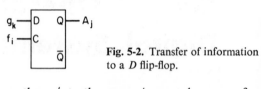

Fig. 5-2. Transfer of information to a D flip-flop.

Since the transfer of another g'_k to the same A_j may also occur for another condition, $f_{i'}$, a *selector gate system* in front of the bit storage A_j is needed. Assume that f_i and $f_{i'}$ do not coincide, and

$$
\begin{aligned}
f_i: \qquad & A_j \Leftarrow g_k \\
f_{i'}: \qquad & A_j \Leftarrow g_{k'} \\
& \cdot \\
& \cdot \\
& \cdot
\end{aligned}
\qquad [5\text{-}3]
$$

A straightforward implementation of Eqs. [5-3] in which $g_k, g_{k'}, \ldots$ do not depend on A_j is shown in Fig. 5-3. The system is described by

$$
A_j = S + \bar{R} A_j \qquad [5\text{-}4]
$$

where

$$
\begin{aligned}
S &= f_i g_k + f_{i'} g_{k'} + \cdots \\
R &= f_i \bar{g}_k + f_{i'} \bar{g}_{k'} + \cdots
\end{aligned}
$$

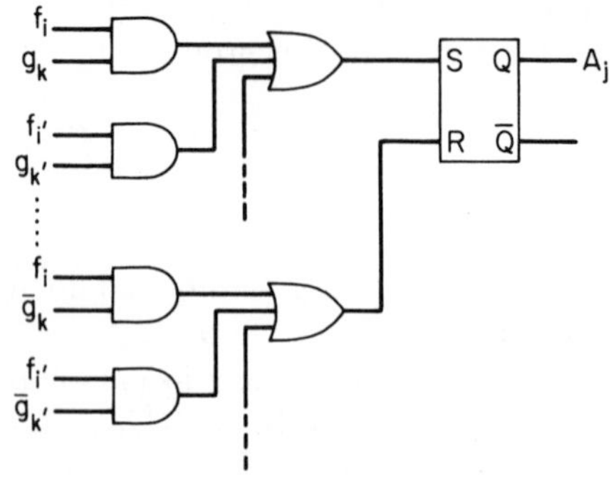

Fig. 5-3. Selector gates for transfer operation.

Since the conditioning signals are usually of the form $f_i = F_i T_i$, some savings in gates may be possible if the clocking signals are separated as in Fig. 5-4. Of course, the right-hand part could be a simpler bistable latch. A selector-gate system can also be designed for clocked flip-flops.

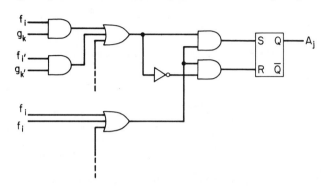

Fig. 5-4. Simpler selector for transfer operation.

In general, the purpose of *control circuits* in a digital system is to produce signals in *space* (at various inputs) and *time*. This chapter is devoted to the discussion of methods by which the control is usually implemented. First some elementary control operations and methods are introduced.

5.2.   Elementary Control Operations

Production of Standardized and Synchronous Control Signals from Asynchronous Commands. Rather many external commands in digital systems are given by switches or electromechanical devices such as relays. The switch seldom closes the circuit in a single operation; even microscopic bouncing usually causes multiple contact operations. To "clean up" such signals from switches for digital purposes, they are usually buffered with direct R-S flip-flops in the way shown in Fig. 5-5. Of course, some logic operations can be combined with the flip-flop. When the switch is thrown up, the first contact with the $\bar{S}$ terminal will set the flip-flop; additional contacts cause no changes in the state of it as long as the lower contact remains open. After turning the switch down, the flip-flop will be reset at the first contact with the $\bar{R}$ terminal. Thus a single pulse is formed at the output during this cycle.

The resistors in Fig. 5-5, which are connected to the switch, are required if the logic value of a floating input is not defined.

To synchronize an asynchronous signal with the clock pulses, any clocked flip-flop (e.g., a D-flip-flop) can be used. The signal is led to its D input. The input information (D) is transferred to the Q output (see Fig. 5-6)

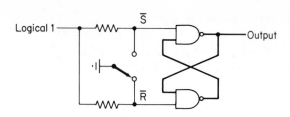

Fig. 5-5. Buffering of electromechanical contacts.

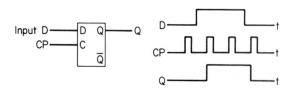

Fig. 5-6. Synchronization of an asynchronous input signal.

at the first clock pulse (*CP*). This circuit alone does not eliminate the effect of bouncing contacts because the switch might be closed at the first clock pulse, open at the second, etc., and, if the clock frequency is high enough, the *D* flip-flop follows the status of the contacts. Thus a buffer is needed anyway. When the input signal ceases, the *D* flip-flop drops the output signal at the next clock pulse.

In some cases an internal control signal must be turned on at the first clock pulse following an asynchronous input (*START*) and automatically turned off after the initiated operation has been completed, which is indicated by a *STOP* signal. An arrangement as shown in Fig. 5-7 might be used. The *START* command is given here by a closing switch ($\overline{START} \rightarrow 0$), which is opened before the arrival of the $\overline{STOP}$ signal.

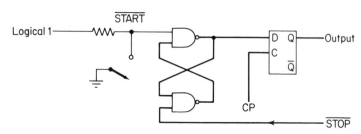

Fig. 5-7. Production of a synchronously started and stopped control signal (*Q*).

Notice that the output signal *Q* is 1 until the next clock pulse after the *STOP* command (i.e., $\overline{STOP} \rightarrow 0$). A *STOP* signal may also be connected to a "direct clear" input, usually provided in the *D* flip-flop. This will always

override all other inputs. In this case the direct *R-S* flip-flop must be controlled by a double-throw switch as in Fig. 5-5.

Production of a Predetermined Number of Pulses. In arithmetic
circuits using *N*-bit shift registers, an often repeated micro-operation is the
production of *N* shift pulses (e.g., at each addition or subtraction command).
Using analog methods, *N* pulses can be produced easily by gating a clock
oscillator for corresponding period. However, digital systems should not be
unnecessarily hybridized with analog ones; for one thing, because the total
system is then no longer subject to uniform electrical specifications, and its
reliability considerations become more complicated. Therefore, digital methods are preferred, and the counting of *N* pulses always requires some kind
of counter with a modulus $m \geq N$. Since *N* clock pulses are always given
during the counting, the required pulse train is derived from the number of
counts.

To stop a counter, two methods are available:

1. *The counter is latched* to some state when a *START* signal is 0, and
   it is released when *START* is made 1 (Fig. 5-8). With a modulo
   N counter the latching state can be arbitrary.

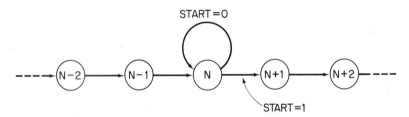

Figure 5-8

2. *The counting input is blocked* by a logic gate. To start, the counter register is cleared and then it begins to count *up* until the state
   N is achieved. This state is decoded and the output of the decoder stops the counting. An alternative method is to load the
   counter with a number corresponding to the predetermined number
   N of pulses and to count *down* until the state zero is detected.

THE LATCHING METHOD

In this method clock pulses are persistently present at the clocking
inputs. Using *J-K* flip-flops in the counter, the latching of the counter is
effected by imposing logic zero on the *J* and *K* inputs; for counting, they are
switched to logic 1. The *J* and *K* inputs are obtained from a logic circuit that

decodes the number N in the counter. The latching condition can also be imposed by the design of a special parallel counter with one latching state. When the counter is counting up and achieves this state, it will latch. To start the counting, the counter is cleared with a temporary *START* command, which is assumed synchronous with the clocking pulse.

Example 5-1. In Fig. 5-9 a binary ripple counter is shown. The first stage is made of a *J-K* flip-flop; arbitrary toggle flip-flops are used in the rest of the stages. The clocking is assumed to happen at the $1 \rightarrow 0$ transition. When a state N is present in the counter, a decoding circuit G_1 produces an output $P_N = 1$ ($\bar{P}_N = 0$), which halts the counter and blocks the *CP* pulses off by the gate G_2.

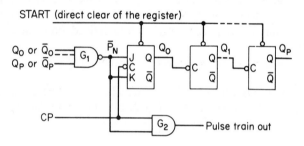

Fig. 5-9. Pulse-train generation by the counter-latching method.

Comment. The decoded signal $\bar{P}_N$ need not be implemented as a normal product. In many cases, incompletely defined truth tables in which states $>N$ are forbidden yield a simpler Boolean function for $\bar{P}_N$.

THE BLOCKING METHOD

In order to exemplify alternative methods for the production of pulse trains, assume that a modulo N counter is available. A train of counting pulses *CP* is given to the counter so that it proceeds once through all the N states and stops at a rest state which is assumed to be 000 . . . 00 in the following. The stopping is due to the blocking of counting pulses in the rest state. To release a new counting sequence, an external synchronous control signal $\overline{START}$, which is applied for a short time, produces an extra count, after which the state in the counter is 000 . . . 01. Let us examine Fig. 5-10. A received count is denoted by T, and the outputs of the modulo N counter are denoted by Q_i ($i = 0$ to p). If the counting is done with the $0 \rightarrow 1$ transition of T, the Boolean function

$$T = CP\left(\sum_{i=0}^{p} Q_i + START\right) \tag{5-5}$$

produces *CP* pulses whenever a state zero is *not* in the counter or when the

$START$ signal is 1. The $START$ signal must have a shorter duration than one counting sequence. The implementation of Eq. [5-5] is shown in Fig. 5-10.

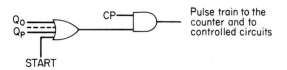

Fig. 5-10. Production of a train of counting pulses by the blocking method.

It is also possible to have several rest states at which the counter stops in succession and from which it can be released by the application of a triggering signal. In this case the input values Q_i or $\bar{Q}_i$ should be selected so that their combination in the logic function uniquely defines all rest states.

Example 5-2. Consider a Boolean function

$$T = CP(Q_2 + Q_1 + Q_0 + START)$$

Regardless of the length of the counter, it will stop whenever the three last bits are zero (i.e., at every eighth of the counting pulses).

The production of pulse trains is often combined with other control functions using common counters for different purposes.

Production of Timing Signals. In Chapter 1 we discussed one type of automatic operation, in which successive *clock phases* were needed. It is common in most automatic computing systems that different clock phases and conditionally initiated *control signal sequences* are needed. These are together called *timing signals*. Timing signals are distributed over the system. They have a definite order in time and each has a particular duration. In some cases a timing signal cannot be produced unless another signal is feasible. All timing signals together comprise the control signals of the system.

Timing signals occur as single commands or as conditionally initiated sequences. A sequence may be a *one-shot cycle* in which a predetermined number of clock phases is triggered by a single command. Sometimes the sequence, when triggered once, is *periodically* repeated until a stop command is given.

In principle there is no difference between the production of one-shot and periodical cycles of clock phases because a periodic operation can be implemented by triggering a new one-shot cycle by the last clock phase of the previous one. The two methods for the production of clock phases are (1) the delay network method and (2) the counter–decoder method.

In the *delay network method*, the circuit elements used to implement pretermined delays may take many different forms. The simplest are *delay lines* (natural or artificial transmission lines), in which, say, a single pulse is propagated with a certain velocity. The line can be tapped at arbitrary points to provide the delay required. Delay lines can be produced over the range 10^{-9} to 10^{-3} sec, but the output signals normally cannot be used without drivers. For large delays the lines become impractically large. *Shift registers* and *ring counters*, in which, say, a single 1 or a train of 1's are propagated, are frequently used in timing, provided that the number of clock phases is not very large. *Analog timing methods* will be discussed in Chapter 8. Sometimes a *chain of monostable multivibrators* triggered in a cascade is used to produce escalated pulses. An advantage of this method is that the length of each pulse can be adjusted individually. A drawback of the use of monostable multivibrators is that they are usually more sensitive to noise pickup than the counter–decoder combinations discussed below, and they need adjustment during use. This author recommends monostable multivibrators only for the formation of proper pulse length when a triggering signal is due. Monostable multivibrators are discussed in more detail in Chapter 8.

In the *counter–decoder method*, as the name implies, an up or down counter or possibly a sequential state counter is used to define a counting cycle. Self-correcting Johnson counters can be often recommended. Each state of the counter register is decoded with a separate decoder, thus providing output signals that are escalated in time. Signals extending over the duration of several states or occurring several times during a counting sequence can be obtained as Boolean functions of the state counter outputs. Also a separate flip-flop may be set by one decoder and reset by another. The state counter method is flexible and reliable. A particular precaution should be mentioned: Although all storage cells of the state counter are controlled in parallel, it is possible that the switching of all of them does not occur quite simultaneously. Some false states may transiently be detected by the decoders. A check for the occurrence of such signal "spikes" should be made and, if they are observed, they should be rejected by filtering, which is usually easily done.

Example 5-3. A *commutator* is a system that periodically selects one of N lines (channels) in succession. If the number of lines is large, the decoding may be implemented in two levels: First a pair of x and y lines are selected and then N items of (x, y) pairs are provided with two-input AND gates. Figure 5-11 shows a system in which two counters are cascaded, thus forming a single synchronous modulo N up counter. The outputs of the two-input AND gates are thus selected in succession.

Review of the Timing Signals. Before the control circuits are discussed in more detail, it is useful to recall what types of control signals are needed in a digital system interacting with its environments. In the following the most common micro-operations are characterized.

Transfer operations or *assignment operations* have the shortest execu-

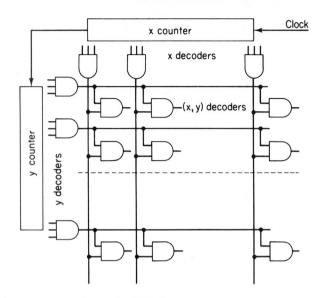

Fig. 5-11. Commutator.

tion times, and their speed is limited only by the propagation delays, which are usually of the order of 10^{-8} to 10^{-7} sec per stage in logic circuits, depending on the circuit type and loading conditions. If the maximum delay through a logic circuit is, say, below 0.1 μsec, the subsequent transfer operations can follow each other with a maximum frequency of 10 MHz. A clock pulse generator with this frequency can be used to control automatic operations, and one period of this clock is usually called a *machine cycle* of the system.

Memory reference operations are also transfer operations but usually involve separate operations for the address selection. They include *writing* of information to memory cells and transferring it to the output buffer register, also called *reading* of memory. There are *active memories* with specifications identical with those of the logic circuits, and the *access time* (the time between the reading command and the instant when the contents are available in the buffer register) is of the same order of magnitude as in a usual transfer operation. On the other hand, most mass memories are based on ferromagnetism and the time needed to magnetize a volume representing a bit is of the order of 10^{-7} to 10^{-6} sec. Therefore, cycle times with this order of magnitude are usual in ferrite memories.

Input/output devices, especially those involving mechanical operations, require control pulses with a duration of 10^{-3} to 10^{-1} sec in order that electromechanical devices (relays, solenoids, etc.) are energized.

Real-time and on-line operations are frequently associated with computing systems. By these operations, signals given by the computer control physical systems (e.g. industrial processes). In principle, some input/output operations could also be considered as real-time operations. On-line operations are usually associated with physical measurements and subsequent analog-to-digital conversion of transducer signals. The duration of real-time control signals may range from 0.1 sec to several hours.

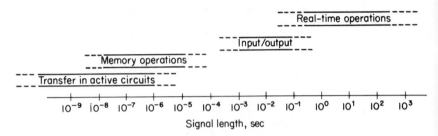

Fig. 5-12. Typical control signal durations.

Figure 5-12 gives an idea of the extension in time of various control signals occurring in a digital system. It is obvious that correct timing of different types of control signals needs synchronizing or at least mutual interaction of several clock systems.

5.3. Sequential Control

Serial and Parallel Processing. By automatic operations in digital systems we usually mean that a completed sequence initiates another sequence, which returns the control to the first sequence or to a third one, and so on. Very often a sequence is interrupted when a particular result has been obtained, at which point the operation is branched to another mode. Branching can also be initiated by an external *interruption request*, which has a higher priority than the internal signals controlling the automatic operation. For example, such an interrupt signal can be given manually.

Quite independent of particular circuit implementations, we can distinguish between two arrangements for the implementation of computing or processing (e.g., input/output) sequences. These are (1) serial processing and (2) parallel processing.

In *serial processing*, only one sequence is active at a time; all other operations are postponed until that sequence has been completed. For example, a running program in a small computer may halt during the printing of results until a feedback signal from the printer permits further execution of the program.

If it is important to speed up digital operations, *parallel processing* of information should be used in several places. For example, some results may be transferred to a *buffer register* by a fast assignment operation, and, while the rest of the automatic computing program is continued, a *printout sequence* using this buffer is simultaneously initiated. Sometimes the main cycle of the automatic computing program is long enough so that a previously initiated simultaneous operation will be completed before its time is due again. In this case there is no need to halt the main cycle. In a more general case it might happen that a simultaneous operation in a parallel system (e.g., satellite computer) should be initiated but that the parallel system is still *busy* (reserved). Another common case is that the main program uses computational results delivered by the parallel system and they are not yet feasible. Thus an *interlocking condition* implies that the main system must *wait* until a parallel sequence has been completed.

Various cases occurring in serial and parallel processing of information are exemplified in Fig. 5-13 by timing diagrams.

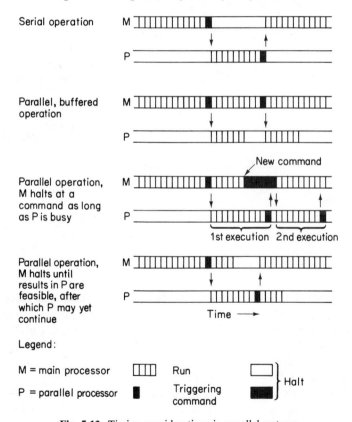

Fig. 5-13. Timing considerations in parallel systems.

Start–Stop Control of Sequential Circuits. A problem that occurs often with the design of sequential control systems is the start–stop control of automatic cycles. For manually and automatically given commands, *start–stop* flip-flops are widely used. These flip-flops can be of the direct *R-S* type, perhaps using combined logic conditions as their Set and Clear commands. More generally, without regard to whether the start–stop flip-flop is asynchronous or clocked, we may describe its effect with two isolated but logically interconnected state graphs, Fig. 5-14. In this method both systems are in the rest states of operation. When an external *START* signal is applied, the flip-flop is turned to its 1 state. The output signal of the flip-flop releases the main system from its latching state. In the subsequent cycle, the decoder of the last state before the latching state (or any other state during the cycle) will return the flip-flop to 0, whereby the main sequential system is able to halt in its rest state.

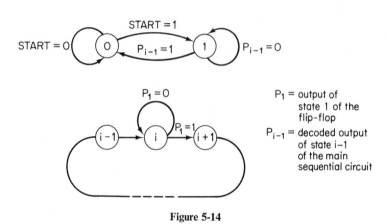

P_1 = output of state 1 of the flip-flop

P_{i-1} = decoded output of state i−1 of the main sequential circuit

Figure 5-14

It should be noted that the start–stop flip-flop is superfluous if the main sequential circuit can be directly released from its latching state by a *START* command. A drawback of this method is that the *START* command must be active until the next clock pulse arrives.

Mutual Control Operations in Interactive Synchronous Sequential Circuits. Only in the simplest automatic devices can the complete time sequence of control signals be delivered by a single sequential circuit. We shall call this kind of circuit an *elementary control circuit*. With increasing complexity of the system, it will be necessary to find groups of operations with almost independent control sequences.

The way in which the operation groups are formed is not unambiguous and might be affected by such points as: which technical solutions are available for the present (especially the degree of integration of circuits), service problems, and whether the control circuits can be designed using computer–aided design methods.

Let us assume that the automatic sequential operation of a digital system can be described by cooperative sequential circuits. A general problem is then to define the way in which the mutual interaction of a pair of sequential circuits, represented by isolated graphs A and B in Fig. 5-15, occurs. To be specific, assume that the states of graph A define a set of *macro-operations*: at a state i, indicated by a decoder output P_i, a *subroutine* represented by graph B is initiated. With the logic condition $P_i = 1$, system B is released from its latching state j, whereby system A *waits* at state i until the cycle of B has been completed. At the last state, $j - 1$, of B the logic condition indicated by a decoder signal P_{j-1} becomes true and the operation of A is continued. Subsequently, B is latched again at state j.

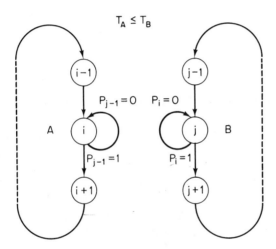

P_i, P_{j-1} = outputs of the decoders of
states i and j−1, respectively

Fig. 5-15. Interactive sequential circuits.

Comment 1. The clock period of system A must be smaller than or equal to the clock period of system B ($T_A \leq T_B$) if the state P_{j-1} is never to be missed by system A. If the clock period of system B should be shorter than the clock period of system A, the branching conditions can be redefined: For system A, use P_j instead of P_{j-1}, and for system B, use P_{i-1} instead of P_i. (Why? Explain this with timing diagrams.)

Comment 2. The total cycle time of system A must be longer than one clock period of system B. Unless this is true, a new cycle may be initiated by A while B is still at state $j - 1$. A modification in which system A waits at state i as long as system B is reserved, is shown in Fig. 5-16. This mode of operation is sometimes met in printout circuits, for example. The systems A and B are not allowed to be at states i and j simultaneously.

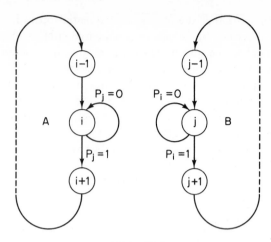

Figure 5-16

5.4. Design Example of an Elementary Control Circuit: Data Logger

In scientific and industrial instrumentation, intermittent acquisition of measurement data is a frequently occurring task. For the recording and processing of information in digital form it is necessary to automatically scan a number of transducers and record their signals in digital form (e.g., on punched tape). The output signals of various types of transducers can be standardized to a common voltage scale, so the problem consists of the execution of a measurement/recording cycle in which every point (called a *channel*) is read and subsequently recorded.

A *digital voltmeter* (DVM) is a device that receives a voltage signal and upon a measurement command in the form of a voltage step or pulse automatically performs one analog-to-digital conversion. After a conversion cycle, usually of the order of 20 to 50 msec, the result is ready in the form of parallel digital signals at the output terminals of the DVM. Very often BCD coding of output values is used. In this example we assume that a DVM is available and that the problem is to design control circuits for *channel selectors* and *punchout operations*.

Definition of the Problem. The design objective is stated as follows: An automatic measurement cycle, scanning m points called *channels*, shall be initiated whenever a control signal is given. Each channel in turn shall be connected to a digital voltmeter through selector relays. The mechanical and electric switching transients need some time before they settle down, which takes, say, 10 msec. After that, a measurement command (step voltage or pulse of a short duration) is given to the digital voltmeter. To eliminate the periodic noise picked up from adjacent power lines, the digital voltmeter forms an average of the transducer signal by integrating it over a time interval equal to one cycle of the line frequency. This time and the interval needed to convert the analog measurement into digital form in the voltmeter takes, say, 50 msec. After this period the digital measurement value has been formed and can be read at the output terminals of the digital voltmeter, in the form of static parallel logic signals. Each digit position has its own BCD signal outputs. These digits and the sign of the number are recorded as separate characters in the usual order on a tape punch, and the solenoids must be driven by, say, 10-msec pulses at least 10 msec apart. Let us take three decimal digits for the representation of the measurement value, after which an *identification mark*, which is used to separate digit groups, is punched.

Character Selection. To each BCD digit there corresponds a unique punch code called a *character*. If a decimal digit is identified in BCD form by four binary signals, D_3, D_2, D_1, and D_0, a *code-converter circuit*, which is a special combinational circuit, forms a proper character of the D_i's. The problem is now to select one decimal digit at a time from corresponding output terminals and substitute it for the D_i's when the punching of this digit is due. This substitution is made with the aid of selector gates and punch control signals $PX, PY, PZ, \ldots$, shown in Fig. 5-17. One of the four-bit groups may represent a special code for the sign and one the identification mark that is to be punched after the decimal digits. The proper codes for the

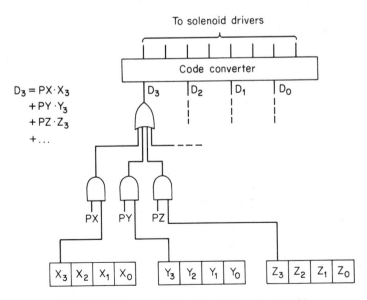

Fig. 5-17. Character selection for tape-punch solenoids.

sign and the identification code shall, of course, not coincide with legitimate decimal number codes. The identification code can be set by manual switches.

It is to be noted that in the selection process the minimum amount of information should be transferred. It is not necessary to select a variable X and its inversion $\bar{X}$ separately; the inversion can be performed on common lines after the selection has been made. Further, it is advantageous to perform the selection in a code in which a minimum number of signal lines are included, before decoding.

Figure 5-17 specifies the selector gates for D_3. Corresponding gates are needed for D_2, D_1, and D_0.

Control Circuits. The whole measuring system with its major parts is shown in Fig. 5-18. The timing signals for it are shown in the diagrams of Fig. 5-19. There are no branchings in the automatic operation and thus each control signal has only one time diagram. In assigning durations for the signal pulses, the time unit has been standardized at 10 msec. The operation of this circuit is explained as follows: Let $P_j(A)$ represent the normal product terms associated with the control counter A ($j = 0, 1, \ldots, 15$) and $P_i(B)$ ($i = 0, 1, \ldots, m - 1$) represent those associated with the channel counter

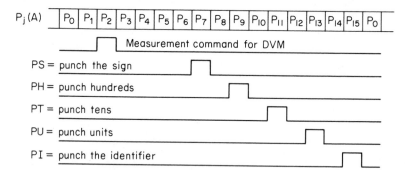

Fig. 5-18. Timing diagram for a data logger.

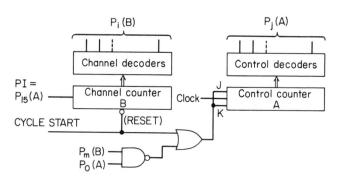

Fig. 5-19. Elementary control circuit for a data logger.

B. The rest states of the counters (i.e., states where these counters halt) are the state 0 of the control counter and the state m of the channel counter. If we assume that $P_m(B) = 1$, which is the case when the channel counter has stopped at state m, and if the "cycle start" signal is 0, we see that the control counter is latched at the state 0 when $P_0(A)$ becomes 1 because the J and K inputs have the value 0. Since no more signals $P_{15}(A)$ are then led to the channel counter, it will also halt at state m, where none of the channels are selected. The operation cycle can now be initiated by the *CYCLE START* signal, which resets the channel counter (whereby the first measuring point becomes selected) and makes the J and K inputs of the control counter equal to 1, thus releasing it from the latching stage. At the first count $P_0(A)$ becomes 0, and inputs J and K remain at the value 1 after the *CYCLE START* signal has ceased. The control counter now repeats cycles (Fig. 5-18). At the end of each, due to the signals $P_{15}(A)$, the channel counter is advanced through the states $0 \cdots m - 1$ so that a similar measurement operation is performed for each measuring channel $0 \cdots m - 1$. Again, when B has counted to the state m, the control counter is latched and the measurement cycle is completed.

5.5. Design Example of Sequential Control Circuits: Auxiliary Arithmetic Unit

The computing load of a central processor in a computer system can be substantially relieved if certain routine tasks, such as multiplication, division, and execution of special algorithms, can be performed in auxiliary arithmetic units. Such a unit is a satellite processor, which is under the control of the main computer. Auxiliary arithmetic units are also needed in simpler data-acquisition systems, in which they perform conversion of measured values into engineering units, calculation of physical quantities from measurements, and so on.

In the present example we assume that two operands are transferred to an auxiliary arithmetic unit by the main processor, and these values are buffered by the operand register R and the accumulator A, respectively. The arithmetic unit shall be capable of performing addition, subtraction, multiplication, or division of these operands upon reception of a corresponding operation code (e.g., 00, 01, 10, or 11, respectively). After a completed operation the result shall be available in the accumulator–MQ register combination, from which it is readily transferred to the main processor. We should point out that this example is of a pedagogic nature and does not necessarily represent circuits realized in practice. Also the design may not be optimal. However, this design reflects features that are common in the design of complete digital systems. In the following we shall exemplify typical steps of a design procedure.

Definition of the Problem. Design an arithmetic unit that operates on two operands stored in an R register and an accumulator, respectively, and which are expressed in 2's-complement representation. Addition, subtraction, and multiplication are to be performed in 2's-complement representation. In division we assume that only positive operands will be present, in order that a signed magnitude division algorithm can be used. A machine cycle corresponding to a clock frequency of, say, 10 MHz is assumed for the system, and all operations are performed in synchronism with the common clock. The accumulator is of the two-phase type, and there is a provision for reading the contents of the R register as 1's complements ($\bar{R}$) when the signal SU ($=$ subtraction) is true. The multiplication and division operations will utilize addition and subtraction subroutines, and a transition to a new micro-operation is to be made as soon as possible.

Instead of using up counters for the definition of clock phases, the execution of arithmetic operations can be speeded up if general conditional sequential circuits are used as state counters. Because sequential systems become complicated if the number of states is large, it is advantageous to organize the control system on hierarchical levels (i.e., to subdivide the system into partial machines which are more easily mastered). Since a standard time unit can be selected for all operations, the control circuitry can be made rather simple. The control signals are derived from the respective states of control counters. In the following, all register operations are conditioned by decoder signals P_i in which the indices i identify different states. Every arithmetic operation is controlled by a separate sequential circuit (program generator) defined by an isolated graph, and the different sequential circuits are mutually interactive. The idea of control hierarchy is illustrated by Fig. 5-20, in which the blocks describe isolated control circuits and the two-way arrows indicate the order of initiation of control sequences.

Macro Control. The five macro states of the arithmetic unit are numbered 0, 1, 2, 3, and 4, and they correspond to the following modes of the system: idling, adding, subtracting, multiplying, and dividing. The macro control circuit is released from its idling state by a short CO ($=$ compute) command, and the branching to various macro states is defined by the operation code O_1O_2 ($=$ 00, 01, 10, and 11, respectively). During the execution of arithmetic operations, the system is latched in macro states. Upon a logic condition RE ($=$ return), which will be defined later, the macro control system returns to its idling state. This is shown in Fig. 5-21.

Addition and Subtraction. The presence of state 1 in the macro control circuit always implies an addition and the state 2 a subtraction. Additions and subtractions are also involved in multiplications and divisions, so we shall introduce two logic variables AD and SU which always initiate an addition

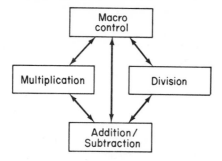

Fig. 5-20. Control hierarchy.

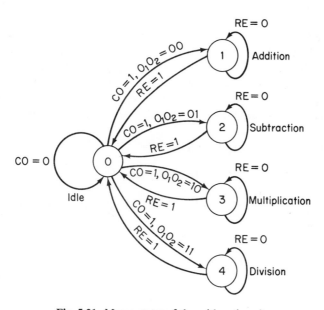

Fig. 5-21. Macro states of the arithmetic unit.

$A \Leftarrow (A) + (R)$ or subtraction $A \Leftarrow (A) - (R)$ when the respective signals are 1. The AD and SU variables will be defined later. Both operations are described by a sequential circuit which is released from the latching state, 5, by $AD + SU = 1$ (Fig. 5-22). In the following we use notations introduced in connection with arithmetic circuits (two-phase accumulator).

P_6: $A \Leftarrow (A) \oplus R'$ where $R' = AD \cdot (R) + SU \cdot (\bar{R})$

P_7: $A \Leftarrow (A) + C_{R'}$ if $AD = 1$ [5-6]

$\qquad A \Leftarrow (A) + C_R + 2^{-N}$ if $SU = 1$

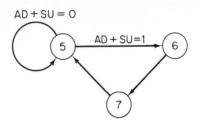

Fig. 5-22. Sequence for addition and subtraction.

Multiplication. A separate state counter for multiplication is latched at state 8 and released by the condition $P_3 = 1$ (Fig. 5-23). Some register operations are performed at states 9 and 11, and state 10 is subject to interactive control conditions with the addition/subtraction operations. At state 10 either AD or SU is made 1 depending on Q_N and Q_{N+1}. The state sequence described by Fig. 5-23 (addition or subtraction) is released and at state 7, $P_7 = 1$, so the multiplication algorithm can proceed to state 11. At this state, according to Booth's algorithm, only cycle counter C is advanced and an arithmetic shift to the right is performed. Depending on Q_N and Q_{N+1}, the successor state will be 11 or 10, until all bits of the multiplier have been used. The stop condition $STM = 1$ ($=$ stop multiplication) is given by the cycle counter. At this condition the multiplication control circuit returns to state 8.

$$P_9: \qquad Q \Leftarrow (A),\ A \Leftarrow 0,\ C \Leftarrow 0,\ Q_{N+1} \Leftarrow 0$$

$$P_{10}: \quad \begin{cases} A \Leftarrow (A) + (R) & \text{if } \bar{Q}_N Q_{N+1} = 1 \\ A \Leftarrow (A) - (R) & \text{if } Q_N \bar{Q}_{N+1} = 1 \end{cases}$$

$$P_{11}: \qquad \text{arithmetic right shift, } C \Leftarrow (C) + 1$$

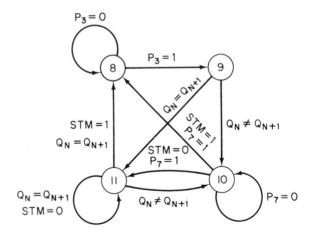

Fig. 5-23. Multiplication sequence.

Division. The sequence starts by the condition $P_4 = 1$, at which point the control circuit is released from its latching state, 12 (Fig. 5-24). The sequence halts at state 14 (subtraction) and 15 (addition) during the execution of one cycle of the addition/subtraction algorithm. State 15 is conditional and is bypassed unless the borrow overflow (BO) signal is 1. At state 16, a left shift is executed and the cycle counter is incremented by one. The successor state of 16 is 14 unless the cycle counter delivers a stop signal $STD = 1$ (= stop division). This signal completes the algorithm by inducing a transition to state 12.

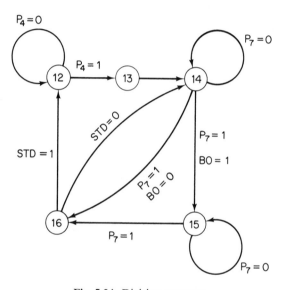

Fig. 5-24. Division sequence.

P_{13}: $C \Leftarrow 0$, shift A and Q left

P_{14}: $A \Leftarrow (A) - (R)$, $Q_{N+1} \Leftarrow 1$

P_{15}: $A \Leftarrow (A) + (R)$, $Q_{N+1} \Leftarrow 0$

P_{16}:   shift A and Q left, $C \Leftarrow (C) + 1$

To complete the interactions in sequential control, we must define the AD and SU signals as well as the return condition RE of the macrocontrol. These are

$$AD = P_1 + \bar{Q}_N Q_{N+1} P_{10} + P_{15} \qquad [5\text{-}7]$$

$$SU = P_2 + Q_N \bar{Q}_{N+1} P_{10} + P_{14} \qquad [5\text{-}8]$$

$$RE = (P_1 + P_2)P_7 + STM + STD \qquad [5\text{-}9]$$

Logical Design of Sequential Control Circuits. As long as the control signals occurring in the state graphs are regarded as external control signals, the four sequential circuits can be designed logically quite independent of each other. This is a standard procedure explained in Chapter 3 and ought to be understood without separate discussion. The isolated graphs can be implemented by $3 + 2 + 2 + 3 = 10$ clocked flip-flops. We shall not question whether material savings could be obtained by merging the graphs.

Micro-operations that are not defined in terms of subroutines are explained in the following:

Micro-operation	*Explanation*
$A \Leftarrow (A) \oplus R'$	executed if $\quad$ HA $= P_6$
$A \Leftarrow (A) + C_{R'}$	executed if $\quad$ CE $= P_7$ and $c_N = 0$
$A \Leftarrow (A) + C_{R'} + 2^{-N}$	executed if $\quad$ CE $= P_7$ and $c_N = 1$
$A \Leftarrow 0$	executed by a direct clear of A by P_9
$Q_{N+1} \Leftarrow 0$	executed by a direct clear of Q_{N+1} by P_9
$C \Leftarrow 0$	executed by a direct clear of C by $P_9 + P_{13}$
$C \Leftarrow (C) + 1$	executed if the counter is enabled by $P_{11} + P_{16}$
$Q \Leftarrow (A)$	executed if the copy gates between A and Q are enabled by P_9

5.6. Microprogramming

If the designer can afford an extra investment on the control circuits, there is an elegant way to systematize the hardware when the complexity of sequential operations might otherwise become overwhelming. This principle is called *microprogramming*. It means in short that all elementary tasks, which are called micro-operations, are given by a memory provided for this purpose only. Since alterations of the *microprogram* are seldom needed, the memory can be a *read-only memory* (ROM), which usually is manufactured, installed, and exchanged as a module. Sequential circuits are also a type of microprogram memory. The memory elements in which the information is stored must be cheap. Microprogramming results in design savings and increase of flexibility when changes are to be made in the program.

The microprogram memory is organized in such a way that *addresses of memory locations correspond to the states of state graphs.* The words in corresponding memory locations contain two kinds of information:

(1) *microinstruction codes* associated with this state, and (2) a *reference or pointer to the next address, which corresponds to the successor state in the graph.* Let us consider the second point first. If there is an unconditional transfer to the next state in the state graph, a unique address of the next state can be given, which is then copied into the address register of the memory, and a new word is fetched from this address. A little more complicated is the situation with branching conditions; the conditioning signals are usually given by external sources. The addresses of the successor states must be derived from the given address by minor modifications (i.e., altering a few bits of it), which thus defines a few alternatives for the next address.

There is a very close relationship between the transfer instructions of a usual computer program and the way in which microinstructions are fetched from the microprogram memory. In fact, a microprogram unit is a kind of minor computer which has fixed contents in the memory and only transfer instructions.

With m bits reserved for the microinstruction code, 2^m micro-operations can be decoded.

When a machine instruction such as multiplication instruction is to

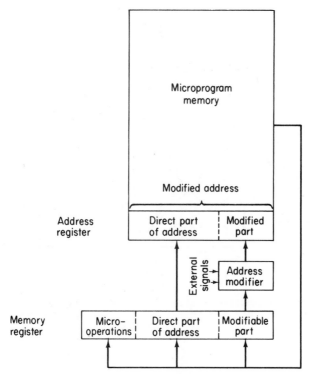

Fig. 5-25. Microprogramming system.

be executed, the beginning address of the multiplication microprogram is copied into the address register of the microprogram memory (Fig. 5-25). After that, the microprogram controls itself under the influence of external conditioning signals, and each microprogram ends in a routine returning the control to the main program.

5.7. Example of Microprogram Control: Arithmetic Unit

The following example has been selected because it allows a comparison of microprogram control with corresponding sequential control. The advantages of microprogramming are more salient in larger systems.

The arithmetic unit described in Section 5.5 is represented by a single state graph in Fig. 5-26, in which different states are identified by indices a to p. There is only one rest state in this system—state a—at which different arithmetic operations are initiated by the application of a CO ($=$ compute)

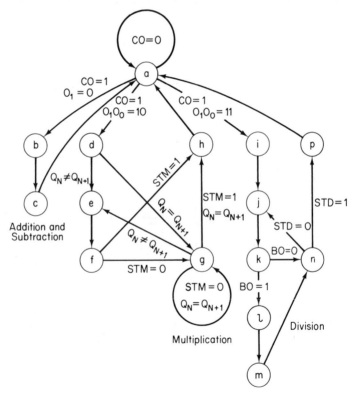

Fig. 5-26. State graph of the arithmetic unit.

command. An operation code is denoted by $O_1 O_0$ ($= 00$ for addition, 01 for subtraction, 10 for multiplication, and 11 for division, respectively), and it determines the branching to different arithmetic operations. In this graph we do not have separate addition and subtraction subroutines, but these operations are incorporated as integral parts in the following sequences:

1. At states b and c, an addition or a subtraction is executed depending on respective operation codes.
2. At states e and f, an addition or a subtraction is executed depending on conditions given in Booth's algorithm.
3. At states j and k, a subtraction is executed.
4. At states l and m, an addition is executed.

More exactly, if a state x is decoded by a Boolean function P_x ($x = a, b, c, \ldots$), the half addition and the addition of carries are always executed at proper clock phases if

$$HA = P_b + P_e + P_j + P_l \qquad [5\text{-}10]$$

$$CE = P_c + P_f + P_k + P_m \qquad [5\text{-}11]$$

Moreover, whether it is an addition or a subtraction is defined by the following conditions:

$$R' = \bar{O}_1 \bar{O}_0 P_b(R) + \bar{O}_1 O_0 P_b(\bar{R}) + Q_N Q_{N+1} P_e(R)$$
$$+ Q_N \bar{Q}_{N+1} P_e(\bar{R}) + P_j(\bar{R}) + P_l(R) \qquad [5\text{-}12]$$

$$c_N = \bar{O}_1 O_0 P_c + Q_N \bar{Q}_{N+1} P_f + P_k \qquad [5\text{-}13]$$

The states d, g, i, and n are equivalent, respectively, to states 9, 11, 13, and 16 of Section 5.6. There are two states h and p which have been added in order to systematize state codes in the microprogram. These states, if wanted, could be used to give a signal indicating that a multiplication or a division, respectively, has been completed.

Address Modification. States a to p can be identified by 4-bit codes of the form $s_3 s_2 s_1 s_0$. Bits s_3 and s_2 are given directly in the microprogram, whereas s_1 and s_0 are conditional. In the case that there are several successor states (maximum of four in this example), bits s_1 and s_0 are defined by selector gates depicted in Fig. 5-27. Here x and y are Boolean functions defined later, and whether it is a pair (s_1, s_0) given in the microprogram or (x, y) is determined by a logic signal MOD ($=$ modify address). If $MOD = 0$ ($\overline{MOD} = 1$), s_1 and s_0 are transferred to the address register without modification.

The assignment of four-bit codes (addresses) to different states of Fig. 5-25 can be made in an arbitrary order. However, to systematize modifi-

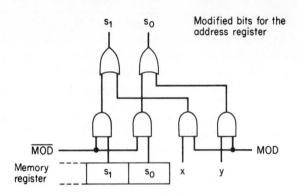

Fig. 5-27. Address-modification circuit.

cation of addresses, we start with states that have several successor states:

Assignment of addresses of the form 00xy:

State	s_3	s_2	s_1	s_0
	Code			
a	0	0	0	0
b	0	0	0	1
d	0	0	1	0
i	0	0	1	1

Addresses of the form 01xy:

g	0	1	0	0
e	0	1	0	1
h	0	1	1	0

Addresses of the form 10xy:

l	1	0	1	1	$\left.\right\}$ successor states to k
n	1	0	0	1	
j	1	0	0	0	$\left.\right\}$ successor states to n
p	1	0	1	0	

Since there are a total of 15 states in the graph, addresses of the form 11xy may now be assigned to the remaining states.

A possible way for the encoding of states is shown in Table 5-1.

The variable bits (x, y), shown as matrix elements in Table 5-2, are defined by branching conditions. When all these conditions are combined, we obtain the circuit equations

$$x = P_a \cdot CO \cdot O_1 + P_f \cdot STM + P_g \cdot STM \cdot E + P_k \cdot BO$$
$$+ P_n \cdot STD \tag{5-14}$$

$$y = P_a \cdot CO \cdot (\bar{O}_1 + O_0) + P_d \cdot \bar{E} + P_g \cdot \bar{E} + P_k$$

**Table 5-1. Encoding of States in the
State Graph**

State	State Code	Successor State	Successor State Code
a	0 0 0 0	a, b, d, or i	0 0 x y
b	0 0 0 1	c	1 1 0 0
c	1 1 0 0	a	0 0 0 0
d	0 0 1 0	e or g	0 1 0 y
e	0 1 0 1	f	1 1 0 1
f	1 1 0 1	g or h	0 1 x 0
g	0 1 0 0	g, e, or h	0 1 x y
h	0 1 1 0	a	0 0 0 0
i	0 0 1 1	j	1 0 0 0
j	1 0 0 0	k	1 1 1 0
k	1 1 1 0	l or n	1 0 x 1
l	1 0 1 1	m	1 1 1 1
m	1 1 1 1	n	1 0 0 1
n	1 0 0 1	j or p	1 0 x 0
p	1 0 1 0	a	0 0 0 0

**Table 5-2. Variable Bits x and y for Different
Branching Conditions**

At state a ($P_a = 1$):

		CO		
O_1 O_0		0	1	
0	0	0 0	0 1	(addition)
0	1	0 0	0 1	(subtraction)
1	0	0 0	1 0	(multiplication)
1	1	0 0	1 1	(division)

At state d ($P_d = 1$):
Denote
$(Q_N = Q_{N+1}) \Leftrightarrow (E = 1)$
$(Q_N \neq Q_{N+1}) \Leftrightarrow (E = 0)$

E	
0	0 1
1	0 0

At state f ($P_f = 1$):

STM	
0	0 0
1	1 0

At state g ($P_g = 1$):

STM		
E	0	1
0	0 1	0 1
1	0 0	1 0

At state k ($P_k = 1$):

BO	
0	0 1
1	1 1

At state n ($P_n = 1$):

STD	
0	0 0
1	1 0

Table 5-3. Complete Microinstructions

			Microinstructions								
			Micro-operations*								
State	State code (address)	Successor state code	HA	CE	$A \Leftarrow 0$	$Q_{N+1} \Leftarrow 0$	$C \Leftarrow 0$	$Q \Leftarrow A$	$C \Leftarrow (C) + 1$	SHR	SHL
a	0 0 0 0	0 0 0 0	0	0	0	0	0	0	0	0	0
b	0 0 0 1	1 1 0 0	1	0	0	0	0	0	0	0	0
c	1 1 0 0	0 0 0 0	0	1	0	0	0	0	0	0	0
d	0 0 1 0	0 1 0 0	0	0	1	1	1	1	0	0	0
e	0 1 0 1	1 1 0 1	1	0	0	0	0	0	0	0	0
f	1 1 0 1	0 1 0 0	0	1	0	0	0	0	0	0	0
g	0 1 0 0	0 1 0 0	0	0	0	0	0	0	1	1	0
h	0 1 1 0	0 0 0 0	0	0	0	0	0	0	0	0	0
i	0 0 1 1	1 0 0 0	0	0	1	0	1	0	0	0	1
j	1 0 0 0	1 1 1 0	1	0	0	0	0	0	0	0	0
k	1 1 1 0	1 0 0 0	0	1	0	0	0	0	0	0	0
l	1 0 1 1	1 1 1 1	1	0	0	0	0	0	0	0	0
m	1 1 1 1	1 0 0 1	0	1	0	0	0	0	0	0	0
n	1 0 0 1	1 0 0 0	0	0	0	0	0	0	1	0	1
p	1 0 1 0	0 0 0 0	0	0	0	0	0	0	0	0	0

* SHR, command pulse for right shift; SHL, command pulse for left shift

The address modification is implemented by the Boolean function

$$MOD = P_a + P_d + P_f + P_g + P_k + P_n \qquad [5\text{-}15]$$

Definition of Micro-operations by the Microinstructions. After the state sequence has been established, the rest of the bits available in the microprogram words can be used to define micro-operations. Each of the individual bits may be reserved for one particular micro-operation which is executed when this bit is 1. In Table 5-3 such *microinstructions* have been exemplified. Note that the two last bits of the successor state code at states a, d, f, g, k, and n are arbitrary and have no effect, because address modification is used.

5.8. Programming

Stored Program Control. The first computers were controlled by mechanically operated code generators (e.g., by continuously running perforated tapes). But *stored program control* was introduced in some of the first electronic computers (e.g., EDSAC at Cambridge University, 1949). This principle, invented by von Neumann, is exemplified in the following.

In most contemporary computers, the stored program control has been implemented at least at the macrolevel. This means that we can define and somehow implement a set of operation sequences called *machine instructions;* included in these are complete arithmetic operations on numbers located in two registers, transfer of the contents of a register or a part of it into another place, etc. Now each sequence corresponding to one machine instruction always proceeds automatically, so usually it is necessary to indicate only some code for the type of operation, the places where the operands are located, and possibly some additional minor specifications (e.g., how long the words are that we are dealing with). The complete computational program is stored in a section of the memory reserved for the program only; this reservation is done during the programming, and almost the whole working memory is usually available for the program allocation except for a few locations in which useful constants or other information is stored and some locations for "working space."

Several machine instructions that are implemented in a sequence may be called a *program block.* Subsequent instructions in computer programs are normally allocated to addresses in numerical sequence. When the program is executed, the instructions are fetched in this order from the memory unless a special instruction, called a *jump instruction*, indicates the place from which the next instruction must be taken. Rather a lot of memory capacity would be wasted if a transfer to an adjacent address had to be indicated every time. Therefore, a convention is made that unless otherwise specified, the next instruction should always be picked up from the next memory location. There is a special counter for this purpose, called an *instruction counter* or *current address register*, which counts up by one each time an instruction is executed. This register then indicates the address of the next instruction. If a skip in the program is necessary, a corresponding new address is automatically copied to it by a special instruction.

It is not necessary to provide a separate memory for the stored programs and another one for the data. A common *working memory* can store both of these, and the following two memory cycles are normally automatically alternated:

1. Fetching the next instruction from the program section.
2. Fetching the data from the data section and executing the instruction that was fetched at item 1.

A typical machine instruction of the simplest type looks like the one in Fig. 5-28(a); the word stored in the memory location is divided into two

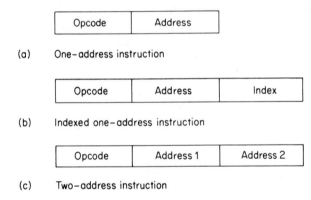

(a)   One–address instruction

(b)   Indexed one–address instruction

(c)   Two–address instruction

Fig. 5-28. Instruction-word types.

parts and one of them contains the *operation code*, shortened to *opcode*. The other part of the word indicates the address of the operand for this operation. Some instructions do not include operands: for example, in a jump instruction an address is indicated in the address part from which the next instruction will be fetched.

The word might also be divided into three or more parts or *fields;* one of these is usually the *index field*, which is used, for example to modify the address or to specify the instructions more closely [Fig. 5-28(b)].

In earlier computers there were two address fields [Fig. 5-28(c)], in which two operands or one operand and the address of the next instruction could be defined. This instruction-word type is now considered uneconomical.

The word length of some computers is so short that the opcode and the address cannot be indicated in the same word. Special provisions are then needed to pick up the complete instruction from the memory in smaller parts. Similarly, a data word might be divided between several locations in the memory. The execution of machine instructions can be described by state graphs, which are closely connected to the electronic operation of the computer; each state is equivalent to a *micro-operation*. In this way, the implementation of machine instructions may be carried out by microprogramming.

All machine instructions together define a *machine language,* in which computational programs can be written. Every computer has its own instruction repertoire.

Most computers have a great number of machine instructions that comes from many operational registers and the way in which information is transferred between them or their portions. Also, there are often numerous types of arithmetic instructions. In principle, there are only a few basic types of machine instructions: *writing* of words from operational registers into specified memory locations; *reading* of numbers from specified memory locations and subsequent implementation of *arithmetic*, *logic*, or other operations between this information and the contents of some operational registers; *branching* instructions, in which a jump to a specified instruction in the program is done unconditionally or conditioned by intermediate results; and the *input/output* instructions, in which information is transferred between operational registers and the input/output devices.

To describe the operation of an automatic computing system, we must first define the machine variables (i.e., all registers and memories) as well as the information transfer paths between them. We shall take a very simple example of a stored-program computer, the *one-address computer*.

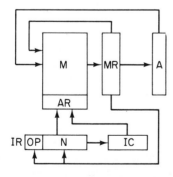

Fig. 5-29. Major blocks of a simple one-address computer.

Its major block diagram is shown in Fig. 5-29. The storage blocks (registers and the memory) are as follows:

1. *Main memory (M).*
2. *Memory address register (AR),* which defines the address of a memory location.

3. *Memory buffer register* (*MR*), to which the contents of a selected memory location are transferred when a read command is given.

4. *Accumulator* (*A*), which is the main arithmetic register of this system, capable of accepting new results of arithmetic operations while using its old contents as one of the operands.

5. *Instruction register* (*IR*), in which a machine instruction is stored in symbolic form as long as its implementation is due.

6. *Opcode part of IR* (*OP*), which is the portion of *IR*, where the machine instruction is specified.

7. *Address part of IR* (*N*), which is the other portion of *IR*, reserved for a number (e.g., a memory location *N* to which *OP* refers).

8. *Instruction counter* (*IC*), which always indicates from which memory location the current instruction must be taken. In a written machine-language program, (*IC*) can be thought to label a machine-instruction statement in the list of instructions.

In Fig. 5-29 the information transfer paths are shown with arrows.

DEFINITION OF A SIMPLE SYMBOLIC MACHINE LANGUAGE

A *symbolic statement* is numbered by a *label* and consists of a mnemonic machine instruction, which is a three-letter operation code, possibly followed by a variable number *N*. This number, written explicitly, may refer to a memory location with this address or to a label. For example, the following is a statement:

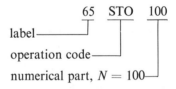

Unless otherwise stated, the instructions are executed in the order shown by the labels.

LIST OF INSTRUCTIONS

1. *No operation* (*NOP*): This operation causes no other changes in the stored data, but the instruction counter *IC* is incremented by one at the execution of this instruction. The purpose of this instruction may be, for example, to reserve a memory location within a stored program for other purposes or to provide a dummy instruction, but it is used rather seldom.

2. *Clear the accumulator (CLA):* A zero is copied into A at this instruction, and the instruction counter is incremented by one.

3. *Store the accumulator (STO N):* With this instruction, the contents of A are written into the memory location N. The instruction counter is incremented by one.

4. *Add to accumulator (ADA N):* The contents of the memory location N are added to the contents of A, and the result is left in A. The instruction counter is incremented by one.

5. *Subtract from accumulator (SUA N):* The contents of the memory location N are subtracted from the contents of A and the result remains in A. The instruction counter is incremented by one.

6. *Add and store to memory (ADS N):* The contents of the memory location N are added to the contents of A, and the result is written back into the memory location N. The instruction counter is incremented by one.

7. *Subtract and store to memory (SUS N):* The contents of the memory location N are subtracted from the contents of A and the result is written back into the memory location N. The instruction counter is incremented by one.

8. *Shift left (SHL):* The word in the accumulator is shifted arithmetically left by one step, whereby the value of the sign bit remains unchanged. The instruction counter is incremented by one.

9. *Shift right (SHR):* The word in the accumulator is shifted arithmetically right by one step, whereby the value of the sign bit remains unchanged. The instruction counter is incremented by one.

10. *Unconditional jump (JMP N):* With this instruction no other operations are executed, but the next instruction to be executed will be the one labeled with N, i.e., a jump to the label N will be made. In other words, the number N is copied into the instruction counter with this instruction. The program then continues by executing the instructions $N, N + 1, N + 2, \ldots$ unless otherwise stated.

11. *Conditional jump (JMC N):* The execution of this program begins by studying the contents of A. If A is zero or negative, the next instruction to be executed will be the one labeled with N, i.e., a jump to the label N will be made. If the contents of A are positive, the program continues from the next label, i.e., the instruction counter is incremented by one.

Instructions describing the transfer of information from input devices and controlling of output devices should be added to the list of instructions. We do not specify input/output operations in this section.

Writing Simple Programs in the Symbolic Machine Language. Consider the algebraic expression

$$X + Y - Z$$

where X, Y and Z are located at the addresses 100, 101, and 102, respectively. This expression must be evaluated and the result stored in address 200, at which point the program has to be stopped. In the symbolic machine language the computations are expressed as follows:

Statement	Explanation
1 CLA	Clear A since we do not know its contents from the foregoing
2 ADA 100	Add X, stored in M⟨100⟩, to A and leave the result in A
3 ADA 101	Add Y, stored in M⟨101⟩, to A and leave the result in A
4 SUA 102	Subtract Z, stored in M⟨102⟩, from A and leave the result in A
5 STO 200	Write A into M⟨200⟩
6 JMP 6	Stop the computation by continuously jumping into this instruction

Routines. There are neither multiplication nor division instructions in our simple machine language, but this does not mean that the instructions cannot be implemented at all. In small computers these operations may be *programmed*, which means that a special program called *routine* or *subroutine* is run every time when operations like these are executed. A routine begins by an unconditional jump instruction to its beginning, and when the routine has been completed, a jump is made back to the original program.

Implementation of Machine-Language Programs. The symbolic machine instructions are translated into binary words and stored in the computer memory, in subsequent memory locations addressed by the labels. Each of the three-letter mnemonic codes of machine instructions has an equivalent binary code, the opcode. The N part of the instruction is expressed as a binary number. If the mnemonic code does not refer to N, the corresponding bits in the N part may be arbitrary, usually 0's.

As a collection of binary words, the stored program is still symbolic, only rewritten from paper on a different medium. It is in the running of the program where the instructions are implemented, each one as a simple sequence of elementary digital operations. The implementation begins by *fetching* the current instruction from the memory and transferring it to the instruction register *IR*. The address of the current instruction is always the same as the current contents of the instruction counter *IC*.

As soon as the opcode resides in the *OP* part of the instruction register *IR*, the output signals of *OP* can be regarded as logic variables that may be used to specify micro-operations needed for the execution of this instruction. We may assume that every opcode is decoded and there is one output for every different opcode. The output signals of this *instruction decoder* are then used as logic conditions for particular micro-operations.

We shall now use the language developed for digital operations to describe the internal operations of the computer in a concise and unique way.

For simplicity we shall assume that the execution of each machine instruction takes the same time. Assume further that a continuously running *control clock* is available. The clock produces seven mutually delayed waveforms, denoted by $P_1, P_2, \ldots, P_7$, such that each one has the value 1 only during the corresponding *clock phase* shown in Fig. 5-30.

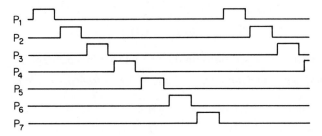

Fig. 5-30. Clock phases for the computer example.

Assume that the program has been stored in subsequent locations starting with the address *n*. In the beginning of computations, *n* has been loaded into *IC*, e.g., using pushbuttons.

FETCHING PHASE

First the current instruction is fetched by the following operations:

$P_1 : AR \Leftarrow (IC)$

$P_2 : MR \Leftarrow (M\langle (AR)\rangle)$ or, briefly, $MR \Leftarrow ((AR))$

$P_3 : IR \Leftarrow (MR)$ $[M\langle (AR)\rangle \Leftarrow (MR)$ or, briefly, $(AR) \Leftarrow (MR)]$

The operation shown in the brackets at P_3 is necessary only with *destructive memories*, in which the reading of a memory location will clear the contents (e.g., in ferrite-core memories). While the address is selected, the contents must be written back to the memory from the buffer register MR. In any case, the current machine instruction now exists in IR. We shall discuss different types of instructions separately. Let us denote the presence of a certain operation code in OP by a Boolean function that bears the same name as the three-letter mnemonic code.

In the following we use the shorter notation for memory locations.

EXECUTION PHASE

No operation (NOP):

$$NOP \cdot P_4: \qquad IC \Leftarrow (IC) + 1$$

Clear the accumulator (CLA):

$$CLA \cdot P_4: \qquad A \Leftarrow 0, \qquad IC \Leftarrow (IC) + 1$$

Store the accumulator (STO N):

$$STO \cdot P_4: \qquad AR \Leftarrow N, \qquad IC \Leftarrow (IC) + 1$$
$$STO \cdot P_5: \qquad (AR) \Leftarrow (A)$$

Notice that the two operations on the first line are independent of each other.

Add to accumulator (ADA N) and *Subtract from accumulator (SUA N):*

$$(ADA + SUA) \cdot P_4: \qquad AR \Leftarrow N, \qquad IC \Leftarrow (IC) + 1$$
$$(ADA + SUA) \cdot P_5: \qquad MR \Leftarrow ((AR))$$
$$ADA \cdot P_6: \qquad A \Leftarrow (A) + (MR)$$
$$SUA \cdot P_6: \qquad A \Leftarrow (A) - (MR)$$
$$(ADA + SUA) \cdot P_7: \qquad [(AR) \Leftarrow (MR)]$$

Add and store to memory (ADS N) and *Subtract and store to memory (SUS N):*

$$(ADS + SUS) \cdot P_4: \qquad AR \Leftarrow N, \qquad IC \Leftarrow (IC) + 1$$
$$(ADS + SUS) \cdot P_5: \qquad MR \Leftarrow ((AR))$$
$$ADS \cdot P_6: \qquad A \Leftarrow (A) + (MR)$$
$$SUS \cdot P_6: \qquad A \Leftarrow (A) - (MR)$$
$$(ADS + SUS) \cdot P_7: \qquad (AR) \Leftarrow (A)$$

Shift left (SHL) and shift right (SHR):

$$SHL \cdot P_4: \qquad A \Leftarrow \rho_L(A), \qquad IC \Leftarrow (IC) + 1$$
$$SHR \cdot P_4: \qquad A \Leftarrow \rho_R(A), \qquad IC \Leftarrow (IC) + 1$$

(ρ_L and ρ_R stand for left and right shifting operations, respectively.)

Unconditional jump (JMP N):

$$JMP \cdot P_4: \qquad IC \Leftarrow N$$

Conditional jump (JMC N): Let a Boolean variable F denote the condition that the contents of the accumulator are zero or negative:

$$JMC \cdot F \cdot P_4: \qquad IC \Leftarrow N$$
$$JMC \cdot \bar{F} \cdot P_4: \qquad IC \Leftarrow (IC) + 1$$

Execution of Machine Instructions. An inspection of the micro-operations of a stored program computer shows that in general they are executed at logic conditions of the type

$$f_k = \sum_{i,j} A_{ijk} F_i T_j \qquad\qquad [5\text{-}16]$$

where F_i is a logic signal telling us that a corresponding machine instruction exists in the instruction register *IR*, T_j is a clock phase, and A_{ijk} (0 or 1) is used to specify the logic condition. Thus the purpose of a control unit in a computer is to produce all the needed micro-operations for each machine instruction during subsequent clock phases T_j. Figure 5-31 shows the principle of the control unit: For every opcode of a machine instruction, a corresponding signal F_i is formed by a decoder. It may be assumed for simplicity that only one F_i is 1 at a time. A clock oscillator drives a timing signal generator, which produces the clock phases T_j. In simpler machines there may be a constant number of T_j's for all machine instructions, whereas in more

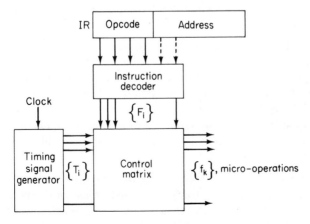

Fig. 5-31. Control unit of a stored program computer.

complicated computers, a more advanced control is needed. Thus the T_j's may also be outputs of a sequential control circuit depending on F_i's.

The F_i's and T_j's are inputs to a combinational circuit, sometimes also called the *control matrix*. This circuit may produce direct control signals for micro-operations or these signals may only enable gates for micro-operations. The control matrix is described by Eq. [5-16].

PROBLEMS

5-1   Design the simplest counter–decoder system that generates the one-shot waveforms G and H depicted in Fig. P5-1 following a synchronous triggering signal F.

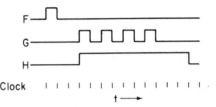

Figure P5-1

5-2   In a channel selector, there are 18 numbered points which are scanned in certain ways depending on external control signals. When $A = 0$, the system is latched at point 1. When A becomes 1 for one clock period, the scanning begins. For $B = 0$, the scanning sequence is 1, 2, 4, 5, 7, 8, 10, 11, 13, 14, 16, 17, 1, whereas for $B = 1$ the sequence is 1, 2, 3, . . . , 17, 18, 1 (all points). Design a control circuit that carries out such channel selection. (*Hint:* Try to group the points in a suitable modulus.)

5-3   Modulo 10 Johnson counters can be cascaded to form counters in modulo 10^n, $n = 2, 3, ...$ Design interconnecting control circuits by which three of such modulo 10 self-corrected counters can be converted into a single 15-bit end-around right-shift register when a signal F is 1 but which work as a modulo 1000 counter in the way mentioned when $F = 0$. (*Hint:* For the J and K inputs of the leftmost flip-flops in every Johnson counter, design suitable combinational circuits.)

5-4   A paper tape reader reads standard eight-channel perforated tape, of which one *character* (fully perforated) is shown in Fig. P5-4. There are photoelectric detectors, one at each hole position, which produce

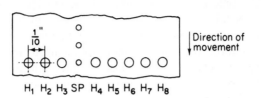

Figure P5-4

logic signals H_1, H_2, ..., H_8 (for data holes), and SP (for the sprocket hole) when a character is positioned in the reader. When the tape is moving, perforated data holes produce logic 1 pulses, whereas for nonperforated positions, the logic signals are 0. Because the sprocket holes are smaller, they produce shorter logic 1 pulses at every character, and this signal is used for synchronizing and sampling purposes (e.g., data are read only during $SP = 1$).

The tape moves when a logic control signal M of the reader is 1 but stops when $M = 0$. There is a small delay in the moving mechanism with respect to the M signal. This delay may be longer than the width of the SP pulse, but it is always shorter than the interval between two SP pulses for a moving tape.

A buffer register for the storage of three characters is made of 24 direct R-S flip-flops. Design suitable selector gates and a control circuit by which the three next characters on the tape are read and loaded into the buffer register at a control signal $P = 1$, after which the reader stops. Discuss separately the cases in which (a) P is longer than the time for loading and (b) P is very short. If the initial position of the tape in the reader is such that $SP = 0$, the three following characters will be read. If $SP = 1$ initially, the reading must begin from the next character. (*Hint:* The only information for the control operations is contained in the SP and P signals. The auxiliary circuits may be usual counters, decoders, and other logic circuits. It is advisable to use the edges of the SP waveform for the counting of characters.)

5-5   In direct digital control of a process, an actuator with control signals M and D moves in one direction as long as $M = 1$ and $D = 1$, and in the opposite direction when $M = 1$ and $D = 0$. In all other cases the actuator stops. A digital computer is used to control the movements of this actuator in the following way: A correction of the position of this actuator is given every second in the form of a rapid burst of short pulses, which, however, are too short to cause any movement. The number of pulses, 0 to 7, given on one signal line indicates the number of clock periods of a 10-Hz clock, which is equal to the time that the actuator will be moving, and a signal pulse on another line tells the direction of movement. Obviously there must be two signal memories for the actuator: one for the time, the other for the direction. These memories remember the orders given by the computer. The memories must be automatically reset when the correction has been made. Using a counter driven by the control signals, a 10-Hz clock, and a flip-flop, construct such a buffer memory for the actuator.

REFERENCES

The logical design of digital systems is a kind of art, comparable to architecture. It is rather difficult to point out detailed treatments of digital-system design. However, we should like to mention the books of Chu [17], [18], Phister [104], and also

Bartee, Lebow, and Reed, *Theory and Design of Digital Machines* [7]

which provide valuable information about systematic design of digital systems. The book

Buchholtz, *Planning a Computer System* [12]

gives a view of the architecture of a large and fast computer. The principles of microprogramming have been described in

Husson, *Microprogramming: Principles and Practices* [236]

The best source for examples of all kinds of control circuits of digital systems and devices can be found in handbooks describing existing commercial equipment.

part two

DIGITAL ELECTRONICS

chapter 6

Fundamentals of
Digital Switching

In this chapter we shall begin a discussion of electronic digital circuits from a physical point of view. It may be supposed that the reader is already familiar with the basic theory of semiconductor components and circuits. The purpose of the chapter, however, is to recapitulate some fundamental results of this theory in a form that is also comprehensible to those who do not wish to search for the necessary background in the semiconductor literature. It has been felt that a review of the basic nonlinear terminal voltage–current characteristics in a simplified analytical form is needed, to explain the large-signal behavior of semiconductor circuits (i.e., the switching concept).

6.1. Voltage and Current Switching

Voltage and Current Dividers. Logic voltage levels are set with *electronic switches*. Any component that exhibits low impedance under certain operating conditions and high impedance in others may be regarded as a switch. The two basic types of elementary switching circuits are called the *voltage switching circuit*, or *voltage divider*, and the *current switching circuit*, or *current divider*. Figure 6-1 shows the two principles of operation that are usual in digital systems. Here v_i and i_i are a constant voltage source and a constant current source, respectively. Both principles make use of a nonlinear two-port voltage–current characteristic:

$$v = f_j(i), \qquad j = 1, 2 \qquad [6\text{-}1]$$

In Fig. 6-1(a), the devices denoted by f_1 and f_2 are connected in series as a *voltage divider*. In Fig. 6-1(b), *current is divided* between the devices f_1 and f_2.

219

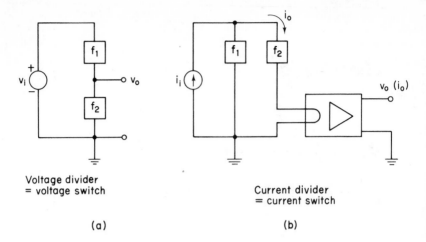

Voltage divider
= voltage switch

(a)

Current divider
= current switch

(b)

Fig. 6-1. Basic types of electronic switches.

Since current levels are not normally defined as logic signals, the output current i_o is usually converted into a dependent output signal voltage $v_o(i_o)$ in a separate output circuit.

Nonlinear Resistive Voltage Divider. To determine v_o in Fig. 6-2, where the supply voltage v_i and the voltage–current characteristics of the nonlinear resistors R_1 and R_2 are known, we may use the well-known *load-line analysis*, modified for nonlinear relations. Without external loads, the current i through R_1 and R_2 is the same, and thus

$$v_o = f_2(i) \qquad [6\text{-}2]$$

$$v_i - v_o = f_1(i) \qquad [6\text{-}3]$$

where f_1 and f_2 are the voltage–current characteristics of R_1 and R_2, respectively. Equations [6-2] and [6-3] can be solved numerically or by a graphi-

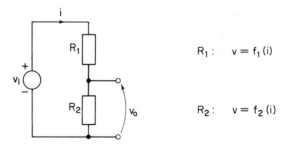

$R_1 : \quad v = f_1(i)$

$R_2 : \quad v = f_2(i)$

Fig. 6-2. Nonlinear resistive voltage divider.

cal analysis. In the latter, they are plotted into v_o-i coordinates, with fixed v_i. Notice that the curve represented by $v = f_1(i)$ is first reflected with respect to the i axis and then translated to the right by an amount v_i, after which it is represented by Eq. [6-3]. This is shown in a series of diagrams, Fig. 6-3.

If R_1 and/or R_2 are modulated (controlled) by external signals, we have a family of curves for f_1 and/or f_2, respectively, and the operating point is found as above for all values of the control signals.[1]

There are also other methods for the analysis of nonlinear voltage dividers (e.g., the graphical elimination method discussed by Chua [19]).

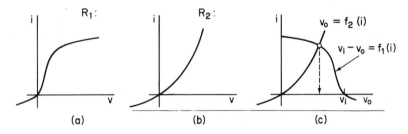

Fig. 6-3. Graphical solution for voltage divider.

Nonlinear Current Divider. If R_1 and R_2 are connected in parallel and fed from a current source as shown in Fig. 6-4, we have a dual of the voltage divider. Similar methods as before can thus be used for the solution of the division of current (Fig. 6-5). Since the voltage v over R_1 and R_2 is the same, we have

$$i_o = f_2^{-1}(v) \qquad [6\text{-}4]$$

$$i_i - i_o = f_1^{-1}(v) \qquad [6\text{-}5]$$

where f_1^{-1} and f_2^{-1} represent the same voltage–current characteristics as f_1 and f_2 but in $i-v$ coordinates.

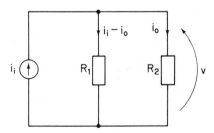

Fig. 6-4. Nonlinear resistive current divider.

---

[1] In the asymptotic cases where R_2 is an open or a closed switch, the transfer characteristics of R_2 have the v axis and the i axis, respectively, as their asymptotes.

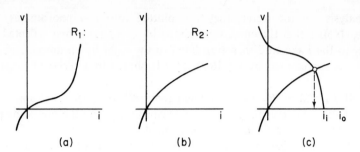

Fig. 6-5. Graphical solution for current divider.

Loading Effects in Nonlinear Networks. The voltage–current charac-teristics over terminals indicate the sensitivity of circuits to loading and are

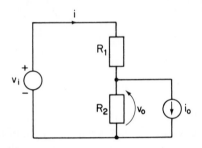

Fig. 6-6. Loaded voltage divider.

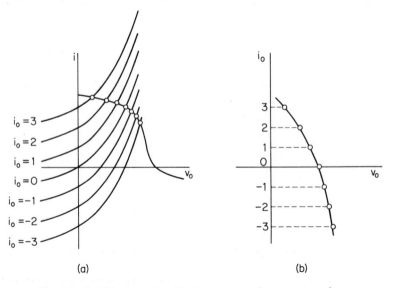

Fig. 6-7. Graphical solution for the output voltage–current char-acteristic (loading curve) of a voltage divider.

usually experimentally established for all practical nonlinear (digital) circuits. Here we shall discuss a nonlinear voltage divider. The current divider is the dual of it and is also treated analytically for diode and transistor networks (Sections 6-4 and 6-5). Let R_2 in Fig. 6-6 be shunted by a load current generator i_o. We have to find the output voltage–current characteristics $v_o(i_o)$. Now we have for R_2 and R_1, respectively,

$$v_o = f_2(i - i_o) \qquad\qquad [6\text{-}6]$$

$$v_i - v_o = f_1(i) \qquad\qquad [6\text{-}7]$$

The shift in the argument of f_2 means that the corresponding curve is shifted *upward* by i_o. Plotting a family of curves according to Eq. [6-6] with i_o as parameter, and searching for intersections with the plot of Eq. [6-7], we obtain the desired relation, as shown in Fig. 6-7.

6.2. Structures of Logic Circuits

Voltage Switches in Inverter Logic. Especially in *large-scale integration* (LSI) of electronic circuits, logic conditions are often set with voltage switches in very much the same way as logic circuits are made of electromechanical contacts. For this purpose the voltage divider must be controllable by logic voltages. The basic element in voltage switching circuits is the *inverter switch*, also called simply the *inverter*. Here we shall use the first name in order to avoid confusion with inverter gates. The inverter switch has a small impedance over its two output terminals when the voltage between the control electrode and the lower output terminal is logical 1, but the terminal impedance is high when the corresponding control voltage is logical 0. Figure 6-8(a) shows a commonly used symbol for an inverter switch (with positive voltages), and the equivalent output circuit is shown in Fig. 6-8(b).

Using a common load resistor R_L, a NOR operation is implemented

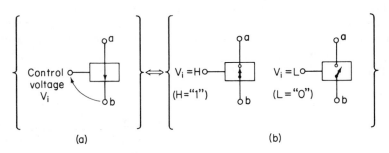

(a)   (b)

Fig. 6-8. Inverter switch.

as in Fig. 6-9(a), by a parallel connection of switches. Also NAND and more complicated combinational operations can be implemented by a series or parallel–series connection of inverters, respectively, although special care is needed in the discussion of input control voltages since they are always referred to the lower terminal of the corresponding inverter switch [Fig. 6-9(b) and (c)]. The logic network principle of Fig. 6-9 is often referred to as *inverter logic.*

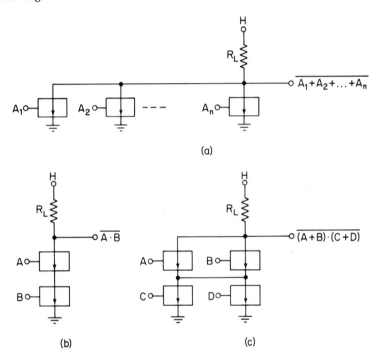

Fig. 6-9. Inverter logic networks.

Classical Structure of Logic Gates. Most bipolar logic families discussed in Chapter 7 are implemented as separate gates by circuits of the type expressed in Fig. 6-7. The input-gate part (CS) is a set of current switches controlled by input voltages (Fig. 6-10). The gate part produces a dependent control current I_o to the output circuit, which usually is of the voltage-switch type (VS). In many logic gates, different output circuits can be combined with different types of input gates.

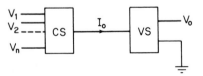

Fig. 6-10. Main parts of logic gates: CS, logic-gate part (current switch); VS, output circuit (voltage switch).

Combined Principles. In medium- and large-scale integrated circuits, circuit topology sets some restrictions on the design of logic networks which can be overcome by using voltage- and current-switching principles in turn. Also some steps of fabrication can be saved and electrical characteristics of circuits simultaneously improved if logic circuits are designed using combinations of current and voltage switches. For example, a two-level AND–NOR circuit is often implemented as shown in Fig. 6-11.

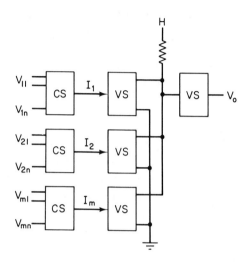

Fig. 6-11. Two-level combined logic circuit implemented by current switches (CS) and voltage switches (VS).

6.3. Voltage-Transfer Characteristics and Noise Margins

Whatever implementation for logic circuits we choose, we may state that electronic logic gates are circuits that have highly nonlinear voltage transfer characteristics. If the input voltages are denoted by V_i, $i = 1$ to n, and the output voltage is V_o, the relation

$$V_o = f(V_1, V_2, \ldots, V_n) \qquad [6\text{-}8]$$

is normally single-valued. (Gates with multivalued input–output relations are sometimes used in special applications, e.g., in tunnel diode circuits.) However, digital circuits should operate on signals for which accurate voltage values are not pertinent but which are confined in certain voltage classes. These classes were called *logic levels* in Chapter 2. In order that digital circuits also operate in the intended way when moderate noise is superimposed on the signals, it is necessary that the normally produced output voltages are confined to narrower limits than the allowed limits of variations of input signals. In other words, the logic circuits must have a *contracting effect* on logic voltages, and the margins that are left between the output voltages and the class limits are called *noise margins*. We shall now give an exact definition for noise margins and clarify the stabilizing properties of logic gates on digital signal voltages.

Noise Margins in an Inverter Chain. Without loss of generality, all the important concepts associated with noise margins can be defined in a chain of cascaded inverters. An extension to multiple-input logic gates will be made in a separate section. First we define *noise sensitivity* (also called *dc noise margin*), *noise margins*, *ac noise margins*, and *noise immunity*.

NOISE SENSITIVITY

Because of variations in temperature, source voltages, and component values, we cannot express a unique voltage-transfer characteristic $V_o(V_i)$ for all inverters. Nevertheless, we assume that over an ensemble of practical inverters, two envelopes of $V_o(V_i)$ can be defined (e.g., by Fig. 6-12).

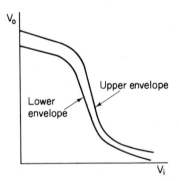

Fig. 6-12. Envelopes of $V_o(V_i)$ for an inverter.

In a chain of cascaded inverters, the output of the former is the input for the latter. Each inverter changes the binary value of the signal, but for each stage, the point (V_i, V_o) shall stay within the envelopes. In Fig. 6-13, let $X(k)$ be the logical signal entering inverter k. The output signal denoted by $Y(k)$ is at the same time an input signal to inverter k', and $X(k + 1)$ is the output of inverter k'. Assuming that the transfer characteristic defined

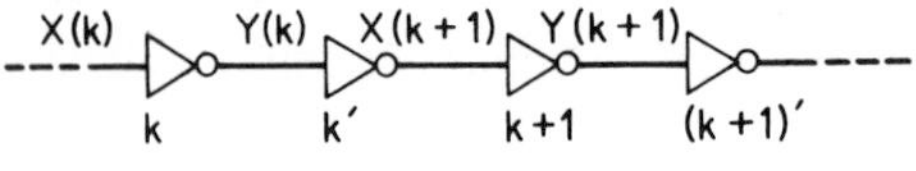

Fig. 6-13. Chain of inverters.

by Fig. 6-12 is valid for all inverters in this chain, it is convenient to use an iterative graphical method for the analysis of signal propagation. In Fig. 6-14 relation R_1 is the same as depicted in Fig. 6-12 with $X(k)$, $X(k + 1)$, ... as the abscissa and $Y(k)$, $Y(k + 1)$, ... as the ordinate. R_1 holds for inverters labeled k, $k + 1$, etc. On the other side, relation R_2 in Fig. 6-14 is otherwise the same as R_1 but with $Y(k)$, $Y(k + 1)$, ... as the abscissa and $X(k)$, $X(k + 1)$, ... as the ordinate (i.e., with X and Y interchanged). Thus relation R_2 holds for inverters labeled k', $(k + 1)'$, etc.

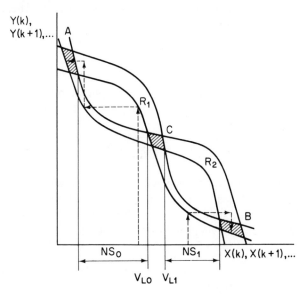

Fig. 6-14. Graphical method for the determination of noise sensitivity.

The areas denoted by R_1 and R_2 have three intersectional areas, A, B, and C. The limits of C in horizontal direction are denoted by V_{L0} and V_{L1}, respectively. Assume now that the voltage $X(k)$ takes an arbitrary positive value. Three cases are now distinguished:

1. $X(k) < V_{L0}$: For any value of $X(k)$ in this range we can show that with increasing length of inverter chain the logic signal finally converges to the region denoted by A: The $Y(k)$ corresponding to $X(k)$ is found from the intersection of the line $X(k) = \text{const.}$ with R_1. Any one of these values of $Y(k)$ is a possible abscissa for R_2 from which a value $X(k + 1)$ is obtained, etc. An arrow in Fig. 6-14 shows a case of signal convergence, and normally the stabilization of logic signals occurs in a few stages.

2. $X(k) > V_{L1}$: In the same way as in case 1, a voltage $X(k)$ from the indicated range will after a few stages converge to region B, as indicated by another arrow in Fig. 6-14.

3. $V_{L0} \leq X(k) \leq V_{L1}$: This is obviously a case in which we cannot state for certain whether it is region A or B to which the signal will converge. Consequently, this range of $X(k)$ must be excluded.

It should now be clear why we call NS_0 and NS_1 defined by Fig. 6-14 the *noise sensitivity of logic* 0 and *logic* 1, respectively. (These values are also called *dc noise margins*.)

NOISE MARGINS

If a small noise component with an amplitude $\Delta X(k)$ is superimposed on $X(k)$, it will cause in the kth inverter an output amplitude

$$\Delta Y(k) = \frac{dY(k)}{dX(k)} \Delta X(k) \qquad [6\text{-}9]$$

and this deviation is further propagated in the later stages. In order that noise should not grow up but rather be attenuated in cascaded stages, it is advisable to define the legitimate logic levels in such a way that for any particular inverter in the ensemble having the input–output relation $V_o(V_i)$,

$$\left| \frac{dV_o}{dV_i} \right| < 1 \qquad [6\text{-}10]$$

over the ranges of logic levels.

The points on a particular characteristic in which $dV_o/dV_i = 1$ are called *unity gain points*, and there are usually two of them, corresponding to logical 0 and logical 1. When we have a family of characteristics, it is natural to define the *worst-case unity gain points* so that for logical 0 this point lies on the lower envelope in the unity gain point and for logical 1 this point is on the upper envelope (obviously giving narrowest or worst-case margins). Now we are in the position to define the *noise margins* NM_0 and NM_1, which are those values usually given in data sheets. From Fig. 6-15 we obtain their accurate meaning.

AC NOISE MARGINS

Because most electronic logic gates act as a sort of low-pass filter for fast signals, the effective noise margins for very short pulses are usually larger than those given before and, of course, depend on the pulse length. It is very

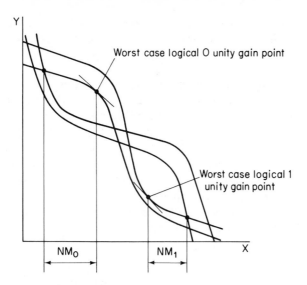

Fig. 6-15. Definition of noise margins.

difficult to give analytical expressions for ac noise margins; they are figures which are usually determined by measurements over a batch of certain logic gates. The typical dependence of ac noise margins on pulse width is depicted in Fig. 6-16.

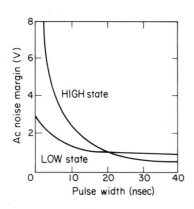

Fig. 6-16. The typical dependence of ac noise margins on pulse width.

NOISE IMMUNITY

The difference between a logic 1 and a logic 0 voltage is called *logic swing*. The smallest logic swing is obviously the smallest difference in the abscissas of the points lying in areas *A* and *B* in Fig. 6-14. A good figure of merit of the *noise immunity* of a logic circuit family is the ratio of noise sensitivity (of logic 0 or logic 1, whichever is smaller) to logic swing. The internally generated noise in a digital system is approximately directly proportional to the logic swings, and therefore it is understood that noise sensitivity (and noise margins, too) should be in the correct proportion to the swing.

Noise Margins of Multiple-Input Logic Gates. Let us discuss the input–output relationships of a logic gate (e.g., NAND) in terms of the transfer characteristic between one of the inputs and the output. This relation depends on the state of the other inputs according to Fig. 6-17(b), where we assume that for NAND, A holds if all other inputs are 1, and B holds if at least one of the other inputs is 0. With a series of cascaded logic circuits, we can proceed in a fashion similar to that for inverters. The only difference is that we have two alternative input–output relations for each stage, and this results in the diagram of Fig. 6-18. The values of noise sensitivity and noise margins of multiple-input gates do not usually differ from the inverter case,

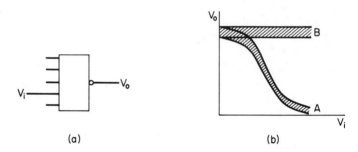

(a)   (b)

Fig. 6-17. Input–output relations of a NAND gate.

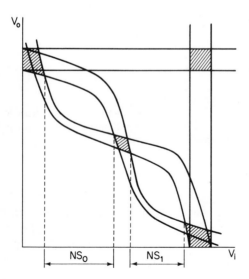

Fig. 6-18. Noise margins of a multiple-input NAND gate.

and the standardizing and noise-attenuating properties of the circuits are also obvious.

6.4.   The Semiconductor Diode as a Switching Component

Diode networks play an important role in many logic gates and they are normally used as current switches. The analysis of diode switching networks starts by setting up an analytical nonlinear model for the diode.

Ideal and Modified Diode Models. Models that are based on a simplified physical theory are called *ideal models*. An example of this kind is the *first-order diffusion-theoretical model for the voltage–current relation $I = I(V)$ of a diode*

$$I = I_s(e^{qV/kT} - 1) \qquad [6\text{-}11]$$

with only one component parameter I_s, the *saturation current* (which depends on temperature). Here q is the electron charge, k the Boltzmann constant, and T the absolute temperature. At room temperature $kT/q = 26$ mV. This model is reasonable for germanium diodes, as is well known, but for silicon devices there are considerable departures. Let us now see whether these departures are essential. With reverse voltages of a few volts, Eq. [6-11] gives values for silicon diodes roughly 1 million times too small, because *pair-generation currents* in the depletion region were neglected. On the other hand, if the diode is used as a digital switch, it is not pertinent whether the reverse leakage current of the diode is 10^{-8} A or 10^{-14} A, as long as it is small. A much more important correction to Eq. [6-11] is to take *recombination in the depletion layer* into account, and thus a formula of the form

$$I = I_s(e^{V/\eta V_T} - 1) \qquad [6\text{-}12]$$

with *two* component parameters, I_s and η ($\eta \simeq 2$ for Si diodes and $\simeq 1$ for Ge diodes), can be established. For brevity, we have denoted

$$V_T = \frac{kT}{q} \qquad \text{(called the *thermal voltage*)} \qquad [6\text{-}13]$$

Equation [6-12] is still physically oriented, and we call it the *modified diode equation*.

Departures from the Diode Equations. At modest forward currents (less than or equal to about 1 mA for signal diodes), or if the reverse voltage is small, Eq. [6-12] agrees rather well with experimental V–I relations. For Ge, the agreement is much better than for Si; in the latter, at least two important types of departures from the ideal model should be taken into account.

1. A voltage drop due to the *series resistance of the bulk volume* is observed with forward currents. With higher currents, the series resistance is decreased, however, as a result of the *high-level injection effect*. This means that the majority charges, when injected over the junction, become minority charges on the opposite side, and if this injection is sufficiently large, an increase in majority charges is induced also, owing to the neutrality condition of the semiconductor. Since the conductivity of the bulk material is mainly determined by the majority charge density, it is seen that at heavy currents the resistance decreases. The high-level injection effect is slower than the usual conduction effects and will be discussed later in more detail. The series resistance is usually a few tens of ohms.

2. With reverse voltages exceeding about 1 V, the leakage currents in a silicon diode are substantially larger than predicted from Eq. [6-12] and increase progressively with increasing reverse voltage. At least two phenomena account for this increase. In order of relative importance they are the *thermal pair generation in the depletion layer* and the *surface leakage currents*. Notice that with increasing reverse voltage, the volume of the depletion layer increases. Since the amount of generated pairs per time unit is proportional to the volume of the depletion layer, this calls for a progressive increase in the reverse current with increasing voltage. The other leakage currents may be described by Ohm's law.

Breakdown Voltages. At higher voltages, ranging from a few volts to hundreds of volts with different components, the *breakdown phenomena* manifest themselves. These are of two kinds: the *avalanche breakdown* and the *Zener effect*. The avalanche breakdown occurs first if the impurity concentration is less than about 10^{18} atoms/cm³, and it is due to impact ionization of the semiconductor atoms caused by electrons that are accelerated in the depletion layer to sufficiently large kinetic energies. The density of mobile charges grows up until ohmic voltage drops formed in the material stabilize it. On the other hand, if the doping of the material is larger than 10^{18} atoms/cm³, it is the Zener breakdown that occurs first. In this phenomenon, the electric field in the depletion layer is so high that it is capable of lifting valence electrons to the conduction band by the quantum-mechanical tunnel effect, thus causing a rapid increase in the conductivity.

This effect is also called field ionization. The avalanche breakdown and the Zener effect manifest themselves in the same way in the experimental $V–I$ relation of the diode; when a certain breakdown voltage in the reverse direction is exceeded, dV/dI (the incremental resistance) has a value of a few ohms.

Experimental $V–I$ Curves of Diode. Figure 6-19 shows how an actual $V–I$ relation departs from the modified diode equation.

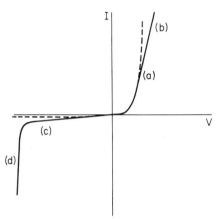

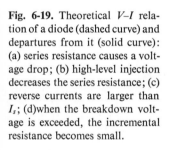

Fig. 6-19. Theoretical $V–I$ relation of a diode (dashed curve) and departures from it (solid curve): (a) series resistance causes a voltage drop; (b) high-level injection decreases the series resistance; (c) reverse currents are larger than I_s; (d)when the breakdown voltage is exceeded, the incremental resistance becomes small.

Transition Capacitance of Junction Diode. For small currents (i.e., mainly for reverse voltages over the junction) it can be shown that the depletion layer between the conductive bulk volumes is equivalent to a capacitor that has the same dimensions and the same dielectricity constant as this layer. This capacitance, called *transition capacitance*, is inversely proportional to the width of the depletion layer, which varies with the applied voltage. To be more accurate, we should always speak of *incremental capacitance*, which is defined as dQ/dV; here dQ is the charge that must be transferred to cause a change dV in the junction voltage. Two types of junctions should be distinguished. In the *abrupt junction*, the doping on both sides is constant up to the junction. In a *linearly graded junction* the difference between donor and acceptor densities is

$$N_D - N_A = bx \qquad [6\text{-}14]$$

where x is the coordinate perpendicular to the junction and b is a constant. On the N side $x > 0$. In a graded junction, the width of the depletion layer is

$$l = \frac{1}{2}\left[\frac{12\epsilon}{qb}(-V + V_D)\right]^{1/3} \qquad [6\text{-}15]$$

where ϵ is the dielectric constant, q the elementary charge, and V_D the diffusion voltage for Si ($V_D \simeq 0.6$ to 0.8 V). In an abrupt junction, a depletion layer with a width l_p on the P side and l_n on the N side is formed. Denoting

$$N_D - N_A = \begin{cases} +N & \text{for } x > 0 \\ -P & \text{for } x < 0 \end{cases} \qquad [6\text{-}16]$$

we have

$$l_p = \left[\frac{2\epsilon}{q} \frac{N}{P(N+P)} (-V + V_D) \right]^{1/2} \qquad [6\text{-}17]$$

$$l_n = \left[\frac{2\epsilon}{q} \frac{P}{N(N+P)} (-V + V_D) \right]^{1/2} \qquad [6\text{-}18]$$

$$l = l_p + l_n$$

For abrupt junctions, V_D is known in closed form:

$$V_D = \frac{kT}{q} \ln \frac{NP}{n_i^2} \qquad n_i = \text{inversion density}, \simeq 10^{10} \text{ cm}^{-3} \qquad [6\text{-}19]$$
$$\text{for Si at room temperature}$$

which yields about the same value as for graded junction. The transition capacitance C_T is now

$$C_T = \epsilon \frac{A}{l} \qquad (A = \text{junction area}) \qquad [6\text{-}20]$$

and as a function of reverse voltage V, C_T has been depicted in Fig. 6-20 for a graded and an abrupt junction, using a relative scale for C_T.

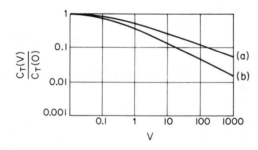

Fig. 6-20. Transition capacitance C_T of a diode, normalized to 1 for small voltages: (a) graded junction; (b) abrupt junction.

Diffusion Capacitance and Diffusion Resistance. In diodes that are intended for high frequencies or fast switching, it is not the transition capacitance that sets a limit for the speed of operation, but another effect, caused by the diffusion of charges, which causes the voltage of a forward-biased diode to always lag the terminal current. This effect is physically due to rearrangement of charge-density distributions. In other words, a minority carrier charge is stored on either side of the junction. The terminal voltage versus the terminal current is described in terms of a capacitance, the *diffusion capacitance* C_D, which depends on the forward current I in the following manner:

$$C_D = \frac{\tau_p q}{kT}(I + I_s) \qquad [6\text{-}21]$$

Here τ_p is a constant, the mean lifetime of holes in the N region. With forward voltages $C_D \gg C_T$ usually.

From the modified diode equation, Eq. [6-12], we obtain the *incremental resistance of a diode*, often called the *diffusion resistance*:

$$r_D = \frac{dV}{dI} = \frac{\eta V_T}{I + I_s} \qquad [6\text{-}22]$$

Notice that

$$r_D C_D = \eta \tau_p = \text{constant} \qquad [6\text{-}23]$$

Turn-on and Turn-off Transients in Diode. Diodes usually operate in circuits which by Thévenin's theorem are described by a voltage generator $V_i(t)$ in series with a resistance R. In the circuit of Fig. 6-21 we should be able to observe the same types of transients as we do in typical practical applications.

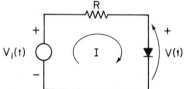

Fig. 6-21. Circuit for the discussion of transients in a diode.

Turn-on Transients. If $V_i(t)$ is a step voltage and its magnitude is much larger than the diode forward voltage, the current through R is approximately a step current. During the turn-on phase the diode voltage $V(t)$ will pass through three transient phases:

1. There is a capacitive or displacement current through the diode caused by the charging up of the transition capacitance C_T and the stray capacitances. This phase is usually very short as compared to the following.

2. When the injection of carriers begins, the current through the junction is primarily determined by the external circuit. After the charge distribution near the junction is built up, $V(t)$ asymptotically approaches a constant value.

3. If the current I is high ($\gg 1$ mA in signal diodes), we would observe an overshoot in $V(t)$ which comes from the high-level injection effect. In the initial phase of conduction, the series resistance of the P and N bulk regions is the same as for neutral homogeneous semiconductors. When the injection of charges over the junction begins, which is a delayed phenomenon, the conductivity of the bulk regions increases, and after the transient the voltage drop will be smaller. Thus the diode behaves as if it would have series inductance proportional to the current (Fig. 6-22).

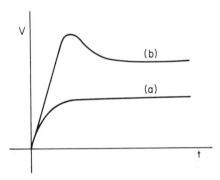

Fig. 6-22. Voltage over a fast diode with step current: (a) low current; (b) high current.

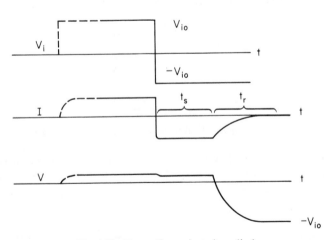

Fig. 6-23. Turn-off transients in a diode.

Turn-off Transients. Figure 6-23 shows the turn-off transients in the voltage and current of a diode. When $V(t)$ has settled down to a constant value, we reverse V_i to a value $-V_{i0}$. On both sides of the junction there is a stored minority carrier charge which sustains a reverse current for awhile. The initial reverse current is determined by the external circuit (i.e., by V_i and R). During this time the voltage over the junction is almost constant.

The stored charge is removed during a time t_s called the *storage delay time*, which depends on the initial reverse current, being roughly inversely proportional to it. After t_s, the incremental conductance of the diode decreases rapidly but there is an extra transient (t_r), during which the transition capacitance C_T is charged up to the negative potential $-V_{i0}$.

Analysis of Current Switching in a Diode Network. The voltage–current relation of a semiconductor diode at low signal currents and voltages is adequately described by the modified diffusion theoretical equation

$$I = I_s(e^{V/\eta V_T} - 1) \qquad [6\text{-}24]$$

which we denote $V = f(I)$ for short.

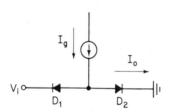

Fig. 6-24. Diode current switch.

Let us now consider the circuit of Fig. 6-24, which is formed of two identical diodes and an ideal current generator. Since V_i must equal the sum of the voltages over D_1 and D_2, we have

$$V_i = f(I_o) - f(I_g - I_o) \qquad [6\text{-}25]$$

From Eqs. [6-24] and [6-25] we have

$$V_i = \eta V_T[\ln (I_o + I_s) - \ln I_s - \ln (I_g - I_o + I_s) + \ln I_s]$$

or

$$\frac{I_o + I_s}{I_g - I_o + I_s} = e^{V_i/\eta V_T} \qquad [6\text{-}26]$$

Then

$$I_o = \frac{I_g}{1 + e^{-V_i/\eta V_T}} + I_s \frac{e^{V_i/\eta V_T} - 1}{e^{V_i/\eta V_T} + 1}$$

$$= \frac{I_g}{1 + e^{-V_i/\eta V_T}} + I_s \tanh \frac{V_i}{2\eta V_T} \qquad [6\text{-}27]$$

Since the latter term is usually negligible unless $-V_i \gg \eta V_T$ (and then its asymptotic value is $-I_s$), we may replace Eq. [6-27] by a very good approximation:

$$I_o = \frac{I_g}{1 + e^{-V_i/\eta V_T}} - I_s \qquad [6\text{-}28]$$

Also, the saturation current I_s is often so small, especially for silicon components, that it can be neglected. Then we can write

$$I_o = \frac{I_g}{1 + e^{-V_i/\eta V_T}} \qquad [6\text{-}29]$$

This equation has been plotted in Fig. 6-25 using normalized coordinates $V_i/\eta V_T$ and I_o/I_g. The two values of V_i denoted by V_{s0} and $-V_{s0}$ correspond to control voltage for which 10 and 90 percent of I_g, respectively, has been switched over. For V_{s0} we now have, from Eq. [6-29],

$$V_{s0} = \eta V_T \ln 9 \simeq 2.2 \eta V_T \qquad [6\text{-}30]$$

For silicon diodes at room temperature, $\eta V_T \simeq 0.05$ V, and then $V_{s0} \simeq 110$ mV.

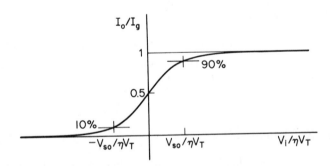

Fig. 6-25. Graphical representation of Eq. [6-29].

Minimum Voltage Selector. The switching circuit of the previous example, slightly modified, can be interpreted in a different way. In Fig. 6-26, two separate input voltages, V_1 and V_2, are connected to the cathodes of the diodes, and the output variable is the middle-point voltage V_o. Analogously with Eq. [6-26] we can set up system equations,

$$V_2 - V_1 = f(I_1) - f(I_g - I_1)$$

$$V_o = V_1 + f(I_1) \qquad [6\text{-}31]$$

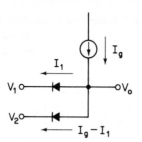

Fig. 6-26. Minimum voltage selector.

Again, after substitutions using Eq. [6-25], we get

$$V_o = \eta V_T \ln \frac{I_g + 2I_s}{I_s} + V_1 + \eta V_T \ln \frac{r}{r+1} \qquad [6\text{-}32]$$

where

$$r = e^{(V_2 - V_1)/\eta V_T}$$

and the terms in Eq. [6-32] have been arranged so that the current I_g is included only in the first term, whereas the signal voltages enter the last two terms.

Let us call

$$\eta V_T \ln \frac{I_g + 2I_s}{I_s} = V_D \qquad [6\text{-}33]$$

the *diode voltage.* Now we have

$$V_o = V_D + V_1 + \eta V_T [\ln r - \ln (r+1)]$$
$$= V_D + V_2 - \eta V_T \ln [e^{(V_2 - V_1)/\eta V_T} + 1] \qquad [6\text{-}34]$$

From Eqs. [6-34] we get some interesting results:

1. *If* $V_1 = V_2 = V$, *then*

$$V_o = V_D + V - \eta V_T \ln 2$$

2. *If* $V_2 - V_1 \gg \eta V_T$ $(V_1 < V_2)$, *then*

$$V_o \approx V_D + V_1$$

3. *If* $V_1 - V_2 \gg \eta V_T$ $(V_1 > V_2)$, *then*

$$V_o \approx V_D + V_2$$

The output voltage V_o *then follows the more negative (minimum) of the voltages* V_1 *and* V_2*, being displaced from it by* V_D*.* Notice that the exponential function in Eq. [6-34] is so strong that the difference in magnitudes of V_1 and V_2 of only a few times ηV_T ($\eta V_T \simeq 50$ mV for silicon) makes approximations 2 and 3 very accurate. We call this circuit the *minimum voltage selector*, because

$$\boxed{V_o \simeq \min \{V_1, V_2\} + V_D}$$ [6-35]

This relation can easily be generalized for several input voltages.

Maximum Voltage Selector. The circuit of Fig. 6-27 is obtained from the circuit of Fig. 6-26 by turning over the diodes and changing all polarities of currents and voltages. The former analysis applies as such, and *the output voltage V_o follows the more positive (maximum) of the voltages V_1 and V_2, being displaced from it by $-V_D$.*

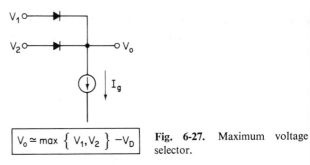

$$V_o \simeq \max \{V_1, V_2\} - V_D$$

Fig. 6-27. Maximum voltage selector.

This circuit is called the *maximum voltage selector*.

$$\boxed{V_o \simeq \max \{V_1, V_2\} - V_D}$$ [6-36]

6.5. Bipolar Transistor as a Digital Switching Component

In Fig. 6-28, the polarities of junction voltages and terminal currents of a bipolar transistor are defined. Stated in words, the currents are positive when flowing toward the device (whether *NPN* or *PNP*), and the junction voltages are positive if the *P* side is positive with respect to the *N* side.

In the illustrative examples we use only *NPN* transistors. The junction voltages V_E and V_C, as defined in Fig. 6-28, cannot be measured directly; however, the voltage drop over the internal base resistance is usually very small, and it can be calculated if the resistance is known.

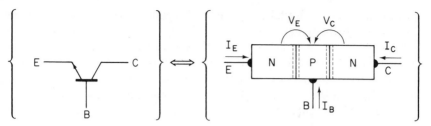

Fig. 6-28. Illustration of a *NPN* transistor and definition of the variables. The dashed lines are boundaries of the depletion layer. (The widths of the base and the depletion layers have been exaggerated.)

States of the Transistors. The three states of a bipolar transistor are the *active, cutoff,* and *saturated states.* In fact, it is the voltage over the *E-B* and the *C-B* junctions that determines the states of operation; these are shown in Table 6-1. To these we should add a fourth state, which occurs in many

Table 6-1

	States of the Junctions	
State of Operation	*E-B* Voltage	*C-B* Voltage
Active	forward	reverse
Cutoff	reverse	reverse
Saturation	forward	forward

expressions: the *inverse active state,* in which the *E-B* junction has a reverse voltage and the *C-B* voltage is forward. The transistor in the *saturated state* has practically no current gain, and the collector current is primarily determined by the external circuit. The voltage drop from the collector to the emitter is mainly determined by the ohmic resistance of all regions in series, which for low-power transistors is usually a few tens of ohms. A transistor in saturation corresponds to a *closed switch.*

When the base–emitter diode becomes reverse biased, the transistor is in the *cutoff state,* where the current gain drops to zero. Only a very small leakage current flows from the collector to the base. This current is practically independent of the collector–base voltage (and also, of the collector–emitter voltage). The *C-E* circuit of a transistor in the cutoff state is equivalent to an *open switch.*

Ebers–Moll Equations. The static relations between the terminal voltages and currents in all states are described by the *Ebers–Moll equations,* which, however, do not take any ohmic effects into account. Although these equations originally have been derived for grown or alloyed junction transistors where the junctions are abrupt, the general form of these equations can be shown to be valid for graded-junction transistors too (e.g., for diffused-junction transistors). These equations are also called *ideal transistor equations*. If η is the same for both junctions, the following relations between the voltages and currents of bipolar transistor are the modified Ebers–Moll equations:

$$I_E = a_{11}(e^{V_E/\eta V_T} - 1) + a_{12}(e^{V_C/\eta V_T} - 1)$$

$$I_C = a_{21}(e^{V_E/\eta V_T} - 1) + a_{22}(e^{V_C/\eta V_T} - 1)$$

[6-37]

where the constants a_{11}, a_{12}, a_{21}, and a_{22} depend on the form of the junctions and the electrical material parameters. An additional result of the diffusion theory is that

$$a_{12} = a_{21}$$

[6-38]

ACTIVE STATE

Since we can assume that $-V_C \gg \eta V_T$, the exponential functions containing V_C vanish:

$$I_E = a_{11}(e^{V_E/\eta V_T} - 1) - a_{12}$$

$$I_C = a_{21}(e^{V_E/\eta V_T} - 1) - a_{22}$$

[6-39]

or

$$I_C = \frac{a_{21}}{a_{11}} I_E + \left(\frac{a_{12} a_{21}}{a_{11}} - a_{22} \right)$$

(by definition)

$$= -\alpha_N I_E + I_{CO}$$

[6-40]

There are no restrictions for biasing the *E-B* junction in the reverse and the *C-B* junction in the forward direction. For this case we obtain the symmetrical equivalent of Eq. [6-40],

$$I_E = \frac{a_{12}}{a_{22}} I_C + \left(\frac{a_{12} a_{21}}{a_{22}} - a_{11} \right)$$

(by definition)

$$= -\alpha_I I_C + I_{EO}$$

[6-41]

Often the manufacturers of devices give average values for I_{CO}, I_{EO}, and α_N over a batch of transistors; then α_I is obtained from Eq. [6-38], which yields

$$\alpha_I I_{CO} = \alpha_N I_{EO} \qquad\qquad [6\text{-}42]$$

In the common-emitter connection we need I_B instead of I_E; denoting

$$\beta_N = \frac{\alpha_N}{1 - \alpha_N} \qquad\qquad [6\text{-}43]$$

and observing that $I_B = -(I_E + I_C)$, we have

$$I_C = \beta_N I_B + \frac{I_{CO}}{1 - \alpha_N} \qquad\qquad [6\text{-}44]$$

Thus in diffusion theory the current gain is constant; Eq. [6-44] is not valid, however, unless $-V_C \gg \eta V_T$. For low collector–base voltages, the experimental β_N is small.

CUTOFF

Now, since $-V_E \gg \eta V_T$, $-V_C \gg \eta V_T$, all exponential functions in [6-37] vanish:

$$I_E = \frac{I_{EO}}{1 - \alpha_N \alpha_I} - \frac{\alpha_I I_{CO}}{1 - \alpha_N \alpha_I}$$

$$[6\text{-}45]$$

$$I_C = -\frac{\alpha_N I_{EO}}{1 - \alpha_N \alpha_I} + \frac{I_{CO}}{1 - \alpha_N \alpha_I}$$

or

$$I_E = \frac{1 - \alpha_N}{1 - \alpha_N \alpha_I} I_{EO}$$

$$[6\text{-}46]$$

$$I_C = \frac{1 - \alpha_I}{1 - \alpha_N \alpha_I} I_{CO}$$

Usually α_N is of the order of 0.99 and α_I may be around 0.50. It is then justifiable to use the approximation

$$I_C \simeq I_{CO}$$

Notice also that $I_E \ll I_C$. We shall use this assumption about the leakage currents when a switching transistor is in the cutoff state, as shown in Fig. 6-29.

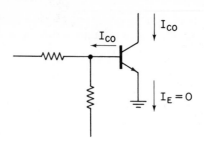

Fig. 6-29. Transistor leakage currents in the cutoff state.

SATURATION

When I_{CO} and I_{EO} are neglected as small quantities as compared with I_C and I_E,

$$V_E = \eta V_T \ln \left(-\frac{I_E + \alpha_I I_C}{I_{EO}} \right)$$

$$V_C = \eta V_T \ln \left(-\frac{I_C + \alpha_N I_E}{I_{CO}} \right)$$

[6-47]

The *collector–emitter saturation voltage* is then (noting the sign conventions for V_C and V_E)

$$V_{CES} = -(V_C - V_E) = \eta V_T \ln \frac{I_{EO}}{I_{CO}} \frac{\alpha_N I_E + I_C}{\alpha_I I_C + I_E}$$

[6-48]

Note that $I_C \simeq -I_E$. Because of the resistance of bulk regions (about 10 to 20 Ω), V_{CES} is usually small compared to the voltage drop in this resistance.

If the transistor is used as a *signal chopper* where the collector current is very small but the collector voltage may have any polarity, we put $I_C = 0$ in Eq. [6-48], and for normal polarity of collector voltages,

$$V_{CES} = -\eta V_T \ln \frac{I_{EO}}{I_{CO}} \alpha_N = -\eta V_T \ln \alpha_I$$

[6-49]

For the inverse polarity of collector voltage (C and E interchanged)

$$V_{ECS} = -\eta V_T \ln \alpha_N$$

[6-50]

For example, for $\alpha_N = 0.99$, $\alpha_I = 0.5$, $\eta V_T = 0.05$ V,

$$V_{CES} \simeq 35 \text{ mV}$$

$$V_{ECS} \simeq 0.5 \text{ mV}$$

Experimental Characteristics of Bipolar Transistors. Although the Ebers–Moll equations are only a first-order mathematical model for a transistor, they are theoretically valid in large-signal conditions and serve as a simple analytical model for transistors in the cutoff, active, and saturated states.

Departures of an actual transistor from its model described by Ebers–Moll equations are partially due to the nonideal properties of *PN* junctions in the way similar to that in a diode. We have, for example, to consider the breakdown effects of both junctions. This behavior is best described in terms of experimental characteristic curves in the common-emitter connection, of which one example is given in Fig. 6-30. Over a rather wide range of V_{CE} and I_C, the transistor model can be linearized. This means that the diagram can be approximated by straight lines that have a constant slope and are equidistant with constant increments in the parameter I_B. For switching applications, however, the transistor must be operated far outside its linear range.

When we inspect the actual characteristic curves of a transistor (e.g., those depicted in Fig. 6-30), we find a region that is normally used for digital operation but for which special pulse-forming devices can be designed. This is the avalanche breakdown region, which occurs at high V_{CE}. The characteristic curves exhibit *negative incremental resistance* and the collector current is a multivalued function of the *C-E* voltage.

The base–emitter circuit can usually be characterized by a diode model in static calculations.

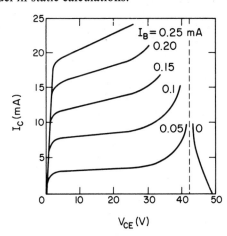

Fig. 6-30. Typical characteristic curves of a bipolar transistor.

Transistor Inverter: Switching Transients. A typical transistor inverter stage is shown in Fig. 6-31. In order that the transistor be in the saturation state, V_i must attain a sufficiently positive value V_{i1}. If the transistor is to be in the cutoff state, the base circuit must be controlled in such a way that V_i has a negative value, $-V_{i0}$.

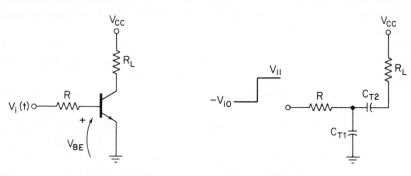

Fig. 6-31. Transistor inverter used for transient analysis.

Fig. 6-32. Model of the inverter in cutoff. The leakage currents have been neglected.

Turn-on Transients. Let us start from the cutoff state, for which we have the approximate equivalent circuit in Fig. 6-32. When the base current is turned on by making $V_i = V_{i1}$, the depletion-layer transition capacitances C_{T1} and C_{T2} of the B-E and B-C junctions are first charged for a short time. When V_{BE} is becoming positive, we must consider the transistor according to the active-state model, for which the diffusion capacitance of the B-E junction is predominant, being shunted by the incremental resistance of the B-E diode. For most cases we can assume that R is much larger than the incremental resistance, and therefore the base current I_B is determined by the external circuit,

$$I_B \simeq \frac{V_{i1}}{R} \qquad [6\text{-}51]$$

Now, considering the approximative equivalent circuit of Fig. 6-33, where

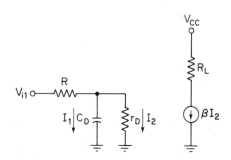

Fig. 6-33. Equivalent circuit for the inverter in the active state.

C_D and r_D are defined in the known way, we can derive a differential equation for the *effective base current* I_2, which is the current actually controlling the collector current.

The other base-current component, I_1, is a parasitic one; in fact, the parasitic current is defined as

$$I_1 = C_D \frac{dV_{BE}}{dt} \qquad [6\text{-}52]$$

and the proportionality constant C_D of Eq. [6-52] depends on the bias current through the diffusion resistance (e.g., on I_2). According to Eq. [6-21], abbreviating τ_p as τ, we have

$$I_2 = \frac{\tau}{V_T}(I_2 + I_s)\frac{dV_{BE}}{dt} \qquad [6\text{-}53]$$

and, by definition,

$$r_D = \frac{dV_{BE}}{dI_2} = \frac{\eta V_T}{I_2 + I_s} \qquad [6\text{-}54]$$

Thus

$$\frac{dI_2}{dt} = \frac{dV_{BE}/dt}{dV_{BE}/dI_2} = \frac{I_1}{\eta\tau} = \frac{I_B - I_2}{\eta\tau} \qquad [6\text{-}55]$$

As long as I_B is constant, Eq. [6-55] is linear in I_2 and its solution is of the known exponential form shown in Fig. 6-34(a). If we can assume β constant,

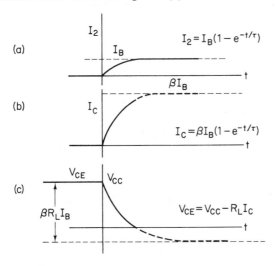

Fig. 6-34. Transient waveforms during the turn-on state of the transistor.

the collector current follows the form of I_2 until the transistor is saturated (i.e., until $I_c \rightarrow V_{cc}/R_L$ and $I_2 \rightarrow V_{cc}/\beta R_L$). When the transistor is saturated, both the B-E and the C-B junction will be forward biased, and the equivalent circuit has the form shown in Fig. 6-35.

The electric charge stored in the two capacitances C_{D1} and C_{D2} after saturation is usually called the *excess base charge*. When reversing the base current, this charge must first be removed from the base before the transistor can become active. Since this system is not of the first order, a simple discussion is no longer possible, especially because of the voltage dependence of the diffusion-model parameters.

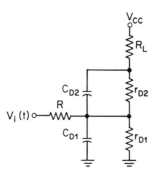

Fig. 6-35. Equivalent circuit of the saturated transistor.

Turn-off Transients. The reversion of base current does not immediately bring the transistor into the cutoff state. The time needed for the removal of the base charge is called t_s. Figure 6-36 shows the various switching waveforms before saturation, during saturation, and after saturation. Notice that

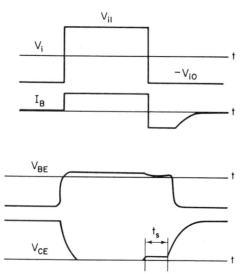

Fig. 6-36. Switching waveforms in a transistor inverter.

I_B in the reverse direction can be assumed constant until the transistor again is in the cutoff state, whereafter it drops off according to the charging of the transition capacitance. Notice also that there are no changes in the collector voltage during the storage delay time except a minor shoulder (usually much less than 100 mV) due to a voltage drop in the internal ohmic resistance of the base region.

Departures from the Previous Model. Most contemporary transistors intended for switching have very short time constants, of the order of 10^{-8} sec. The diffusion capacitance alone can no longer account for the observed total switching times. Transport delays of charge carriers in the base region, lead inductances, and stray capacitances must be taken into account. Thus the switching waveforms of currents are no longer purely exponential.

Current Switching in Active Circuits. According to the Ebers–Moll equations, we can introduce an approximative transistor model, shown in Fig. 6-37, which consists of a diode input circuit. The output circuit is a cur-

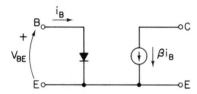

Fig. 6-37. Switching transistor model for $I_B > 0$.

rent generator the value of which is linearly proportional to the input current. The current gain β is assumed to be $\gg 1$. The model equations are

$$I_B = I_s(e^{V_{BE}/\eta V_T} - 1)$$
$$I_C = \beta I_B$$

[6-56]

The usual active current switching circuit is shown in Fig. 6-38.

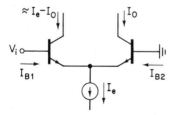

Fig. 6-38. Active current switch.

$$I_{B1} \simeq \frac{I_e - I_o}{\beta_1}$$

[6-57]

$$I_{B2} \simeq \frac{I_o}{\beta_2}$$

Denote

$$V_{BE1} = f_1(I_{B1})$$
$$V_{BE2} = f_2(I_{B2})$$

[6-58]

In a manner similar to that of passive circuits, we set up the system equation

$$V_i = f_1\left(\frac{I_e - I_o}{\beta_1}\right) - f_2\left(\frac{I_o}{\beta_2}\right)$$
$$= \eta V_T\left(\ln \frac{I_e - I_o + \beta_1 I_{s1}}{I_o + \beta_2 I_{s2}} - \ln \frac{\beta_1}{\beta_2} - \ln \frac{I_{s1}}{I_{s2}}\right)$$

[6-59]

where we can denote

$$V_f = -\eta V_T\left(\ln \frac{\beta_1}{\beta_2} + \ln \frac{I_{s1}}{I_{s2}}\right)$$

[6-60]

Solving for I_o and neglecting $\beta_1 I_{s1}$ and $\beta_2 I_{s2}$ we obtain

$$I_o \simeq \frac{I_e}{1 + e^{(V_i - V_f)/\eta V_T}}$$

[6-61]

6.6.   Field-Effect Transistors in Digital Switching

Junction FET. In its original form, the *field-effect transistor* (FET) consists of a resistive bar of lightly doped semiconductor material onto which some electrodes are fitted. A device of *P*-type material is called a *P-channel FET*, and one of *N*-type material is called an *N-channel FET*. Two electrodes, called *source* (*S*) and *drain* (*D*), are fitted to each end of the bar. Let the bar consist of *N*-type silicon. On the opposite sides of it, heavily doped *P*-type regions (*P⁺*) are diffused (Fig. 6-39). These regions are connected with an external circuit and serve as the *gate* (*G*) electrode. A depletion layer is formed along the *PN* junction. The thickness of this layer depends on the voltage over the junction. Since the gate is more heavily doped than the rest of the bar,

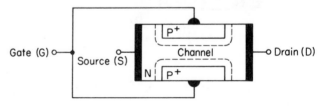

Fig. 6-39. Structure of a field-effect transistor. (Dashed lines indicate the boundary of the depletion layer.)

the depletion layer extends mainly into the N region. We can regard the depletion layer as an insulator, the thickness of which is controlled by the gate-to-bar voltage. Therefore, the average cross section of the remaining conductive N region (the *channel*) and, accordingly, the source-to-drain resistance are electrically controlled by the gate voltage. The bar is usually very thin; typical dimensions are thickness 2 mil, length 10 mil. This type of FET is called a *junction FET* (JFET). There are also planar constructions of JFET's.

In the same way as the characteristic curves for bipolar transistors are usually given in the common-emitter connection, we can represent the characteristic curves for a FET in the common-source connection, Fig. 6-40, showing the drain current I_D as a function of the drain-to-source voltage V_{DS} and with the gate-to-source voltage V_{GS} as parameter.

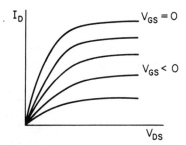

Fig. 6-40. Characteristic curves of JFET in common-source connection.

Metal–Oxide–Semiconductor (MOS) Construction. A special technology has been developed to produce field-effect transistors on silicon wafers. The usual construction is depicted in Fig. 6-41. Onto a high-resistivity P-type silicon substrate (denoted P^-) low-resistivity N-type electrode areas (denoted N^+) are diffused. On the surface, an accurately controlled silicon oxide layer is formed by oxidizing, and this layer acts as a dielectric. An aluminum overlay is deposited over the oxide. Electrodes are connected to the N^+ areas as well as to the overlay, and they form the source, the gate, and the

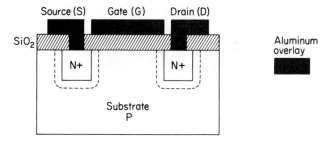

Fig. 6-41. Metal–oxide–semiconductor construction of a FET. (Dashed lines indicate depletion layers.)

drain, respectively. The gate is thus isolated and has a typical leakage resistance of about $10^{14}\ \Omega$.

**Enhancement-Mode Metal–Oxide–Semiconductor Field-Effect Tran-
sistor.** The two basic principles of FET's are called the *depletion mode*
and the *enhancement mode* of operation. The depletion mode of operation
was discussed at the beginning of this section. The main disadvantage of
this principle is that the gate-control voltage must attain opposite polarity
with respect to V_{DS}. There are additional problems in the mutual dc coupling
of such devices.

Next we shall describe the *enhancement mode* of operation. The control
voltage of this device has the same polarity as V_{DS}, and such devices can be
interconnected directly.

The basic structure of a MOS-type FET (or MOSFET) was depicted
in Fig. 6-41. If the gate potential is floating, a depletion layer is formed
around the source and the drain electrodes. There are, in fact, two opposed
PN junctions in series between the source and the drain, and one of these is
always reverse biased. The only current flowing in the *S-D* circuit is leakage
current.

If the gate electrode is now made positive with respect to the substrate
as shown in Fig. 6-42, the electric field will repel positive charge carriers
and draw negative charges onto the upper silicon surface by electrostatic
induction. These charges, being conduction electrons, are minority carriers
in *P* silicon. At a certain gate voltage V_{GS}, called *gate threshold voltage V_o*,
the concentration of induced electrons will equal the hole concentration,
and, with increasing V_{GS}, the semiconductor type at the surface will be
inverted. Thus an *N* channel is induced which ties the source and the drain
together. The thickness of this conductive channel depends on the applied
voltage V_{GS}. Figure 6-42 shows the channel formation. Let us discuss the
operation of the enhancement-mode MOSFET more closely in the following.

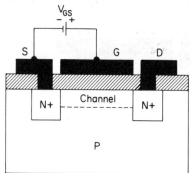

Fig. 6-42. Channel formation in an enhancement-type MOSFET.

Surface Charge Density with $V_{DS} = 0$. With V_{DS} equal to zero, the surface charge per unit area, or the *surface charge density* Q induced in the channel, is assumed to be a linear function of V_{GS} in the same way as the surface charge density in a capacitor:

$$Q = \frac{\epsilon}{d}(V_{GS} - V_o) \qquad V_{GS} \geq V_o \qquad [6\text{-}62]$$

where V_o is the threshold voltage at which the semiconductor type near the surface is inverted, and ϵ and d are the dielectric constant and the thickness of the SiO$_2$ layer, respectively.

MOSFET Equation for $0 \leq V_{DS} \leq V_{GS}$. If a voltage V_{DS} is applied between the drain and the source, a potential distribution $V(x)$ along the surface is observed (Fig. 6-43). The potential does not fall linearly because the charge density at x depends on $V_{GS} - V(x)$, but $V(x)$ again depends on the distribution of channel resistance.

The total current in the channel is constant and is denoted by I_D. The charge density, according to Eq. [6-62], is now

$$Q(x) = \frac{\epsilon}{d}[V_{GS} - V(x) - V_o] \qquad [6\text{-}63]$$

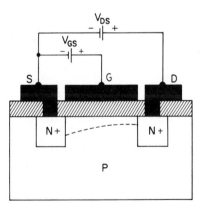

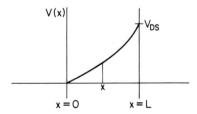

Fig. 6-43. Surface charge distribution.

V_{DS} causes a longitudinal electric field $K(x) = -dV(x)/dx$ in the channel, which again drives electrons with an average velocity

$$v(x) = -\mu_n K(x) = \mu_n \frac{dV(x)}{dx} \qquad [6\text{-}64]$$

Here μ_n is the mobility of electrons.

Denoting the width of the channel by W, we have for the drain current

$$I_D = WQ(x)v(x) \qquad [6\text{-}65]$$

From Eqs. [6-63] to [6-65] we get a differential equation for $V(x)$ by substitution,

$$I_D = \frac{\epsilon \mu_n W}{d}[V_{GS} - V(x) - V_o]\frac{dV(x)}{dx} \qquad [6\text{-}66]$$

which can be integrated from $x = 0$ to $x = L$ with the boundary conditions $V(0) = 0$, $V(L) = V_{DS}$:

$$I_D = \frac{\epsilon \mu_n W}{Ld} V_{DS}\left(V_{GS} - \frac{V_{DS}}{2} - V_o\right) \qquad [6\text{-}67]$$

or

$$I_D = 2I_{DSS}\left[\left(\frac{V_{GS}}{V_o} - 1\right)\frac{V_{DS}}{V_o} - \frac{1}{2}\left(\frac{V_{DS}}{V_o}\right)^2\right] \qquad [6\text{-}68]$$

where

$$I_{DSS} = \frac{\epsilon \mu_n W V_o^2}{2Ld}$$

MOSFET Equation for $V_{DS} > V_{GS} - V_o$. From the experimental $V_{DS} - I_D$ curves we can see that they level off when V_{DS} exceeds a certain value which depends on V_{GS}. This is due to the fact that if the gate-to-surface voltage for some x is less than the threshold voltage V_o, there can be no surface charge at this place, and the surface resistance becomes very high. Now, if V_{DS} exceeds $V_{GS} - V_o$, it is just near the drain, where with increasing V_{DS} the channel begins to vanish. The increase in channel resistance will oppose the increase in current. We can then assume that the current I_D attains a value obtained from Eq. [6-67] with $V_{DS} = V_{GS} - V_o$. Thus we get the MOSFET equation for the constant-current region $V_{DS} > V_{GS} - V_o$,

$$I_D = I_{DSS}\left(\frac{V_{GS}}{V_o} - 1\right)^2 \qquad [6\text{-}69]$$

Departures from the Ideal MOSFET Model. The most important departure from simplified MOSFET equations is that instead of the constant-current relation postulated in Eq. [6-69], the $I_D(V_{DS})$ curves also have a nonzero slope in this region. Figure 6-44(a) gives empirical characteristic curves for MOSFET transistors, and Fig. 6-44(b) defines the *transfer curve* $I_D(V_{GS})$ with V_{DS} = constant.

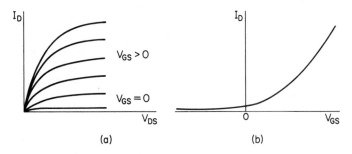

(a)   (b)

Fig. 6-44. (a) Characteristic curves of an enhancement-mode MOSFET; (b) transfer curve.

Capacitances of the MOSFET. The MOSFET has a lower switching speed than bipolar transistor due to gate capacitances. The gate capacitance is actually distributed along the channel but can be approximated by a gate-to-source and a gate-to-drain lumped capacitance (C_{GS} and C_{GD}, respectively). Further, the source and the drain have their capacitances (C_S and C_D, respectively) with respect to the substrate, as a result of the depletion layers formed at the *PN* junctions. A high-frequency equivalent circuit of the MOSFET is shown in Fig. 6-45. The transfer relations of the MOSFET are described by an equivalent current generator $g_m V_{GS}$ and an internal resistance r_{DS} according to the previously discussed analytical model. The internal source and drain resistances R_S and R_D can still be taken into account.

Notice that the capacitance C_{GD} gives rise to a Miller effect and is more important than C_{GS} in limiting the speed of operation.

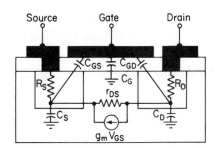

Fig. 6-45. High-frequency equivalent circuit of the MOSFET.

Depletion-Type MOSFET's. There are also MOSFET's which work in the depletion mode of operation. By diffusion, a permanent impurity channel is doped on the surface between source and the drain. The depletion of majority carriers from this channel is controlled by the gate voltage, which is opposite to the drain-to-source voltage.

Symbols for FET's. The JFET where the gate forms a junction with the channel is usually described by the symbols in Fig. 6-46(a) or (b). Other ways of symbolizing have been developed for the MOSFET's. If the polarity and the substrate have to be shown explicitly, the symbols in Fig. 6-46(c) and (d) can be used to denote the *P*-channel and the *N*-channel device, respectively. In more complex diagrams the simplified symbols in (e) and (f) can be used. The even simpler symbols of Fig. 6-46(g) and (h) can be used if the type of the MOSFET and the identification of the drain and the source are known from the context.

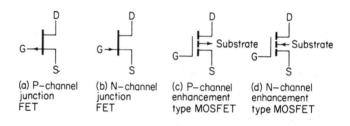

(a) P–channel junction FET (b) N–channel junction FET (c) P–channel enhancement type MOSFET (d) N–channel enhancement type MOSFET

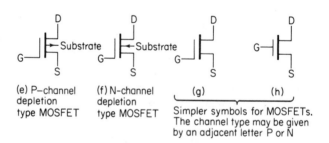

(e) P–channel depletion type MOSFET (f) N–channel depletion type MOSFET (g) (h)

Simpler symbols for MOSFETs. The channel type may be given by an adjacent letter P or N

Fig. 6-46. Symbols for field-effect transistors.

6.7. Transistor Inverters and Output Circuits

It is an objective of circuit design to implement the characteristic nonlinear input–output relations $V_o(V_i)$ of Section 6.3 using signal-limiting properties of components and elementary circuits. Diode current-switching circuits were found to have saturable voltage-to-current transfer relations, and a transistor is a current-controlled device. Therefore, diode logic gates (minimum or maximum selector gates) followed by transistor output circuits are natural combinations for logic circuits. Bipolar transistor stages alone can also be used as inverter switches, as discussed in the following section. The field-effect transistors, especially MOSFET's, are voltage-controlled devices, and inverter logic is the natural mode of their use. In the following paragraphs we shall discuss the setting of logic levels with various types of inverter switches. More details about circuit engineering of logic gates will be found in Chapter 7.

Resistor-Transistor-Logic and Direct-Coupled-Transistor-Logic Inverters. The setting of logic levels can be implemented by a series connection of a transistor and a load resistor R_L as indicated by Fig. 6-47(a). There is in this circuit a base resistor R_B in which the input voltage V_i causes a proper control current. According to the Ebers–Moll equations, the circuit is approximatively equivalent to the model of Fig. 6-47(b), where D is a semiconductor diode described by the modified diode equation, Eq. [6-12], and h_{FE} can be regarded constant as long as $I_B \geq 0$. (In any case, if $I_B < 0$, the generator current can be assumed very small.) Also, it is assumed that V_o cannot become more negative than the collector–emitter saturation voltage V_{CES}, which usually is of the order of 100 mV.

Figure 6-48(a) shows a graphical solution for I_B as a function of V_i. Here R_B defines a *load line*

$$V_{BE} = V_i - I_B R_B \qquad [6\text{-}70]$$

which together with the diode equation $V_{BE} = f(I_B)$ yields a solution for I_B. The output circuit is correspondingly described by the equation of another load line,

$$V_o = V_{CC} - h_{FE} I_B R_L \geq V_{CES} \qquad [6\text{-}71]$$

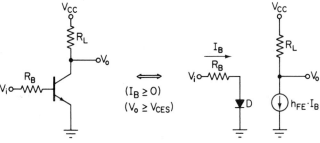

Fig. 6-47. RTL inverter.

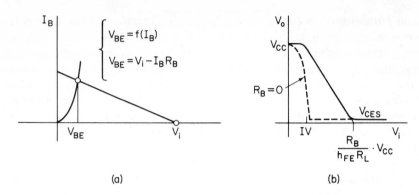

(a)   (b)

Figure 6-48

to which I_B is substituted from Fig. 6-48(a). Thus Eq. [6-71] is represented by the graph of Fig. 6-48(b). The dashed curve in Fig. 6-48(b) represents $V_o(V_i)$ with $R_B = 0$, and the transistor in this connection is called a *direct-coupled-transistor-logic* (DCTL) *inverter*. The circuit with an external base resistor R_B is called a *resistor-transistor-logic* (RTL) inverter.

Loading Effect of Input Circuits. Especially in DCTL, the input diode D will cause a problem in the matching of stages to each other. This is discussed in terms of Fig. 6-49.

Let the loading input circuit be described by a linearized equivalent circuit according to Thévenin's theorem, as shown in Fig. 6-49(a). The switching transistor driving this circuit is represented by a switch SW. When SW is closed, it will act as a *current sink*; with SW open, the collector resistor R_L acts as a *current source*. The internal impedance of the transistor

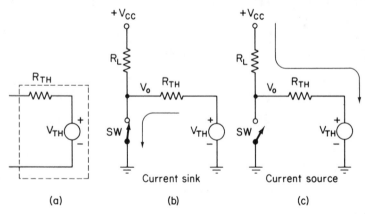

(a)   (b)   (c)

Fig. 6-49. Loading an inverter.

stage is thus very different in the two states of SW and this manifests itself when there is a capacitive load C in parallel with the switch. The rise time of logic signals is relatively long, with a time constant of $C(R_L \| R_{TH})$; the signal fall time is mainly determined by the transistor as a current generator.

Inverter States with High Sink Currents. There are cases in which the asymmetry between the source and sink currents does not matter but a large sink current must be drawn. On such occasions reinforced inverter stages, as shown in Fig. 6-50, can be used. In the former, Q_1 acts as an emitter follower for Q_2, the collector voltage of Q_1 being reduced by R_1. The case of Fig. 6-50(b) could be approximatively described by a single-transistor inverter of Fig. 6-47(b), in which the h_{FE} of the virtual transistor Q is

$$h_{FE} = h_{FE1}(h_{FE2} + 1) \qquad [6\text{-}72]$$

with subscripts 1 and 2 as for Q_1 and Q_2.

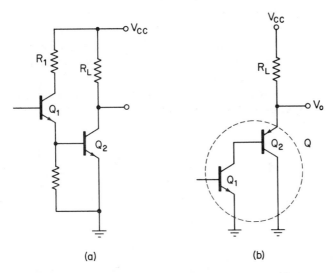

(a)   (b)

Fig. 6-50. High-current inverter stages: (a) using an emitter-follower amplifying stage; (b) using a common-emitter amplifying stage.

Active Pull-up Circuits. When high source and sink currents must be delivered by the output circuit, and when large signal amplitudes are desirable, the collector resistor R_L of a usual inverter stage should be replaced by another transistor switch, which is made to operate in the opposite phase with respect to the lower transistor. (Fig. 6-51). This is often termed an *active pull-up operation*, in contrast to *passive pull-up* by a resistor. This circuit has the additional advantage that there is no unnecessary power

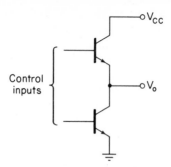

Fig. 6-51. Principle of an active pull-up output circuit.

dissipation due to R_L during $V_o = $ LOW[1]. Precautions must be taken in order that both transistor switches are not closed during any period of operation; otherwise a heavy surge current from V_{CC} to ground is drawn. In addition to the danger of damaging the circuit, the surge current will cause heavy transients in the supply voltage, reflected as "logical noise." Some possible active pull-up circuits are shown in Fig. 6-52, where Q_1 always acts as a phase splitter.

The most popular of these, depicted in Fig. 6-52(c), has certain advantages over the others: low power consumption and simple circuit design. This circuit is now analyzed more closely.

Using the same simplified equivalent circuit as before, the stage is adequately described by Fig. 6-53, where the Thévenin equivalent voltage and

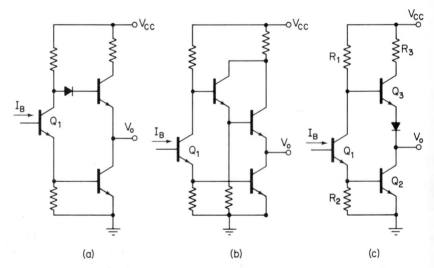

(a)   (b)   (c)

Fig. 6-52. Active pull-up output circuits.

[1] According to a common usage, we shall denote the more positive logic voltages by HIGH (= H) and the more negative logic voltages by LOW (= L) in the following.

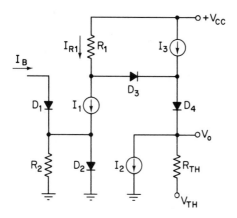

Figure 6-53

impedance V_{TH} and R_{TH} for the next input circuit are used. Equations [6-73] are derived from this circuit. For further simplicity, we have assumed that R_2 can be neglected. The error resulting from this assumption is not large.

$$I_1 = h_{FE1}I_B$$

$$I_2 = h_{FE2}(1 + h_{FE1})I_B = (1 + h_{FE3})(I_{R1} - I_1) - \frac{V_o - V_{TH}}{R_{TH}}$$

$$I_{R1} = \frac{V_{CC} - V_{D3} - V_{D4} - V_o}{R_1}$$

[6-73]

$$I_3 = h_{FE3}(I_{R1} - I_1)$$

where V_{D3} and V_{D4} are diode voltages of D_3 and D_4, respectively, and the h_{FE}'s have the same indices as the transistors.

For stationary logic signals we can assume two extreme cases:
$I_B = 0$: Now $I_1 = I_2 = 0$, and the *source current* due to Q_3 is

$$I_3 = h_{FE3}I_{R1} \qquad \text{[6-74]}$$

$I_B \geq I_{BS}$, where I_{BS} is the current needed to saturate Q_1: Since $V_o > 0$, we can assume that D_3 is reverse biased and $I_3 = 0$. Then

$$I_2 = h_{FE2}(1 + h_{FE1})I_B \qquad \text{[6-75]}$$

is the available sink current.

The output impedance of this active pull-up circuit is usually of the order of or less than about 100 Ω; in the HIGH and LOW states this is due to saturated Q_2 or Q_3 in series with R_3. In the active state during switching, Q_3 acts as an emitter-follower and its output impedance is then approximately $R_1(h_{FE3} + 1)$, plus the impedance of D_4.

Field-Effect Transistor as an Inverter. It is a significant advantage for the design and manufacturing of large-scale logic circuits that there are available components which can be interconnected without auxiliary matching circuits. The most important of these is the enhancement-mode metal–oxide–semiconductor field-effect transistor. Another MOSFET (one with different geometrical dimensions) can be used as a current generator by tying the control electrode (gate) and the upper electrode (drain) together. In automatically produced integrated systems, it is easier to produce an extra MOSFET than a resistor. An inverter circuit using the MOSFET technology is shown in Fig. 6-54. We shall discuss MOS circuits in more detail in Chapter 7.

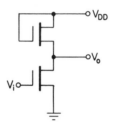

Fig. 6-54. Enhancement-mode MOSFET inverter.

PROBLEMS

6-1 Indicate by means of illustrations what are the true directions of bias currents, leakage currents, and bias voltages for a *PNP* and *NPN* transistor

(*a*) in the active state.

(*b*) in the cutoff state.

(*c*) in the saturation state.

6-2 Transistors are used as inverter switches in the way depicted in Fig. P6-2. What is the maximum number of such switches that can be connected in series for the implementation of a NAND function in order that each transistor would attain either a cutoff or a saturated state for all possible input logic voltages? (LOW = 0 to 1 V; HIGH = 2.2 to 6 V.) The power-supply voltage (= H) is +6 V and the load resistor R_L is 1 kΩ. The following facts about the transistors are known:

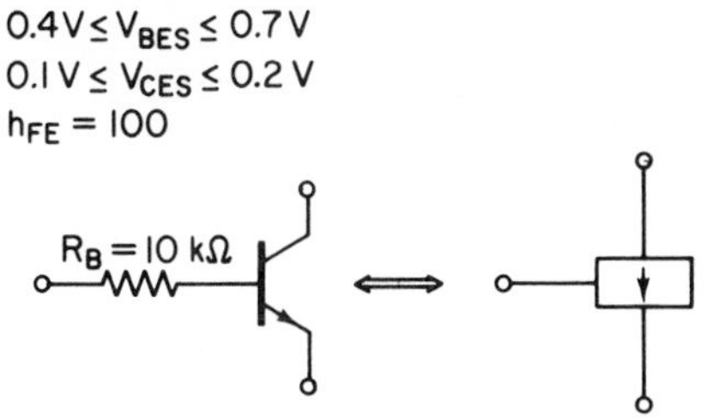

$0.4V \leq V_{BES} \leq 0.7V$
$0.1V \leq V_{CES} \leq 0.2V$
$h_{FE} = 100$

$R_B = 10 \text{ k}\Omega$

Figure P6-2

$$0.4 \text{ V} \leq V_{BES} \leq 0.7 \text{ V}$$

$$0.1 \text{ V} \leq V_{CES} \leq 0.2 \text{ V}$$

$$h_{FE} = 100$$

6-3 Derive the Boolean function X for the logic circuit of Fig. P6-3 using positive logic convention. (The input impedances of the inverters are assumed to be very high.)

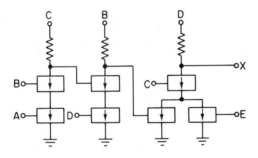

Figure P6-3

6-4 The input–output voltage relations of an inverter family are represented graphically by the envelopes of Fig. P6-4. Determine graphically the HIGH and LOW noise sensitivities and noise margins.

6-5 (*a*) For a minimum voltage selector (Fig. 6-26) plot V_o versus time when V_1 and V_2 are given by Fig. P6-5. Use the modified diode equation ($\eta V_T = 50$ mV, $V_D = 650$ mV).
(*b*) Repeat this problem for a maximum selector (Fig. 6-27).

6-6 In the active current switch depicted in Fig. P6-6 the input voltage V_i varies from -2.4 to -1.6 V. Derive the maximum values of R_1 and R_2 for which neither of the transistors is saturated (i.e., V_{CB} does not get negative). The component values are assumed to be

$$V_{BE} = 0.6 \text{ V}$$
$$h_{FE} \gg 1 \quad\Big\} \text{ for an active transistor}$$

$$R_3 = 1.2 \text{ k}\Omega$$

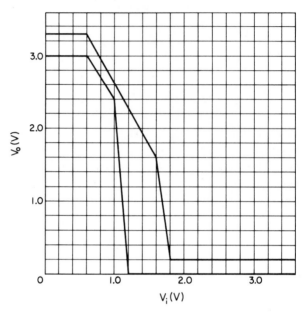

Figure P6-4

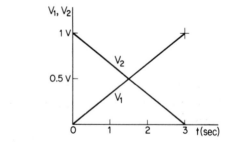

Figure P6-5

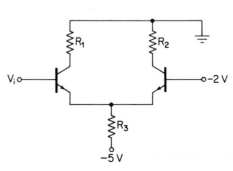

Figure P6-6

6-7 Calculate the saturation voltage V_{CES} according to the Ebers–Moll equations for the transistor of Fig. P6-7 using the following values:

$$I_{E0} = 25 \text{ nA} \qquad I_{C0} = 50 \text{ nA}$$

$$\alpha_N = 0.99 \qquad \eta V_T = 50 \text{ mV}$$

$$I_B = 100 \ \mu A \qquad R_L = 3.3 \text{ k}\Omega$$

$$V_{CC} = 3 \text{ V}$$

$I_{E0} = 25 \text{ nA} \qquad I_{C0} = 50 \text{ nA}$
$\alpha_N = 0.99 \qquad \eta V_T = 50 \text{ mV}$
$I_B = 100 \ \mu A \qquad R_L = 3.3 \text{ k}\Omega$
$V_{CC} = 3 \text{ V}$

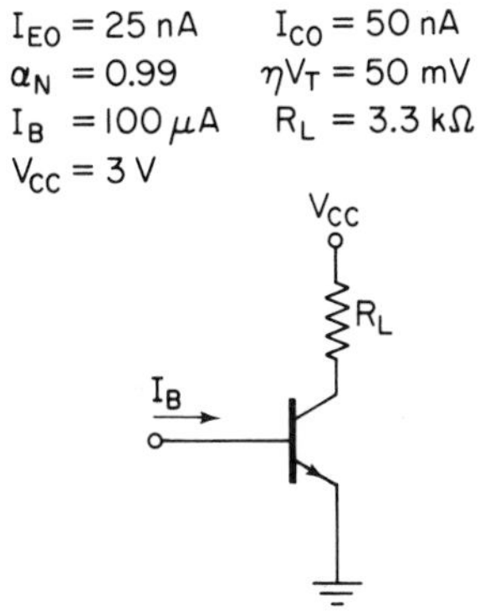

Figure P6-7

6-8 Derive the voltage–current relation, corresponding to Eq. [6-29], for a diode current switch in which the saturation currents of the diodes (I_s) are not equal.

REFERENCES

The basic ideas of nonlinear voltage and current dividers (i.e., the theory of switching) can be found in the book

> Chua, *Introduction to Nonlinear Network Theory* [19].

A rather complete view of the theory of bipolar transistor circuits may be obtained from

> Lynn, Meyer, and Hamilton (eds.), *Analysis and Design of Integrated Circuits* [86].

Field-effect transistors, especially MOSFET's, are described in

> Crawford, *MOSFET in Circuit Design* [22]

and in numerous applications reports circulated by semiconductor manufacturers; to mention a particular one,

Motorola Semiconductor Products, Inc., *Developments in LSI* [233].

Also, the following book serves as a classical textbook of semiconductor theory:

Gibbons, *Semiconductor Electronics* [40].

Other references can be found in the Bibliography.

chapter 7

Logic Circuit Families

7.1. General

Logic gates can be implemented by electronic, optical, hydraulic, pneumatic, or mechanical components. In this book we shall be dealing only with electronic circuits. In logic design, the general structure of logic networks and the interconnections for logic signals can be defined in a rather unique way; the selection of electronic implementation and the partitioning of individual logic gates on component packages, circuit boards, and mounting frames belongs to the field of detailed circuit design, in which the number of solutions is usually infinite. Optimal circuit-design practice depends on the state of the art of semiconductor components, especially on the degree of integration. A circuit is called *integrated* if it is fabricated, tested, and vended as an indivisible unit. *Monolithic circuits*, which are fabricated by photolithographic techniques on single silicon crystals (chips), comprise the most important integrated circuits. The degree of integration is expressed by the terms *medium-scale integration* (MSI) and *large-scale integration* (LSI). There is no sharp distinction between these two, but it is customary to call circuits with less than 100 gates on a chip MSI circuits, whereas circuits with about 100 gates or more are LSI circuits.

Up to about the middle 1950s, practically all digital circuits, including logic gates, were assembled of discrete components, although in the production of systems, printed-circuit-board technology allowed automated production of rather large circuit assemblies. At that time the electrical design of all circuits was left completely to the user of semiconductor components. The circuit boards comprised the smallest units that could be directly interconnected by back-panel wiring. With the advent of integrated circuits shortly before and around 1960, a big change in the roles of the circuit designer and the manufacturer of semiconductor components took place. Large portions of electronic circuits together with their intraconnections are produced on silicon chips by photolithographic techniques. These circuits can be interconnected with ease, and the manufacturers of semiconductor components assume the responsibility for the electrical characteristics of individual logic circuits. Logic gates, as well as assemblies of different logic circuits, are available as standardized packages. For economic reasons and convenience of handling, several separate elementary circuits may be mounted in a single package.

In this chapter we shall concentrate primarily on circuit engineering with integrated circuits. There are several different principles for the construction of such electronic circuits, which possess the required nonlinear voltage-transfer relations needed in the classification of digital signals. Some of these methods can be mixed. However, to ensure uniform electrical specifications over a digital system because of requirements for standardization, a small number of well-established *logic circuit families* have been developed, each with characteristic properties and compromises between price and performance. Every family ought to contain a complete choice of standard logic circuits so that arbitrary digital systems could be realized within each family. Some of the families were developed in the era of discrete-component logic circuits and transferred almost as such to integrated circuits; other families have been directly designed for integrated circuit production. The integration of a large number of components on a single chip brings about problems that were not met with discrete components: leakage currents, the creation of parasitic components (e.g., extra diodes) generated by the fabrication steps, and packaging problems.

The two main lines of semiconductor components used in digital circuits are (1) *bipolar* transistors and diodes (2) *unipolar* or field-effect transistors.

The best-known logic circuit families nowadays are the following:

WITH BIPOLAR TRANSISTORS

Resistor-transistor logic (RTL)
Diode-transistor logic (DTL)
Transistor-transistor logic (TTL or T^{2}L)
Emitter-coupled logic (ECL)

WITH UNIPOLAR TRANSISTORS

Usual MOSFET logic
Complementary symmetry MOSFET (COS-MOS) logic

In addition, there are some variants of these principles bearing different tradenames and having particular characteristics. For example, there are special circuits with low power dissipation in many bipolar families. A variant of DTL with high noise immunity is called HTL (high-threshold logic) or HLL (high-level logic). Also, a certain family called complementary-transistor logic (CTL) will be discussed here.

7.2.   Bipolar Logic Circuit Families

7.2.1. Resistor–Transistor Logic

It was mentioned in Chapter 6 that digital circuits are based on voltage or current switches or both. In bipolar logic circuits, only two types of voltage switches (inverter switches) are used for logic gates: inverters of direct-coupled transistor logic (DCTL) and of resistor-transistor logic (RTL). The former has almost totally lost importance, mainly because of its small signal voltage levels and poor stability. However, DCTL has the best speed/power figure of merit of all bipolar logic types. This figure alone is not sufficient to compensate for its shortcomings. In this section, we consider the RTL inverter logic only.

Figure 7-1 shows a typical NOR gate with three RTL inverters. Positive logic is used throughout the following, and all signal voltages are positive. If the input to an inverter is HIGH, the corresponding transistor

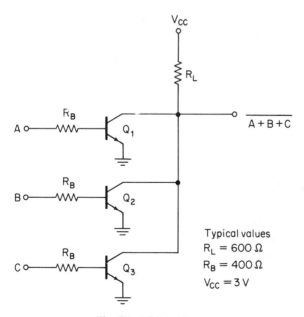

Fig. 7-1. RTL NOR gate.

is saturated. The output is LOW if any of the transistors are saturated. It is obvious that similar circuits connected to the output of this gate form a current sinking load, and therefore the saturation of a single transistor must be guaranteed with no external load (which is the worst case). The ratio of the base resistor R_B and the collector resistor R_L must be such that the base current under the worst conditions exceeds the saturation base current I_{BS}; a sufficient condition is

$$\check{h}_{FE} \frac{\check{V}(1) - \hat{V}_{BES}}{\hat{R}_B} \geq \frac{\hat{V}_{CC}}{\hat{R}_L} \qquad [7\text{-}1]$$

where $\check{h}_{FE}$ = minimum h_{FE} over a batch of transistors, evaluated at a V_{CE} near to the saturation voltage V_{CES}

$\check{V}(1)$ = minimum input signal voltage still classified as a logic 1

$\hat{V}_{BES}$ = largest base–emitter voltage of saturated transistor occurring over the batch

$\hat{\ }, \check{\ }$ = in general, notations for the largest and smallest allowable values, respectively, for components and variables

Another restriction imposed on the circuit is that the maximum allowable number of similar circuits loading the output (fan-out) is limited. For simplicity, assume that these circuits can be described by a current sinking resistor R_{LS} connected to a ground. The minimum allowable R_{LS} is determined by the requirement that the output voltage, if classified as a logic 1, must not fall below the class limit. Neglecting leakage currents we have

$$\frac{\check{R}_{LS}}{\hat{R}_L + \check{R}_{LS}} \check{V}_{CC} \geq \check{V}(1) \qquad [7\text{-}2]$$

Comment. The fan-out of RTL logic can be increased by selecting a smaller R_L than the nominal value.

The number of inputs in RTL inverter logic can be increased by tying outputs of NOR gates together ("Wired AND", "Dot AND"). For this reason, the load resistor R_L of the circuit modules is usually left unconnected in order that only one R_L of parallel gates could be used as the common-collector resistor. The maximum number of parallel inverters (*fan-in*) is limited only by the fact that the total collector leakage current must not cause a voltage loss in R_L, which would make the logic 1 output voltage smaller than $\check{V}(1)$. Thus Eq. [7-2] reads, more accurately,

$$\frac{\check{R}_{LS}}{\hat{R}_L + \check{R}_{LS}} \check{V}_{CC} - \hat{N}_i \hat{I}_{co} \cdot \hat{R}_L \| \hat{R}_{LS} \geq \check{V}(1) \qquad [7\text{-}3]$$

where $\hat{N}_i$ is the maximum fan-in and $\hat{I}_{co}$ is the maximum collector leakage current per transistor.

The RTL family is the cheapest of all bipolar logic circuit families, because of its simplicity. On the other hand, its noise margins are not very large. Typical specifications are

Fan-in (of integrated gates)	$N_i = 4$
Fan-out	$N_o = 5$
Noise margin	150 to 500 mV
Propagation delay	25 to 35 nsec
Power dissipation	3 to 30 mW

Typical input–output voltage-transfer relations $V_o(V_i)$ for an RTL gate, for three different fan-out numbers (F.O.), are shown in Fig. 7-2. (All data in Figs. 7-2 to 7-21 refer to circuits made by Motorola, Inc.)

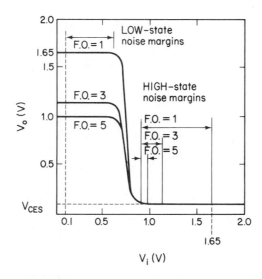

Fig. 7-2. RTL input–output voltage relations.

Multiple-Input RTL Gate. The name RTL was originally coined for a multiple-input transistor gate shown in Fig. 7-3. In this circuit, a NOR gate, the resistors R_1, R_2, and R_L are dimensioned so that if one of the inputs is HIGH and the other inputs are LOW, the transistor is still saturated, and more HIGH inputs only increase saturation base current. Because of its low price, this principle was used in earlier discrete-component logic circuits. Its worst shortcomings are low speed, small noise margins, and the need for a separate negative voltage.

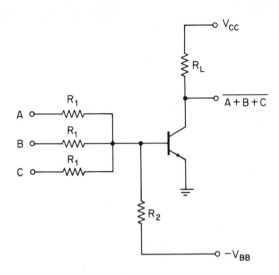

Fig. 7-3. Multiple-input RTL gate.

7.2.2. Diode–Transistor-Logic and High-Threshold-Logic Families

Logic circuits in which the logic operation is executed by a diode logic gate and which contain a separate amplifying output circuit for the compensation of signal attenuation and standardization of logic levels are called diode–transistor-logic (DTL) circuits. The most common of the DTL circuits is the basic NAND circuit, consisting of a diode AND gate and a transistor inverter. Also, more complicated (e.g., two-level) diode logic networks may be used in the gate part. Different DTL families are distinguished by the methods by which the standardized signal levels and the noise margins are defined. We shall discuss the basic DTL, high threshold logic (HTL), and noise-filtered DTL's.

Basic DTL. Let us consider the circuit of Fig. 7-4(a), the basic DTL gate, and decompose it into two parts as in Fig. 7-4(b). If all input voltages are HIGH, the input diodes are reverse biased, and Q is saturated by the current through R_1 and R_2. For a saturated transistor we can approximately assume a fixed value for V_{BE}, called V_{BES}, at least in the worst-case analysis. Then V_G, the control voltage of the output inverter, attains a value $V_G(1)$ through the resistor voltage divider,

$$V_G(1) = \frac{R_2 V_{CC} + R_1 V_{BES}}{R_1 + R_2} \qquad [7\text{-}4]$$

The condition for all input diodes to be reverse biased is obviously that

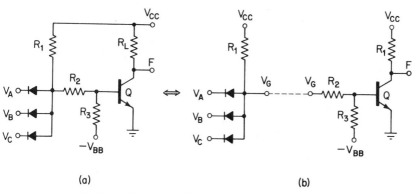

Fig. 7-4. Basic DTL (a) and its decomposition into two parts (b).

$V_A, V_B, V_C, \ldots$, etc. $> V_G(1)$, given by Eq. [7-4]. On the other hand, if any of the input voltages is less positive than $V_G(1)$, the diode gate will act as a minimum voltage selector. For input voltages which are logic signals and at least one of which has the value 0, the value for V_G, now denoted by $V_G(0)$, is

$$V_G(0) = \min (V_A, V_B, V_C, \ldots) + V_D \qquad [7\text{-}5]$$

where V_D is the diode voltage discussed in more detail in Chapter 6. Thus, in order that the output transistor be cut off when at least one of the input voltages is LOW, we must require that

$$V_{BE} = \frac{R_3 V_G(0) - R_2 V_{BB}}{R_2 + R_3} \leq -|V_{BEX}| \qquad [7\text{-}6]$$

where V_{BEX} is an arbitrarily selected, negative base-to-emitter voltage which assures cutoff and a reasonable noise margin against disturbances coming from the ground line.

The purpose of R_2 is only to provide a certain drop in signal voltage levels, to compensate for the diode voltage V_D and to provide a reasonable reverse V_{BE} voltage for the transistor in cutoff. This resistor is often bypassed by a speed-up capacitor which is large enough to transfer a sufficient base charge to and from the transistor due to the voltage steps $\pm(V_G(1) - V_G(0))$ at the switching. The resistor R_2 may also be replaced by other level-shifting circuits. For example, since an extra negative supply voltage $-V_{BB}$ is not desirable in integrated circuits, where the layout of extra supply lines may be a problem, the $-V_{BB}$ voltage can be avoided and noise margins are still guaranteed if the levels are shifted by an extra transistor and a series diode in place of R_2 (Fig. 7-5). The first transistor simultaneously provides for current amplification. The point denoted by E is an *extender* (or *expander*) input to which external diodes can be connected to increase the fan-in.

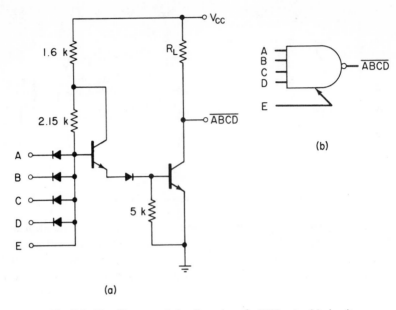

Fig. 7-5. Usual integrated circuit version of a DTL gate: (a) circuit; (b) symbol. Typical $V_{CC} = 5 \pm 0.5$ V.

The inputs of DTL act as sources of current when connected as loads to other output transistors. Accordingly, to increase the current sinking capability of the output transistor, it is advantageous to select the collector resistor R_L as large as possible; in many cases, R_L may even be abandoned. An input at the logical 1 potential draws only leakage current of a reverse biased diode. Figure 7-6 shows a typical input voltage–current relation. The control current of the output circuit is independent of the input voltages,

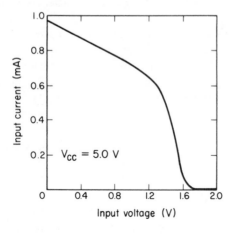

Fig. 7-6. Typical input voltage-current relation of a DTL gate

as long as they are classified as HIGH, and this results in better stability of the output circuit than, say, in RTL.

Comment. An unconnected diode input attains the logic value 1.

DTL families are rather common in digital technology. Their most obvious advantages are high fan-in and fan-out, reasonable noise margins, and rather safe construction of circuits. They have a medium speed of operation. Typical specifications of a basic DTL NAND gate (integrated circuit version) are as follows:

Fan-in	Up to several tens; in integrated gates, typically 4 inputs + extender
Fan-out	Typically 8
Noise margin	650 to 800 mV
Propagation delay	25 to 30 nsec
Power dissipation	8 to 10 mW

The input–output voltage-transfer relation does not depend to a first approximation on the fan-out. A typical curve is shown in Fig. 7-7.

Where speed/power requirements are stringent, special heavy-duty output circuits can be used. Figure 7-8 gives an example of these. This circuit has a high current gain and accordingly a high fan-out capability, up to $N_o = 25$. As a result of the complementary (push-pull) switching action of the output transistors, this circuit may act as a source and a sink current generator, being able to charge and discharge capacitances very rapidly. The emitter-follower action of Q_4 in the logic 1 state of the output provides a small output impedance and attenuates noise picked up capacitively.

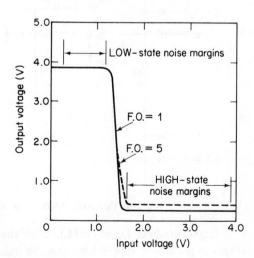

Fig. 7-7. Input–output voltage–transfer relation of a typical DTL gate.

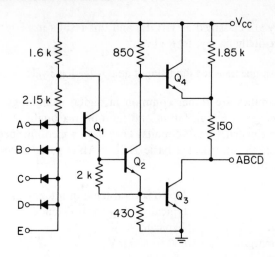

Fig. 7-8. DTL NAND gate with high fan-out. Typical $V_{CC} = 5 \pm 0.5$ V.

Two-Level DTL Circuits. A two-level AND–NOR package, of which one example is shown in Fig. 7-9, was commonplace in digital systems using discrete-component circuits, but it has lost importance in integrated circuits. If all inputs are HIGH, the output transistor is driven to a deep saturation and long saturation delays result.

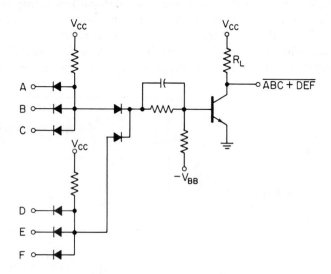

Fig. 7-9. Two-level AND–NOR DTL package.

High-Threshold Logic (HTL). With the presence of excessive noise voltages in ground or supply lines (e.g., in industrial environments) there is a danger that the output transistor, if intended to be in the cutoff state, is occasionally turned on if the emitter potential becomes sufficiently negative. The safest protection against noise is a large $|V_{BEX}|$. A large noise margin (high threshold for conduction) in the cutoff state can be provided if a Zener diode is connected in series with the base lead of the output transistor. A typical HTL NAND gate is shown in Fig. 7-10.

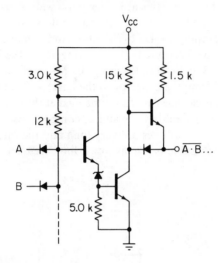

Fig. 7-10. HTL NAND gate. Typical $V_{CC} = 15 \pm 1$ V.

The supply voltage of the HTL circuit must be selected rather high, typically 15 V, in order to produce signal voltages that are in the correct proportion to the threshold voltages. A typical input–output voltage-transfer relation of HTL is shown in Fig. 7-11. For comparison, the corresponding relation of the usual DTL circuit is depicted by a dashed curve. Typically, the HTL noise margins are of the order of 5 V.

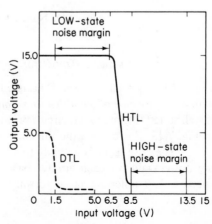

Fig. 7-11. Input-output transfer relations of HTL and DTL.

Noise-Filtered DTL Circuits. Some disturbances (e.g., those generated in supply lines by surge currents) may have a very short duration. Since most realistic electrical circuits, as mentioned in Section 6-3, act as low-pass filters for very short noise pulses, it is to be expected that the maximum tolerable amplitude of a noise pulse is a function of its width. Such relations can be found experimentally and they are called *ac noise margins*.

Enhanced ac noise margins at longer pulse widths can be achieved by designing the logic gates as special low-pass filters with a specified low limit frequency. Especially in DTL circuits using discrete components, a couple of methods have been used: (1) replacing R_2 of the basic DTL by a low-pass T filter as in Fig. 7-12(a), (2) using the Miller effect (i.e., connecting a feedback capacitor from the collector to the base) as in Fig. 7-12(b).

Noise-filtered logic gates are recommended at the interface of computing units and their peripheral devices and at computer-process interconnections. Especially when signal lines are very long and the rate of information transfer is low, it is advisable to use as low a limit frequency for filters as is possible.

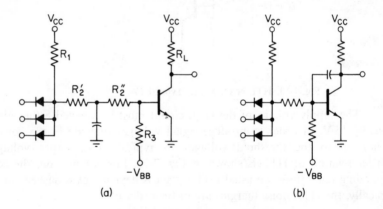

Fig. 7-12. Noise-filtered DTL circuits.

7.2.3. Transistor–Transistor Logic

Closely related to the DTL is a logic circuit called TTL, in which the gate part is formed of a *multiemitter transistor*. The latter is a component that is well suited for integrated circuit production; its diffusion profile is depicted in Fig. 7-13. For clarity, the thickness of the base is exaggerated.

The base layer is actually thin, as in normal transistors, thus providing power gain for each emitter–base–collector section, as long as the respective *E-B* junction has a forward bias and the *C-B* junction is reverse biased.

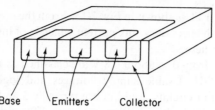

Fig. 7-13. Diffusion profile of a multiemitter transistor (schematic).

In a logic gate, the operation of a multiemitter transistor is roughly equivalent to the operation of a diode gate of DTL, as shown in Fig. 7-14.

As the output circuit for this gate, an active pull-up circuit is usual, except when the gate is included as a part of larger integrated networks. The logic circuit family using multiemitter transistors is called *transistor–transistor logic* (TTL) and it is one of the most popular bipolar logic families. The TTL family was developed directly for integrated circuits.

If at least one of the input emitters has a LOW voltage, the "partial transistor" corresponding to the lowest emitter voltage has a forward B-E bias voltage. The reverse biased C-B junction is then able to draw current from the output circuit, turning it off. On the other hand, if all emitters attain a HIGH voltage, the B-E junctions of the multiemitter transistor become reverse biased while the C-B junction has a forward bias. The current of the base resistor R_B will now be diverted to the output transistor, which will be turned on. To be more specific, we should now consider the multiemitter transistor as an emitter follower, with its collector in the role of the emitter (inverted operation).

From the previous discussion we can infer that the power gain of the multiemitter transistor in the normal as well as the inverted mode of operation provides a faster turn-off and turn-on than in DTL, although the logic-gate operation is similar. Notice that due to the transistor action, the circuit driving a TTL input must supply current toward the TTL gate when the input voltage is HIGH. This was not the case in DTL, where the input diodes were reverse biased in the corresponding situation. A typical input voltage–

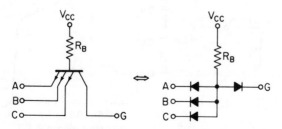

Fig. 7-14. Multiemitter transistor and its junctions, represented by diodes.

current relation of TTL is shown in Fig. 7-15. From the point of view of a design problem, it is the necessary sink current which is of interest and which is defined as positive here.

All voltages from 0 to about 0.8 V are classified as LOW and voltages over 2.0 V as HIGH. Usable supply voltages and logic 1 output levels depend on the output circuits used but are typically 5 and 3 V, respectively.

Because of the high switching speed of TTL circuits, transmission of digital signals from one circuit board to another may already present a problem as to the proper termination of signal lines. We shall return to this problem in Chapter 10. Let it suffice to mention that the dynamic input impedance for positive as well as negative input voltages ought to be reasonably small, and for this reason extra diodes for the clamping of negative voltages are fabricated at each input in most TTL circuits, as shown in Fig. 7-16.

The TTL gate as a separate logic circuit needs a high-level output circuit. The Phoenix gate shown in Fig. 7-17(a) is rather popular. Another circuit [Fig. 7-17(b)] uses more active components, which provide a higher current gain and faster operation but also dissipate more power. The output voltage–current relations of these circuits are shown in Fig. 7-18. Typical specifications for TTL circuits are as follows:

Fan-in	Typically 4 inputs or more on one gate
Fan-out	Typically 10 to 20
Noise margin	0.7 to 1.0 V
Propagation delay	6 to 10 nsec
Power dissipation	10 mW (Phoenix gate)
	25 mW (high-gain output circuit)

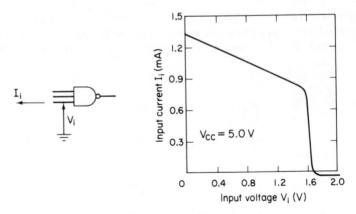

Fig. 7-15. Typical input voltage–current relation in TTL.

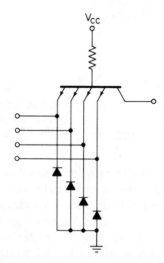

Fig. 7-16. Input clamping by diodes.

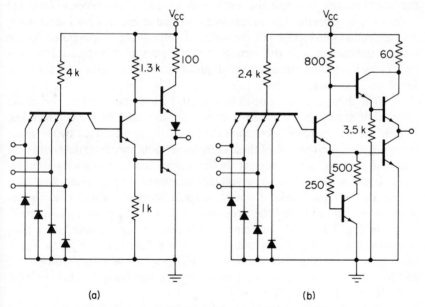

(a)

(b)

Fig. 7-17. TTL gates with typical resistor values: (a) TTL with a Phoenix-gate output; (b) TTL with a high-gain output.

7.2.4. Emitter-Coupled Logic

The nonlinear input–output voltage-transfer relations for the standardization of logic levels were in all previous circuits implemented by driving output transistors from cutoff into saturation. One shortcoming of this

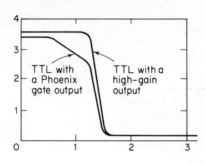

Fig. 7-18. Comparison of input–output voltage transfer relations of the TTL gates shown in Fig. 7-17.

method is that the propagation delay of a logic signal, when a saturated transistor is turned off, is longer than the propagation delay when the transistor is turned on. In many cases these delays are still sufficiently short and do not cause any harm. In TTL this difference in delays has been reduced or almost totally eliminated by the active operation of the multiemitter transistor. Another drawback in the use of saturated switches is that the switching currents that are taken from the supply lines have short rise times, often of the order of nanoseconds. The supply lines may have appreciable inductances, and self-induction voltages of the order of volts might be generated by the circuits themselves. For this reason it is necessary to use shunt capacitors near every integrated circuit or circuit group, in order to have a local supply for short-term surge currents.

The difficulties mentioned before can be overcome and the switching speed simultaneously increased if transistors are used as *current switches*. In ideal current switches, a constant current is diverted between two or more current paths, and the sum of these currents remains approximately constant. Thus the drain of supply current by the system and by every subsystem remains constant, too, and noise voltages generated by the system are eliminated. In the basic mode of operation, transistors as current switches are never saturated, and thus the propagation delays of signals remain small.

In Fig. 7-19(a), the principle of a two-input current switching circuit is shown. If $V_A = V_B$, the output current I_o is a function of V_A (and V_B) in the way shown in Fig. 7-19(b). If $|V_A - V_B| \gg \eta V_T$, the *more positive* of V_A and V_B will determine the output current, the relation being roughly the same as depicted in Fig. 7-19(b).

If the asymptotic levels of output current are classified as logical 0 and logical 1, the more positive being 1, and an input voltage that is sufficiently positive with respect to the bias has the logical value 1 while a sufficiently negative voltage is a logical 0, the output current I_o is obviously a logical NOR function of the input voltages.

As a realistic implementation, a current switching logic gate called *emitter-coupled logic* (ECL), sometimes called *current mode logic* (CML), is shown in Fig. 7-20. A bias voltage is set by an emitter-follower and a voltage

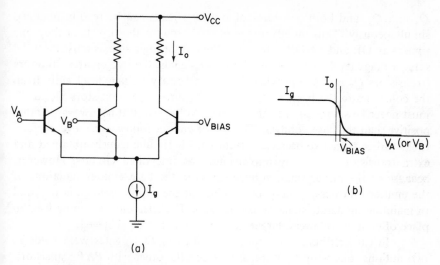

Fig. 7-19. (a) Two-input current switching circuit; (b) voltage–current transfer relation.

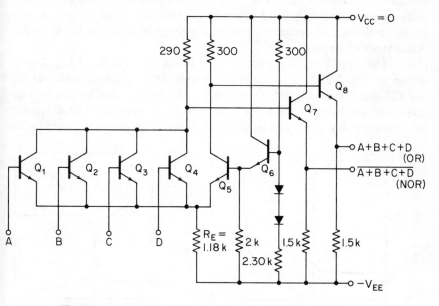

Fig. 7-20. Emitter-coupled logic (ECL) gate with typical component values. $-V_{EE} = -5$ V (typically).

divider. The constant-current generator I_g is replaced by a resistor R_E. Notice that the circuit may be grounded at the V_{CC} terminal, whereas the supply voltage is usually brought to the emitter supply line. The emitter current (neglecting the base currents) is diverted between the collectors of Q_1 to

Q_4 and Q_5, and both polarities of output signals can be readily obtained simultaneously. Thus an advantage of this circuit is that it acts at the same time as an OR and a NOR gate. Since the output logic signals must meet the same voltage-level standards as the input signals, the two emitter-followers (transistors Q_7 and Q_8) provide for the necessary voltage-level shift from the collectors to the bases of the next transistors. The transistors Q_1 to Q_5 must never be saturated, and the collector voltages will thus always be more positive than the base voltages. The use of emitter-followers as level shifters looks rather attractive because current gain is simultaneously obtained and extra transistors are easily implemented as integrated circuits. However, because of the rather small voltage drops in the base–emitter junctions of the emitter-followers, quite good stability of the supply voltage is required to maintain moderate signal noise margins. If Zener diodes can be used in place of emitter-followers, larger noise margins can be achieved.

In earlier discrete-component versions of ECL, gates were made in two options: one with *NPN* transistors and the other with *PNP* transistors. When these types are used alternately in successive stages, a direct connection from a collector to the base of the next stage is possible. Such *complementary stages* are an electrically good solution, but there are some restrictions imposed on the logical design because logic voltage levels are different in different types of stages.

The input voltage–current relation of the ECL circuit of Fig. 7-20 is shown in Fig. 7-21(a), and the input–output voltage-transfer relations for the OR and NOR outputs are shown in Fig. 7-21(b). The logical voltage levels are nominally -1.5 ± 0.4 V in this circuit example. The difference between them, 0.8 V, may be regarded as being rather small compared to the

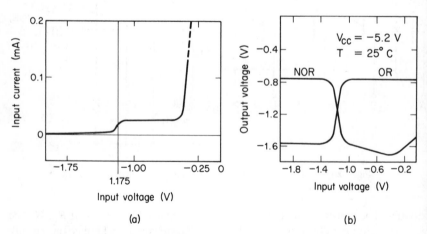

(a)   (b)

Fig. 7-21. (a) Typical input voltage–current relation of an ECL gate; (b) typical input–output voltage transfer relation of an ECL gate.

other logic families. On the other hand, as was mentioned before, the noise signals generated by the circuit itself remain small, and thus the rather small noise voltage margin of this circuit is not a serious shortcoming.

Typical specifications for this circuit are as follows:

Fan-in	Of the order of 20
Fan-out	Typically 25
Noise margins	0.35 V
Propagation delay	3 to 4 nsec (down to 1 nsec in special circuits)
Power dissipation	40 mW

7.2.5. Complementary Transistor Logic

An active equivalent of the voltage-selector diode gate discussed in Chapter 6 is a current switching transistor circuit in which an output voltage signal is taken from the emitter point. In the circuit shown in Fig. 7-22, the output of an emitter-coupled *PNP* transistor gate is, in fact, a minimum voltage selector of *A*, *B*, and *C*, and an output emitter follower performs a level-shifting and current-amplifying operation. This gate is called a *complementary transistor logic* (CTL) circuit, for obvious reasons.

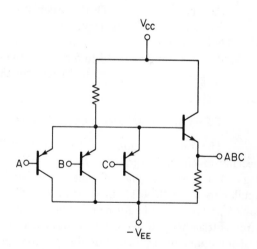

Fig. 7-22. Logic gate of voltage–selector type in CTL.

A logic circuit such as the one depicted in Fig. 7-22 cannot standardize signals, because the voltage levels of output signals depend on the voltage levels of input signals. Therefore, after a few minimum voltage-selector gates, logic signals must be standardized by special saturating logic gates of the type shown in Fig. 7-23. The CTL system thus necessarily involves both types of gates discussed. Of course, a saturating gate has a larger propagation delay, but the relative number of saturating gates in CTL systems can be kept small.

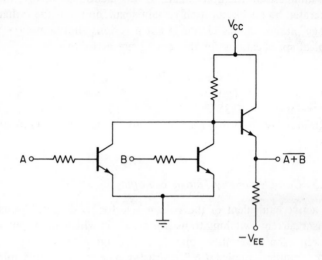

Fig. 7-23. Standardizing logic gate in CTL.

Typical specifications of CTL circuits are as follows:

Fan-in   Of the order of 8
Fan-out   Typically 12
Noise margins   0.25 to 1.0 V
Propagation delay   3 to 5 nsec for the selector gate,
                    12 to 15 nsec for the standardizing gate

7.2.6. Wired AND as a Logic Operation

The implementation of combined (multilevel) logic functions, especially in larger integrated arrays, may be based on the idea of inverter logic. The output circuits of many logic families using resistor pull-up are similar to a simple RTL inverter switch. Consequently, certain combined logic operations can be formed by tying the collectors of two or more output circuits together. To avoid unnecessary current drain, only a single pull-up resistor ought to be used. For this purpose the pull-up resistors, although incorporated in the circuit, are often disconnected. (In DTL circuits, pull-up resistors are not mandatory because a disconnected diode input attains the logic value HIGH. Pull-up resistors are used only to define larger voltage swings at the inputs.) The logic operation implemented by wiring, as will be easily seen, is a two-level AND–NOR. Another way of describing the new operation is to say that the original gate operations (NAND's) are followed by AND. [This is shown by a "phantom" gate in Fig. 7-24(b).] This method is called *Wired AND* (also *Dot AND*). Figure 7-25 gives another example of

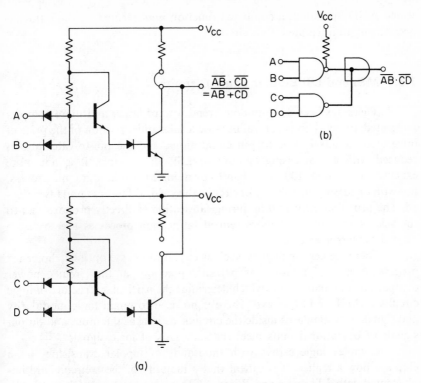

Fig. 7-24. Wired AND in DTL: (a) circuit connection; (b) symbolic representation in which the load resistor is shown separately.

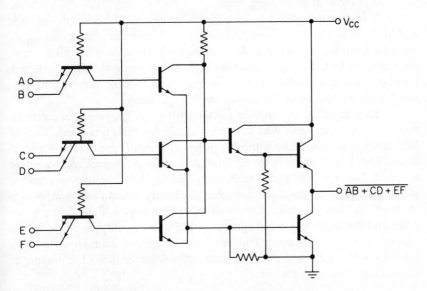

Fig. 7-25. Wired AND within a TTL circuit.

Wired AND operation, a combined function implemented by the intraconnections in an integrated TTL circuit.

7.3. MSI and LSI Bipolar Logic Circuits

There has been a continuous trend toward larger and more complex
integrated arrays. This is natural because a substantial portion of the price of
integrated circuits in due to pin connections, and the pin-to-gate ratio is
reduced with an increasing amount of LSI. In principle, it is not more
expensive to produce 100 gates than 10 by a single photolithographic process,
but with increasing complexity of circuits the yield of faultless units is reduced. The probability for failure during fabrication is directly proportional to
the size of the unit, and more refined fabrication processes are therefore
needed for larger arrays.

There are certain factors, such as circuit topology, optimal choice for
processing steps, formation of parasitic junctions at the circuits, power
dissipation, and many more, which determine the particular structure of logic
circuits in MSI or LSI arrays. There is no need to retain logic modularity
and signal-level standards inside the circuits, as long as the input and output
signals of operational units meet the standards of logic signals.

Complex logic arrays with thousands of bipolar transistors on a
chip are now a reality. It is salient that a majority of these circuits are implemented with TTL logic and Wired AND intraconnections. Obviously the
main arguments for this selection are a well-established technology and a
minimal number of processing steps.

It is almost hopeless to give typical representative examples of MSI
or LSI circuits. The fabrication technology is changing at a pace that would
make an extensive review out of date before publication. Bearing this in mind,
the examples in this chapter must be regarded as being randomly selected.
However, the basic solutions for circuits do not differ very much from
small-scale integration to the most complex LSI arrays.

Such repeated and regular circuit patterns as are found in counters,
registers, memories, and decoders are suitable for the mass production of
MSI and LSI cicuits. There is another practical point of view for the introduction of LSI in these units: Since interconnections from one package to another
are made through pins and discretionary wiring, it is desirable to keep the
pin-to-gate ratio as small as possible, as already stated. For example, in
shift registers access is usually necessary to only a few flip-flops, and selector
gates and decoders may be integrated to the memory chip in memories. In
general, the communication between large integrated arrays ought to be
made by as few signals as possible, even if this would increase the number of
interfacing circuits on the die.

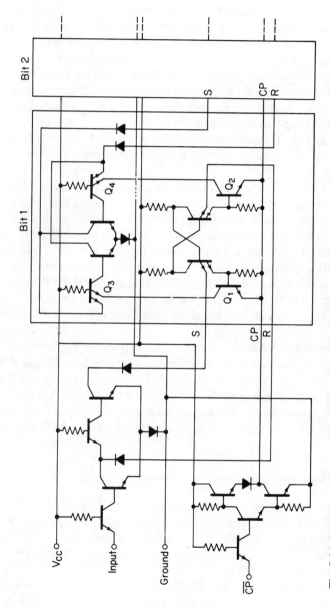

Fig. 7-26. Integrated bipolar shift register. (Simplified representation of a circuit as made by Texas Instruments, Inc.)

Example of MSI Registers: Shift Register. Since the internal signals within logic arrays need not be standardized, a shift register using master–slave R-S flip-flops, for example, can be implemented with fewer logic gates than are normally needed. The minimum distinguishable difference in logic signal levels is that which guarantees that a certain flip-flop can be turned on or off. In the TTL logic version of Fig. 7-26, a special method for the production and coupling of clock pulses is used. The bistability of the lower flip-flop, the master, is implemented by a direct cross-connection of the multiemitter transistors. This is allowable because there are no collector loads. Notice also that the master flip-flop has no ground reference potential exept that provided by CP, S, and R. The collector currents are diverted between the emitters depending on the relative potentials of the CP signals and the S and R inputs.

In the normal state, CP is LOW and one of the multiemitter transistors of the master flip-flop is conducting through the respective emitter. The LOW potential of the S and R inputs is higher than the LOW value of CP due to the input diodes and an extra series diode in the driver of S and R signals. Thus the effect of S and R on the master flip-flop is blocked out, because corresponding emitter junctions are reverse biased. At the same time, either Q_1 or Q_2 is conducting, depending on the side of the master flip-flop that is conducting. Thus Q_1 or Q_2 is holding the corresponding emitter of Q_3 or Q_4 LOW, respectively, and the state of the slave flip-flop follows the state of the master.

When CP is made HIGH, the inputs to the slave flip-flop are blocked out. The state of the master flip-flop is now determined by the S and R inputs. The S and R signals in this circuit are always complementary and thus one side of the master flip-flop is always conducting current to a LOW potential.

The operation of this master–slave combination is thus easily recognized as logically equivalent to the operation of a normal master–slave flip-flop.

7.4. MOS Logic Circuits

Although numerous possibilities for the implementation of logic gates are available in bipolar families, there is practically only one logic principle that has been applied in MOS logic circuits. The basic mode of operation of unipolar transistors in MOS logic circuits is *voltage switch*, and this is what we call *inverter logic*.

The enhancement-mode MOSFET is very suitable for an inverter switch because its input signal can be directly coupled to the gate, without need for any impedance matching or voltage-level-shifting circuits. Accordingly, a MOSFET inverter switch is nothing other than the MOSFET itself. In inverter logic, a network of inverter switches' is connected to the supply voltage through a *pull-up resistor*. In integrated arrays, it is more economical and electrically superior to fabricate a special MOSFET as a current generator in place of the pull-up resistor.

The circuit configuration of a simple MOS inverter is shown in Fig. 7-27. The gate of the upper MOSFET may also be connected to a separate voltage $-V_{GG}$. MOS logic circuits are formed of serial and parallel connections of inverter switches. For example, the NAND and NOR circuits, as well as the implementation of a full adder, are shown in Figs. 7-28 and 7-29.

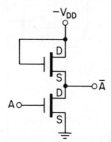

Fig. 7-27. MOS inverter.

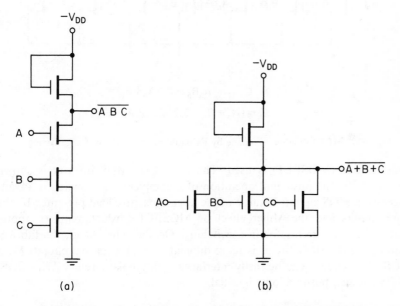

(a)   (b)

Fig. 7-28. MOS logic gates: (a) NAND; (b) NOR. (Negative logic.)

Because of negative supply voltage, *negative logic* definition is used in MOS circuits unless otherwise stated.

In *complementary symmetry MOS circuits* (COS-MOS), an active

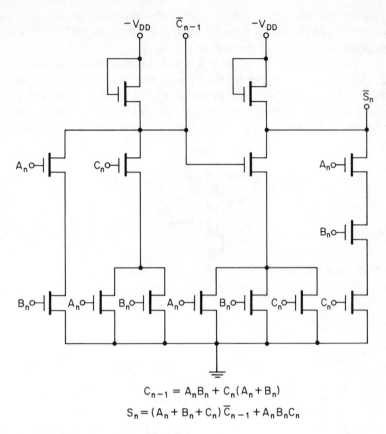

$$C_{n-1} = A_nB_n + C_n(A_n + B_n)$$
$$S_n = (A_n + B_n + C_n)\overline{C}_{n-1} + A_nB_nC_n$$

Fig. 7-29. MOS full adder, as made by Fairchild Semiconductor. (Negative logic.)

pull-up load (MOSFET) is used. The basic COS-MOS inverter is shown in Fig. 7-30. The push–pull operation of a complementary switch is advantageous in MOS circuits because of the large capacitive load presented by the gate circuits and the Miller effect of MOSFET's, which requires appreciable control currents for fast switching. On the other hand, the fabrication of a COS-MOS circuit is more difficult than of a normal circuit. Most COS-MOS circuits can be easily interfaced with bipolar circuits. With COS-MOS circuits, positive logic is usual.

Clocked MOS Logic Circuits. One shortcoming of MOS circuits is their lower speed as compared to bipolar circuits. For to achieve smaller time constants, the transconductances could be selected larger, but this results in increased power dissipation, which may become a problem, especially if the high packing density of MOS circuits is fully utilized.

The pull-up MOSFET used in place of a load resistor offers a possi-

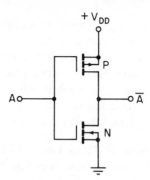

Fig. 7-30. Complementary symmetry MOS inverter.

bility to reduce the average power dissipated in the circuit by lowering the duty cycle of the circuit. In many applications, logic signals may be strobed at prescribed intervals. The output voltage of a logic circuit can then be sampled or gated by applying a short clocking signal ϕ to the the gate of the pull-up MOSFET according to Fig. 7-31. Thus the duty cycle of the current generator remains small.

Notice also that the isolation resistance of gate inputs is extremely high (of the order of $10^{14}\ \Omega$ typically), and their stray capacitances are of the order of a few picofarads. The charges stored at the gate capacitances thus have their only leakage paths through the reverse biased drain-to-substrate and source-to-substrate junctions, described by the leakage currents I_{DSS}, usually of the order of 1 μA. Thus a logic voltage level applied at the gating is sustained typically for a period of microseconds. Especially if the logic levels are changed often, this "sample-and-hold" arrangement of logic circuits is useful in MOS logic. Switching power is needed only when the values of logic signals are changing. However, clocked gates are not practical in static logic circuits.

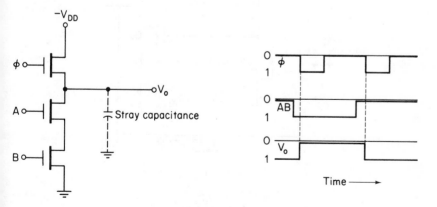

Fig. 7-31. Clocked MOSFET gate. (Negative logic.)

The clocking of logic circuits serves other purposes, too; for example, since the feedback of amplified signals through gate capacitances might present problems in larger arrays, unidirectional flow of information through the system can be guaranteed if, in cascaded stages, signals are coupled in succession, enabling only one stage at a time. This method is analogous to the familiar master–slave flip-flop operation. We shall discuss later similar problems with four-phase circuits.

Special Features of MOS Logic: Bistable Latch. Since MOSFET's can be used as logic switches as well as linear gates for signal voltages, it is particularly advantageous to implement sequential operations in MOS circuits. An electric charge stored in the gate capacitance represents stored information. In the circuit of Fig. 7-32, which is a bistable latch, information is handled in this way. When the clocking signal CP is made 1, Q_1 switches the input information to Q_2 while the switch presented by Q_3 is open. The logic value of the input signal is also stored at the gate of Q_4 in inverted form. When CP is made 0 and $\overline{CP}$ becomes 1, the input gate is disabled and the feedback path of the flip-flop is closed through Q_3. Thus the binary information transferred to this flip-flop in the clocking operation is retained as long as CP is 0.

MOS Shift Registers. Let us recall the problem of temporary storage in shift registers, as discussed in Section 3.5. In order that a bit storage location would be able to accept new information while still emitting the old

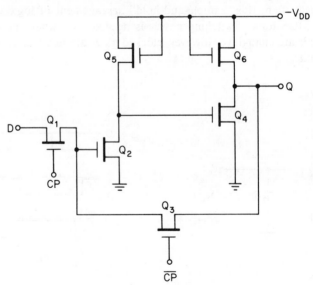

Fig. 7-32. Bistable latch implemented in MOS logic. (Negative logic.)

information, the gate capacitance can act as a temporary charge storage which holds binary information during a sequential shift operation. This charge is gradually leaking off through the reverse biased PN junction formed by the driving circuit. In order to hold the binary information in the storage cell for indefinitely long periods, the storage must be made bistable by a logical feedback.

One stage of *static MOS shift register* is shown in Fig. 7-33(a), and the timing diagrams of two clocking periods of its control signals are depicted in Fig. 7-33(b). Negative logic convention is used in the discussion of this circuit. Input data are copied to the gate capacitance of Q_2 at $\phi = 1$. The old information is stored in the gate capacitance of Q_5 while Q_3 and Q_4 are open, and it can be copied to the next stage simultaneously. When ϕ goes off, the new information from Q_2 is transferred to the output circuit (Q_5) through Q_4 at the "fast" clocking signal ϕ_F while the "slow" clocking signal ϕ_S is still 0 and Q_3 is open. Thus there is no interference with the old information of Q_5 on Q_2 through the feedback path. When the shifting of binary information has been completed, ϕ_S is made 1, and the whole system (except Q_1) now acts as the usual static logically coupled flip-flop. The small notches in the gate voltage of Q_2 are exaggerated to show the effect of feedback signals through Q_3.

A drawback of this circuit is the need for three carefully timed clock phases.

Dynamic MOS shift registers are simpler than the static registers. They usually need two clocking signals, ϕ_1 and ϕ_2, which are not allowed to overlap. A shortcoming of such circuits, of which a typical example is shown in Fig. 7-34, is that information is not permanently stored in any storage cell. In order to preserve the contents in this register, a continuous shifting at a certain frequency must be made. The circuit of Fig. 7-34 operates in a fashion similar to that of a two-phase master–slave flip-flop except that gate capacitances are used in place of storage flip-flops. At the clocking phase ϕ_2, the state of Q_4 is copied to the output terminal through Q_5. In the same way, input data are copied to Q_1 from the previous stage. This information is temporarily stored at the gate of Q_1 until the state of Q_1 is copied into Q_4 at the other clocking phase ϕ_1 through Q_2.

Notice that the binary information at the drain electrode of Q_2 is a negation of the input signal; at the drain of Q_5, the original logic value is restored.

This example of dynamic shift registers is selected for illustrative purposes only; some modifications of this principle require more clock phases and are correspondingly faster and more reliable.

Four-Phase MOS Circuits. The four-phase (4ϕ) principle combines features of gated static logic circuits and dynamic shift registers. Information

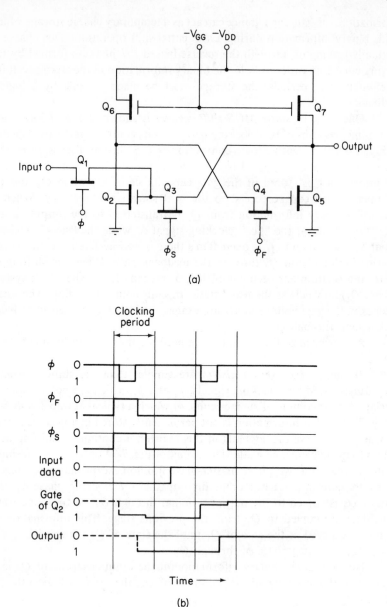

Fig. 7-33. One stage of a static MOS shift register: (a) circuit; (b) timing diagrams. (Negative logic.)

is transferred as an electric charge between circuit nodes (data points), and a complex clocking and gating system is needed for the transfer of signals through the circuitry, at the same time guaranteeing unidirectional flow of

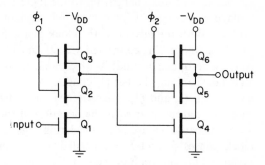

Fig. 7-34. One stage of a dynamic
MOS shift register.

information. It was pointed out recently by Boysel and Murphy [201] that 4ϕ
logic is also applicable to complete digital systems, such as general-purpose
computers, and that the clocking system actually is no problem. The main
advantages of 4ϕ logic are a large speed/power ratio and low cost in large
systems because of the simplicity of basic logic gates.

The basic operation of a 4ϕ gate can be seen from Figs. 7-35 and
7-36. Here Q_1, Q_2, Q_3, and Q_4 are MOSFET's which are gated by negative
logic signals. The circuit of Fig. 7-35 as a whole acts as an inverter; if more
complex logic conditions are needed, Q_4 may be replaced by MOSFET
inverter switches connected according to the static inverter logic. The
capacitance indicated by C is the total capacitive load of stages connected
to the output. Only two of the clock phases, ϕ_1 and ϕ_2, are needed at this
stage; the other two are used elsewhere in the system. When ϕ_1 becomes 1
(which in negative logic means that Q_1 and Q_2 start conducting), the capa-

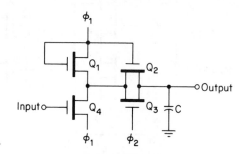

Fig. 7-35. Four-phase inverter.

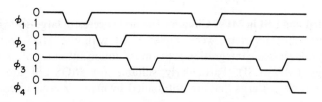

Fig. 7-36. The four clock phases.

citance C is at first charged up to the logic 1 level through Q_1 and Q_2. This voltage excursion always occurs independent of the value of input signal voltage. With the advent of the clock phase ϕ_2, which does not overlap ϕ_2, the capacitance C is connected via Q_3 and Q_4 to a potential represented by ϕ_1 (now zero, i.e., ground). With $\phi_2 = 1$, Q_3 is conducting, whereas the input signal determines whether Q_4 conducts or not. If the input is 1, C is discharged through Q_3 and Q_4, whereas for input equal to 0, Q_4 is off. Thus in both cases the output after ϕ_2 is the negation of the input.

Because we are not considering logic voltages but charges, the other clock phases ϕ_3 and ϕ_4 are needed to transfer a charge signal through the system. This is accomplished in the way similar to that of master–slave flip-flops, by opening transfer paths in succession. Since the output of the circuit of Fig. 7-37 is stable during the other two clock phases, ϕ_3 and ϕ_4, we can divide the four clock phases cyclically for different parts of the system so that "information waves" are propagated through the system.

Another variant of MOSFET connections, here implementing one stage of a shift register, is shown in Fig. 7-37.

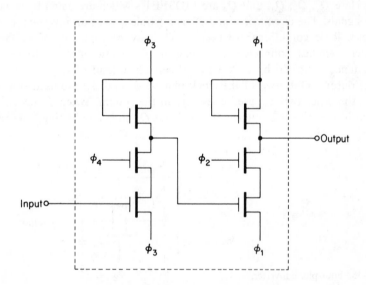

Fig. 7-37. One stage of a 4ϕ shift register.

MSI and LSI in MOS Circuits. The first successful large-scale integrated circuits, shift registers, were implemented with the MOS technology. Nowadays, large-scale integration is the most important mode of using MOS technology (MOS/LSI). Particularly suitable for MOS/LSI are regular, repeated circuit patterns that are not loaded by many external circuits. Such circuits are found in memories (random access and read-only memories), integrated accumulators, serial/parallel and parallel/serial converters, code converters, channel selectors, and digital-to-analog and analog-to-digital converters.

Typical integrated memory circuits of MOS/LSI will be discussed in Chapter 9.

7.5. Comparison of Various Logic Families

Recapitulation of Bipolar Logic Families. *RTL.* As an independent family, RTL has almost vanished from the market, but in combined circuits its basic solutions are still found. The main reason for the introduction of *RTL* is that it is possible to implement reasonably fast circuits using low-quality transistors. There are two shortcomings of RTL: the need for resistors in the base leads (and therefore much space on an integrated circuit) and low noise margins. The advantages of RTL are simple construction, low price, and availability from many vendors. For special interfacing purposes, RTL circuits made of discrete components can be used.

DTL. Since DTL has been adopted from discrete-component technology, it does not have all the advantages that are offered by integrated active components. One of the drawbacks of DTL is that if diodes and other junctions are used in series with the base lead for the necessary voltage drop, the saturation base charge has no external path for leaking off. This results in very unsymmetrical propagation delays of the logic signals (i.e., the switching from the LOW to HIGH state is much slower than the opposite transition). On the other hand, this feature may sometimes be a remedy for noise problems, because short noise spikes generated by capacitive and inductive stray coupling between signal lines are filtered out during saturation. For this reason DTL circuits might be used at the interface of logic units (e.g., in LSI circuits in which the internal structure is made by TTL logic).

There is another advantage of DTL circuits which is helpful in interfacing applications: In the HIGH state of inputs, the input diodes have a large reverse bias and the operation of gates becomes practically independent of the input voltage levels. This isolation is also effective in attenuating propagated noise signals. Also, there is usually no danger of causing damage to logic gates by having input signals that exceed supply voltages by moderate amounts. The DTL also has high fan-in and fan-out figures.

There is a special feature in DTL—that the inputs are controlled by logic 0 voltages by switching the inputs to ground. An unconnected diode thus has a state that is equivalent to the state caused by a HIGH signal.

Consequently, pull-up resistors in the output circuits of DTL are not mandatory, but they are used for faster and larger voltage swings.

HTL. Being a modified version of DTL, this circuit has approximately the same advantages and shortcomings as DTL except for the very large noise margins. This feature is needed in industrial equipment.

TTL. This is the most widespread logic family for the present, and it seems to be extremely suitable for LSI. With many TTL logic-gate modules, some problems are caused by the high switching speed. Mechanisms for the internal generation of noise are discussed in more detail in Chapter 10.

One special problem in the use of TTL is the coupling of logic 1 signals to the inputs. Precautions must be taken in order not to exceed the allowed emitter–base voltages. Unused inputs of NAND gates ought to be connected to a well-stabilized logic 1 potential which is lower than the supply voltage to protect for voltage transients during the turning on of power supplies. Other methods are the connection of unused inputs to V_{CC} through, say, a guarding resistor (of the order of 100 Ω typically), or tying the unused inputs to other inputs.

In the active pull-up output circuits of many TTL families, there is a possibility that both complementary switches conduct simultaneously during the driving of a capacitive load. A surge current spike like this calls a heavy current drain from the supply line and is reflected as a noise signal.

The TTL is electrically superior to many circuits, but, owing to easily produced logic noise, its use needs special care.

ECL. As already mentioned in the introduction to logic families, ECL is the fastest logic family available in the form of standard components, and the noise generated by the circuit itself is small. The fan-out of this family is very large. A drawback is a rather high power dissipation and the need for well-stabilized power supplies. Also, the logic signals do not have a fixed reference potential such as is formed by the ground potential in saturated circuits.

The ECL has the special advantages that NOR and OR outputs are available simultaneously. On the other hand, the interfacing of ECL with other logic families is more difficult than between other families.

CTL. The complementary-transistor logic was introduced for applications in which high speed at a low cost must be attained. Because of the two kinds of logic circuits needed, the use of this family needs some special expertise. The CTL is favorable if high capacitive loads are driven.

Survey of the Relative Merits of Bipolar Versus Unipolar Devices in Digital Technology. Most logic circuits are still built by bipolar technology, and the fabrication processes of bipolar logic circuit families are well established. However, much work has also been done in recent years on MOS circuits, and it seems that the MOS technology would have many advantages over the bipolar components, especially in LSI circuits. The following is a brief listing of the most characteristic features of the relative merits of these technologies.

ADVANTAGES OF BIPOLAR CIRCUITS

1. Good long-term stability.
2. High switching speed and small propagation delays of signals (typical minimum switching and delay times of 5 to 20 nsec).
3. Rather low resistances in the conducting state of transistors (of the order of 10 Ω) and, accordingly, good matching of signal lines.
4. Small interelectrode capacitances (of the order of 10 pF).
5. High current-handling capability.

SHORTCOMINGS OF BIPOLAR CIRCUITS

1. Power dissipation per stage appreciable (however, speed/power characteristics satisfactory).
2. Complex structure, requiring many fabrication steps (typically 130 steps per circuit).
3. Relatively large area necessary, because of the electrode structure and because several bonding pads are needed; typical minimum area 24 mil^2 per transistor.

In the appraisal of MOS technology, the assertions are roughly inverse to those mentioned before. However, some specific features are listed below.

ADVANTAGES OF MOS CIRCUITS

1. Very simple structure (typically requiring 38 fabrication steps).
2. Small area per transistor, typically 1 mil^2; fewer bonding pads than with bipolar transistors.
3. No need for resistors in the circuits because MOSFET's can be connected as pull-up resistors.
4. Isolation of circuits on a chip better than with bipolar devices; no extra isolation diffusion steps necessary.
5. Current flow bilateral.
6. High-input impedance, accordingly high fan-out capability at low switching speeds.

SHORTCOMINGS OF MOS CIRCUITS

1. Long-term stability of gate insulation questionable; however, improvements are taking place.
2. Although input capacitances are not much larger than with bipolar devices, they may form charging-time constants of considerable size with the generator impedances of active stages. Thus bipolar transistors are still superior if high speed at high fan-out is necessary.
3. MOS circuits cannot be easily combined with bipolar circuits. Thus MOS circuits need special matching circuits when interfaced with bipolar circuits.

Some Hints for the Selection of Logic Families. No one of the various logic families is clearly superior over the others: in fact, the optimal selection for the logic family depends on the degree of integration, speed, cost, noise margins, etc. In small-scale integration, in which individual logic gates are interconnected through pins, many of the electrical properties of the complete system are determined by the circuit designer. Consequently, there can be numerous possible choices for logic families for different purposes. We must state that the need for small-scale integrated circuits is still there, mainly because it is very difficult to find large standardized arrays for all applications. This is more so in customized design, whereas in long production lines standard dies can be used.

The first question that usually arises is whether bipolar or MOS technology should be used. Although the same question has been asked for several years, answers to it have changed. As of this writing, there is still too little experience with MOS circuits to make definite assertions. The question is therefore left open, and the following discussion applies mostly to bipolar circuits.

Unless there are special reasons to favor a particular logic type, the designer should start with the most common families. First he ought to make clear whether he is aiming at the fastest possible system or whether he could do with a switching rate in the megahertz, or slower, region. If the latter, he could choose a slow family, such as DTL, with which no special precautions need be taken. But if the environment is noisy, he might do much better to select HTL. If high switching speed has a high trade-off in the system, TTL might be the right answer. If selecting TTL, the designer must be aware of the many precautions necessary when switching times approach signal propagation times in interconnecting lines. Some of these difficulties will be discussed in Chapter 10.

If ultimum speed is important and cost is not a primary factor, ECL is the right selection. With the fastest circuits, however, open interconnections are no longer possible except at the shortest distances (much less than 1 foot). The use of transmission lines becomes mandatory at longer distances. The electronic design of circuits in the nanosecond region can be mastered only by experienced and skilled designers.

7.6. Integrated Circuit Packages and Their Use

Standardized Packages for Integrated Circuits. Silicon dies must always be connected to the outside world through pin connections. For this purpose, several standardized packages have been developed. A good package must have the following features:

1. It must allow effective layout of printed circuit boards.
2. Insertion of packages onto circuit boards must be easy when made by hand or by special tools; also, socket mounting may be necessary.
3. The package must provide mechanical protection for the circuits and the leads must form a sturdy footing for the package.
4. If the total power dissipation in the circuit is high, heat convention from the package to the surroundings must be good.

The first commercial package for integrated circuits was the round, transistor-outline metal can in which pin connections were made through 8 to 10 leads arranged in a circle. This package was developed for discrete transistors, and this type of can forms a good housing for integrated circuit amplifiers and other linear circuits. In digital circuits this package has not been very popular because its pin arrangement does not allow the complicated wiring patterns typical for digital circuits.

The *flat package* is another early-developed package which has survived, especially for linear circuits. The envelope of flat packages consists of metal, glass, or ceramic, and the connections are made through 10 to 14 metal strips. The mounting of this package onto printed circuit boards, especially in digital circuits, is still a problem.

The most widespread package for digital integrated circuits is the *dual in-line package*, which usually consists of a plastic encapsulation, although heat-resistant materials such as ceramics are also used. One reason for the popularity of dual in-line packages is their low price and easy installation on printed circuits. The same type of pins can be used as soldering leads as well as socket pins. As small-scale and medium-scale integrated circuits, 14-lead and 16-lead packages are the most common. Dual in-line packages can be used for LSI circuits, too.

Figures 7-38 through 7-45 show some packages for integrated circuits. (All measurements are in inches).

Temperature Ranges of Integrated Circuits. One of the most important specifications imposed on electronic devices is the intended range of temperatures in which reliable operation is guaranteed. This implies that the vendor of integrated circuits must specify the temperature range in which all electrical specifications for semiconductor circuits are valid. The reliability is, of course, a concept of statistical quality control, and very often components obtained from the same batch are divided into different quality classes after testing. Accordingly, there is always a compromise between the price and the quality of a circuit.

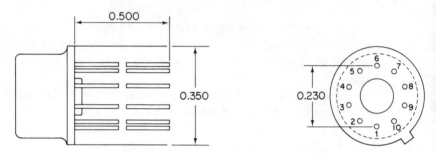

Fig. 7-38. TO-100 metal can outline.

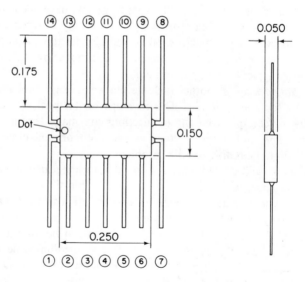

Fig. 7-39. Flat package outline.

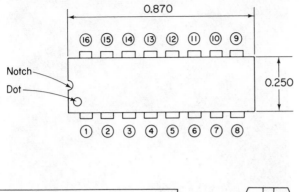

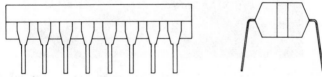

Fig. 7-40. Sixteen-pin dual in-line package outline.

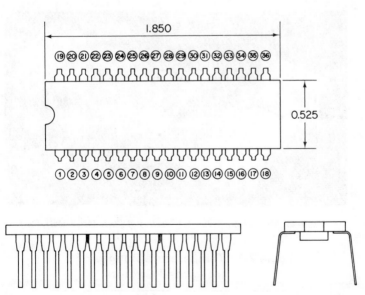

Fig. 7-41. Thirty-six-pin dual in-line package outline.

The most usual temperature ranges of integrated circuits are shown in Table 7-1. A word of caution is necessary in order to give the proper meaning to these terms. *Industrial use* actually means industrialized pro-

Fig. 7-42. Flat-pack, plastic dual in-line, ceramic dual in-line, and metal-can integrated circuit packages. (*Courtesy of Texas Instruments, Inc.*)

Fig. 7-43. MOS 24-pin package. (*Courtesy of Texas Instruments, Inc.*)

Table 7-1. Temperature Ranges of Integrated Circuits

Military range	−55 to +125°C
Military midrange	0 to 100°C
Professional range	0 to 75°C
Industrial range	15 to 55°C

Fig. 7-44. Artist's rendering of a medium-scale integrated package detailing inner connections. (*Courtesy of Texas Instruments, Inc.*)

duction of, say, such mass-produced devices as TV sets, radios, etc. If any integrated circuits are intended for reliable long-term use *in industrial environment,* a *professional line* of components ought to be selected. Sometimes even this might not be enough. Taking into account the high value of raw materials in a production line that is directly controlled by electronic devices, and special environments, it might sometimes be a wise policy to select components of the best military range for such applications.

Design of Integrated Circuits. The layout of integrated circuits is often made using computer-aided design (CAD) techniques. Figs. 7-46 and 7-47 show equipment used in CAD.

System Packaging. Almost invariably, integrated circuit packages are mounted on printed circuit boards. The most usual method for the creation of circuit patterns is to use both sides of circuit boards for wiring. Printed circuits are usually made of copper foils cemented on epoxy-glass boards, but the type of plating of the conductors varies. It is usually necessary to interconnect circuits located on opposite sides of the board, and this

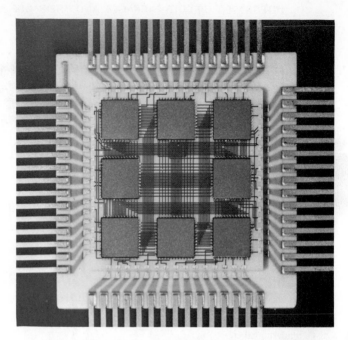

Fig. 7-45. MOS memory-chip package utilizing the "beam-lead" approach in packaging. (*Courtesy of Texas Instruments, Inc.*)

Fig. 7-46. Computer-aided graphic design of an integrated circuit takes place at a cathode-ray-tube terminal using a light pen. (*Courtesy of Motorola Semiconductor Products, Inc.*)

Fig. 7-47. Automatic drafting machine for computer-aided integrated circuit mask master preparation. (*Courtesy of Motorola Semiconductor Products, Inc.*)

is made by plated-through holes. A good practice is to let the connector strips on opposite sides cross each other perpendicularly. Unused portions of circuit boards ought to be used as ground planes, and in order to reduce lead inductances, large conducting areas instead of narrow conductors are to be preferred as supply lines. But in complicated circuits, especially in LSI, it might happen that the interconnections require more space than provided by two conducting layers. In this case, multilayered boards may be used. Several one-sided circuit boards are fabricated separately and bonded together using adhesives. Interconnections between layers are made by plated-through holes. In multilayered boards, some layers may serve as ground planes which have openings at all places where component leads or plated-through holes must pierce them.

Very large portions of electronic devices can be manufactured on printed circuit boards. However, since servicing must always be taken into account, it is advantageous to have a high degree of modularity in the

Fig. 7-48. Small general-purpose computer showing printed circuit boards (front) belonging to it. Notice that printed conductors of opposite sides cross at right angles, and a ground plane of copper foils surrounds the boards. Conductor strips, ending at the small dots, are plated-through connections between the top and bottom sides. Notice also the capacitors which are used to bypass noise generated inside or outside the system. (*Courtesy of Raytheon Co.*)

system. Therefore, professional systems are often assembled of several smaller circuit boards interconnected by back-panel wiring. Admittedly, the introduction of several printed-circuit connectors brings about extra problems as to the reliability of electromechanical contacts. In automated production, back-panel wiring may be implemented by a printed circuit

Fig. 7-49. Sockets for dual in-line integrated circuits and discrete components are mounted in planes and drawers. Socket pins extend through the frames, positioned accurately for automatic wiring. (*Courtesy of Electronic Engineering Company of California.*)

Fig. 7-50. In this figure, components are mounted on printed circuit cards which plug into connectors mounted in drawers and cabinets. (*Courtesy of Electronic Engineering Company of California.*)

board, called the *mother board,* on which circuit connectors are mounted. Mother boards are very often interconnected by flexible printed cables.

The LSI technology will certainly affect conventional system packaging, but a detailed discussion is beyond the scope of this book. Instead, some general problems regarding the interconnections are discussed in Chapter 10.

Some examples of mechanical packaging are given in Figs. 7-49 through 7-51.

Fig. 7-51. Automatically wired assemblies. (*Courtesy of Electronic Engineering Company of California.*)

7.7. Future Trends in System Packaging

Some improvements in the existing logic circuit families are to be expected in the near future, but the basic semiconductor technologies seem to be well established for the present. However, the largest changes in circuit technology during recent years have occurred in system packaging. While the size of monolithic circuits is increasing year by year, many users of integrated circuits have started to make their own integrated circuit modules by *hybridizing* (i.e., buying the semiconductor circuits as chips and interconnecting them into larger assemblies by *thin-film* or *thick-film technology*). In both these methods, metalized conductor strips and some passive components, primarily resistors, but also capacitors, are deposited on a ceramic substrate on which monolithic chips are also mounted. After fabrication, the circuit assembly is encapsulated as a whole. Reliability is increased because there are fewer junctions than with separate integrated circuit packages mounted on printed circuit boards. On the other side, the initial expenses of hybrid circuits are higher than with the conventional technology, so the fabrication of such modules does not become economical until the production line consists of some thousands of similar units or more.

Thin-Film Circuits. If the bonding pads of semiconductor chips are interconnected with vacuum-deposited (evaporated) metal conductors, resistors, and capacitors, we speak of thin-film technology. Using SiO_2 layers between the strips, thin-film networks can be deposited in several layers. Figure 7-52 shows a typical example of a thin-film hybrid microcircuit before encapsulation.

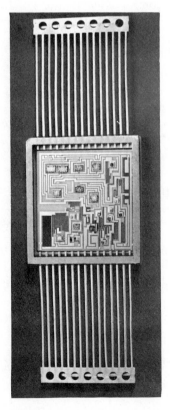

Fig. 7-52. Integrated circuits mounted on a substrate by the thin-film technique. Light strips are metalized conductors, dark strips are resistors. (*Courtesy of Collins Radio Co.*)

Thick-Film Circuits. A cheaper but not as stable interconnecting method as the thin-film circuit is the thick-film circuit, in which conductor strips and resistors are deposited by *printing* (i.e., using screening masks and inks which consist of colloidal metal and metal oxides mixed with organic binders and solvents). After painting, the substrate is fired at temperatures of 800 to 1000°C, after which the circuit is encapsulated. In most thick-film circuits, planar conductor networks must be used. If crossing conductors are needed, they are made by discretionary wiring (welding), which is also used with thin-film circuits.

PROBLEMS

7-1 Which family of integrated circuits, and for what reasons, is best suited
(*a*) for the logic control of an elevator?
(*b*) for a general-purpose computer having machine cycles of about 0.2 μsec?
(*c*) for ultrafast time measurements?
(*d*) for a toy?

7-2 Using the input–output voltage relations and output voltage–current relations given in the text, point out which of the logic families indicated in Table P7-2 can be mutually interconnected. Also discuss restrictions that are imposed by fan-in and fan-out when interconnecting these circuits.

Table P7-2.

	Driving Circuit	Receiving Circuit
(*a*)	RTL	DTL
(*b*)	DTL	TTL
(*c*)	TTL	DTL
(*d*)	ECL	TTL
(*e*)	HTL	TTL
(*f*)	RTL	HTL

7-3 Choose a suitable (not too fast) logic family to implement the half adder shown in Fig. 4-13 when the time allowed for addition is
(*a*) 10 nsec.
(*b*) 100 nsec.
(*c*) 1 μsec.
Search for practical circuit diagrams from literature.

7-4 Search for typical HIGH- and LOW-state noise margins for some integrated circuits of the following families, using available brochure material distributed by the vendors:

(*a*) RTL

(*b*) DTL

(*c*) HTL

(*d*) TTL

(*e*) ECL

7-5 Simplify the circuit shown in Fig. P7-5 using Wired AND Gates belong to the DTL family.

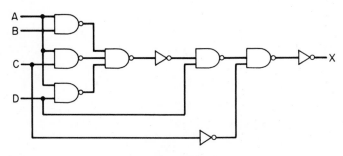

Figure P7-5

REFERENCES

Bipolar logic circuit families are extensively analyzed in

Lynn, Meyer, and Hamilton (eds.), *Analysis and Design of Integrated Circuits* [86].

The operation of MOSFET families can be found in application reports and handbooks distributed by semiconductor manufacturers, and also in separate articles. See the article of Boysel and Murphy [201] for references on four-phase MOSFET logic. A more complete list of references will be found in the Bibliography.

chapter 8

Timing Circuits

If enough time is reserved for each assignment operation in digital computing, the exact length of subsequent clock phases is of no primary importance. However, there are special digital systems in which the mutual distance in time of two or more clock pulses must be defined with a relative accuracy ranging, say, from $1:10^3$ to $1:10^8$. Such applications are usual in measurements of time or of related variables such as frequency, or with other measurements using time bases.

The timing of signals, regardless of whether it is made with high or low accuracy, is implemented by dynamic circuits. A dynamic system used in timing is usually described by a differential equation of a low order ($n = 1$ or 2):

$$f\left(\frac{d^n x}{dt^n}, \frac{d^{n-1} x}{dt^{n-1}}, \dots, \frac{dx}{dt}, x, t\right) = 0 \qquad [8\text{-}1]$$

where x is a voltage, current, or charge variable.

A rather typical timing system is a first-order linear dynamic system consisting of a resistor and a capacitor. In Fig. 8-1, the switch denoted by SW

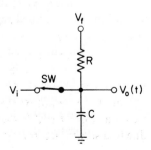

Fig. 8-1. Dynamic system used in timing.

317

represents a general method by which the initial voltage V_i of the capacitor at time $t = t_i$ is set; SW is subsequently opened. The differential equation describing $V_o(t)$ and its solution with the initial condition $V_o(t_i) = V_i$ are given in Eqs. [8-2] and [8-3].

$$C\frac{dV_o}{dt} = \frac{V_f - V_o}{R}$$

or

$$\frac{d(V_o - V_f)}{dt} = -\frac{V_o - V_f}{\tau} \qquad [8\text{-}2]$$

where $\tau = RC$ is the *time constant*. The solution of Eq. [8-2] is

$$V_o(t) = V_f - (V_f - V_i)e^{-(t-t_i)/\tau} \qquad [8\text{-}3]$$

At time t_d, assume that V_o has achieved a value V_d. The time interval $T = t_d - t_i$ is

$$\boxed{T = t_d - t_i = \tau \ln \frac{V_f - V_i}{V_f - V_d}} \qquad [8\text{-}4]$$

where $_i$ = initial, $_d$ = discrimination, and $_f$ = final. This timing relation occurs in the following circuits.

8.1.   Timing with Logic Gates

Pulse-forming Circuits. In the definition of logic voltage levels an undefined region was left between the HIGH and LOW voltages in order to avoid ambiguous classification. It is necessary, however, that every signal voltage used in digital systems attain either of the logic values 0 and 1 at all times; other possibilities are excluded by definition. Therefore, a voltage that is changing monotonically in time shall change its logic value when crossing a certain *discrimination level*. We may assume that the exact location of this level is somewhere in the undefined region.

Let us examine the circuit of Fig. 8-2, in which two cascaded inverters are shown. Assume that the output circuit of the former inverter can be described by a single transistor and that the input impedance of the latter gate is infinite or can be combined with the collector resistor R. A timing capacitor C is used to shunt the output of the first inverter. Above a certain level V_d, $V_o(t)$ is detected as logic 1 and below that level as logic 0. The discharging time of C due to a saturating transistor is usually much shorter than

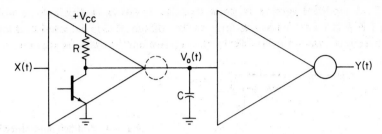

Fig. 8-2. Pulse stretcher.

the time for it to charge up through R. If X was initially 0, $V_o(0)$ was equal to V_{CC}. During a short excursion of X to 1, C is discharged but after that C starts charging toward $V_f = V_{CC}$. After a time T given by Eq. [8-3], V_o crosses the level that separates the logical 0 and 1 voltages. Accordingly, $Y(t)$ is 1 during the interval which starts at the discharging of C and extends for a time T after the end of the X pulse. This operation is called *pulse stretching*. The complete waveform of $V_o(t)$ (and the response for a new pulse) are shown in Fig. 8-3, in which the waveform of $Y(t)$ has been idealized.

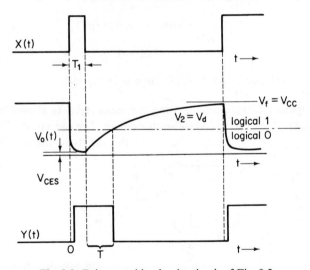

Fig. 8-3. Pulse stretching by the circuit of Fig. 8-2.

Example 8-1. When using DTL logic, typical pulse lengths of the circuit of Fig. 8-2 are (as indicated by Motorola, Inc., for their MDTL family)

$$T(\text{nsec}) = 0.55C(\text{pF}) + 60$$

$$T_1(\text{nsec})_{\min} = 0.01C(\text{pF}) + 30$$

A modified version of pulse stretcher is shown in Fig. 8-4, in which Y is 1 if X is 1 and further as long as the voltage of C is detected as a logic 1. This circuit has a much wider range of pulse lengths than the previous one,

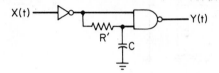

Fig. 8-4. Modified pulse stretcher.

without danger of dissipation of too large discharge energies by the feeding gate. On the other hand, since C must be discharged through R', a certain minimum trigger pulse width is needed. This difficulty can be circumvented by using a pulse stretcher of the first type in front of this circuit. The modified pulse-stretcher principle is applicable for any type of logic circuit, also for those with high-level output circuits.

Example 8-2. The pulse length of the modified pulse stretcher given in Fig. 8-4 is $T(\text{nsec}) = 0.55C(\text{pF}) + 60$ (for the Motorola MDTL). If $R' = 68 \,\Omega$, the limiting conditions are

$$T_1(\text{nsec})_{\min} = 0.5C(\text{pF})$$

$$C_{\max} = 15 \,\mu\text{F}$$

The meaning of T and T_1 is the same as in Fig. 8-3.

A logical feedback circuit using capacitor delay is shown in Fig. 8-5. This circuit may be called a *pulse-forming circuit*. The output Y of this circuit in the steady state is 0, because either of the inputs to the last NOR (G_2)

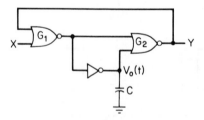

Fig. 8-5. Pulse-forming circuit.

is 1. The input X is normally 0, whereby C attains the LOW voltage. When X is made 1 and the voltage of C starts rising, both inputs of gate G_2 are 0 for a time T given by the earlier discussion. During this time Y makes an excursion to logic 1. After T, Y returns to 0 and cannot depart from this value until X has been returned to 0 and C has been discharged. The waveforms of this circuit are depicted in the timing diagram of Fig. 8-6. This circuit also produces

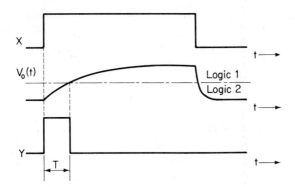

Fig. 8-6. Waveforms for the pulse-forming circuit.

pulses with a length T from short pulses. The proof is left to the reader as an exercise.

The shunt capacitor timing method is in general best applied to RTL circuits. However, it has some disadvantages. The worst of them is that V_o crosses the discrimination level rather slowly, and if any noise is superimposed on V_o, multiple pulses at the trailing edge of Y may be observed.

Digital circuits ought to be insensitive to noise, and several methods can be used to remove the indeterminacy associated with the trailing edges of pulses. These are the use of the trigger circuit as a level detector and flip-flop buffering. The direct R-S flip-flop of Fig. 8-7 is a trigger circuit in which

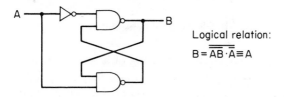

Logical relation:
$$B = \overline{\overline{AB} \cdot \overline{A}} \equiv A$$

Fig. 8-7. Flip-flop trigger circuit.

the output signal is logically equivalent to the input signal. Because of a regenerative action (positive feedback) by the flip-flop and differences in logic levels between input and output, however, any changes in the output signal always occur with the maximum possible switching speed, even when the input signal A changes rather slowly. In other words, the output voltage has a small hysteresis with respect to the input signal, and such phenomena will be discussed later. The circuit of Fig. 8-7 can be inserted between a timing capacitor C and the corresponding gate input circuit. This method is best applicable to DTL circuits.

Several flip-flop buffering methods make use of the fact that a trigger pulse initially sets a flip-flop; after a certain delay provided by a passive dynamic circuit or a special *delay element*, the flip-flop is reset. If there are multiple pulses in the feedback signal, it is the first one of them which performs the resetting. The pulse shortener discussed before implies this kind of set–reset action. Thus a pulse stretcher–pulse shortener combination might constitute a possible solution for a pulse-forming circuit. In what is to follow, we shall prefer clocked flip-flops as buffers, because these flip-flops are usually completely symmetrical with respect to their half-sections, and it can be expected that they are also the most insensitive to noise picked up from power-supply lines or from the ground. Moreover, the timing circuits ought to have uniform specifications with other sequential control circuits, which also recommends to us the use of clocked flip-flops.

In the circuit of Fig. 8-8, a clocked *J-K* flip-flop is set to 1 by a signal having the usual clocking specifications. A direct $\bar{R}$ input is able to override all other functions (resetting the flip-flop when $\bar{R} = 0$), and this signal is

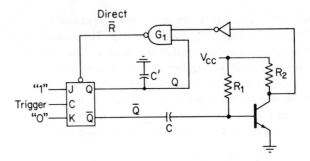

Fig. 8-8. Practical pulse-forming circuit.

obtained from a delay circuit consisting of a *RC* differentiating circuit and an external transistor. The transistor is normally saturated, whence $\bar{R} = 1$. After triggering, $\bar{Q}$ becomes 0, the base of the transistor attains a negative voltage, and the transistor current will be cut off. The purpose of C' is to delay the other input of G_1 for a time which is larger than the propagation delay of signals through the external circuit. The direct R signal remains 0 ($\bar{R} = 1$) until the transistor again conducts current. At this moment both inputs of G_1 become 1 and the flip-flop will be reset.

Example 8-3. A possible set of component values for the practical pulse-forming circuit given in Fig. 8-8 is

$$R_1 = 13 \text{ k}\Omega$$

$$R_2 = 1 \text{ k}\Omega$$

$$C = 470 \text{ pF}$$

$$C' = 330 \, \text{pF}$$

When using these values, the output pulse length is $T = 1 \, \mu\text{sec}$.

The special timing method discussed before has an additional advantage over previous methods—that the timing voltage passes the discrimination level faster than in previous examples. Thus the trailing edge of pulses is better defined.

There are numerous other possibilities for making pulse-forming circuits of standard logic gates. Various implementations utilize special properties of particular logic circuit families and will not be discussed in this book.

Monostable Multivibrator. A *monostable multivibrator* (MMV), or a *one-shot multivibrator*, as it is also called, is a circuit that has one stable state, from which it can be triggered to a metastable state. The duration of the metastable state is defined by a dynamic circuit. After that, the circuit automatically returns to its stable state. An ideal timing diagram showing the triggering and the output pulses is as shown in Fig. 8-9. Actually the two pulse-forming circuits discussed in the previous paragraph are monostable multivibrators, too.

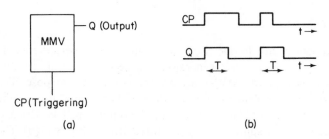

Fig. 8-9. Ideal monostable multivibrator: (a) symbol; (b) waveforms.

We shall discuss the operation of several types of monostable multivibrators in what is to follow. Each one has its advantages and drawbacks. Using standard logic gates, a MMV can also be built. For this purpose, DTL NAND gates with a gate extender input (E) are usually selected (Fig. 8-10). *In this circuit the triggering signal must be shorter than the output.* The principle of operation is roughly given in what follows: Gate G_1 forms a NAND function of the trigger input and of the extender input E, which are both normally 1. When a 0 signal is applied to the trigger input, the output of G_1 jumps HIGH, and the output of G_2—a negative-going step—will cause a negative voltage, interpreted as a logic 0, at the extender input. After triggering, the voltage of E is approximately $-V_{CC}$ and starts rising toward a

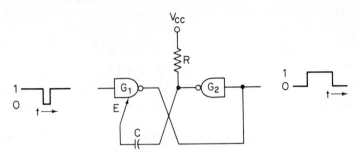

Fig. 8-10. Monostable multivibrator.

positive value defined by the DTL gate. After a time T given by a timing relation similar to Eq. [8-3], the extender input voltage becomes logic 1, the output jumps LOW, and the original stable state is restored. A more detailed analysis would be very similar to that given in connection with the collector-coupled monostable multivibrator of Section 8.2.

Example 8-4. The output pulse length of the monostable multivibrator of Fig. 8-10 is approximately

$$T(\text{nsec}) = 2.75C(\text{pF})$$

for DTL.

Astable Multivibrator. For the production of periodic clock pulses, continuously running pulse generators are needed. Various circuits, called *astable multivibrators* (AMV's), are applicable for this purpose. The circuit of Fig. 8-11 is an implementation of AMV by DTL logic circuits which have gate extender inputs. The operation of this circuit will be discussed later in more detail. Let is suffice to show the waveforms in Fig. 8-11(b). The output circuits of G_1 and G_2 switch in opposite phases.

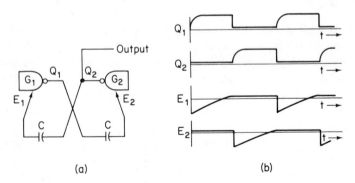

(a)   (b)

Fig. 8-11. Astable multivibrator: (a) circuit; (b) waveforms.

Example 8-5. The output frequency of the astable multivibrator, given in Fig. 8-11, is approximately

$$f(\text{MHz}) = \frac{1}{0.005C(\text{pF}) + 0.15}$$

when using DTL. The maximum value of the capacitors is $C = 0.3 \ \mu\text{F}$ and the maximum frequency is 5 MHz.

The principle of another AMV, which is made of logic gates and a delay element, is shown in Fig. 8-12. The output waveform of an ideal delay

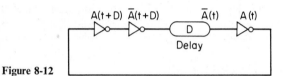

Figure 8-12

element is similar to the input waveform but is delayed by a time D with respect to it. Assuming that the inverters have a negligible delay, we have for the system equation

$$A(t + D) = \bar{A}(t) \qquad\qquad [8\text{-}5]$$

which obviously has no static solutions for $A(t)$. On the other hand, the waveform of Fig. 8-13 represents a solution that satisfies Eq. [8-5].

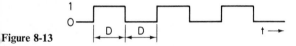

Figure 8-13

To avoid impedance matching needed with delay elements, a simpler "ring oscillator," shown in Fig. 8-14, may be constructed for the same purpose. A delay in logic values of the propagated signal is implemented by a capacitor C, which acts approximately in the same way as a Miller capacitor in an integrator. (Notice that three inverters connected as a ring without C would oscillate with a very high frequency as a result of their inherent transport delays and internal capacitances. This method can be used to measure gate delays.)

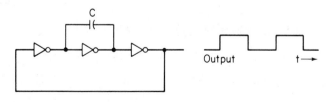

Fig. 8-14. Astable multivibrator.

Example 8-6. The approximate output frequency of the astable multivibrator (DTL) of Fig. 8-14 is

$$f(\text{MHz}) = \frac{6.7}{0.015C(\text{pF}) + 1}$$

The maximum value of the capacitor is 30 nF.

Crystal Oscillator as a Clock-Pulse Generator. The temperature stability of pulse lengths in all previously discussed multivibrators, except those implemented with ideal delay elements, is poor to fair. This is due to the fact that discrimination levels depend on semiconductor junction voltages which are functions of temperature. It is quite typical that temperature coefficients of pulse widths are of the order of 2 percent per 10°C. If piezoelectric crystals are used in timing circuits, astable multivibrators with a relative stability of frequency of 1: 10⁴ over 50°C can be built.

A circuit with DTL gates in which C_v is a trimmer capacitor is shown in Fig. 8-15.

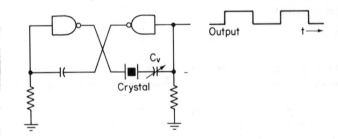

Fig. 8-15. Crystal oscillator.

8.2. Analysis and Design of Some Basic Multivibrators

Collector-Coupled Astable Multivibrator. As a classical example, we consider the collector-coupled astable multivibrator shown in Fig. 8-16. This circuit also approximately represents the AMV of Fig. 8-11, which was made of standard logic gates. Assume the following starting state: Q is cut off and Q' saturated, the left side of C' attains some negative value $-V_{B1}$, and C has been discharged. To sustain saturation on Q', the current through R_2 (which now is equal to the base current of Q') must exceed the necessary saturation base current I_{BS}.

In the beginning, the only dynamic phenomenon is the discharging of C' through R'_2 and the saturated Q'. The voltage over Q' is assumed constant, $= V_{CES}$. The discharge time constant is R'_2C' and the voltage V_{BE} is rising toward the final value V_{CC}. However, as soon as V_{BE} becomes $\geq V_{BEA}$, where

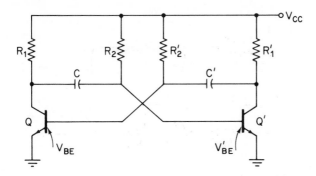

Fig. 8-16. Collector-coupled astable multivibrator.

V_{BEA} is a value of V_{BE} above which Q is active, Q starts to draw collector current through R_1 and C.

At first the base charge of Q' is pumped off by the current flowing through C, and Q' becomes cut off. Immediately Q is saturated by the rise in the collector potential of Q' and the charge transferred through C' to the base of Q. This switching occurs very fast, owing to positive feedback during the transition phase. After switching, the currents and voltages attain well defined levels; the voltage swing of V'_{BE} is the same as the change in collector voltage of Q, $V_{CC} - V_{CES}$. (The voltage over C cannot be changed instantaneously.) The voltage over C' is $V_{CES} - V_{BEA}$, and since V_{BE} can be approximated by a constant $= V_{BES}$, the collector voltage of Q' has an initial value $V'_{CES} - V_{BEA} + V_{BES}$ from which it is rising toward the final value V_{CC} with a time constant R'_1C'. Usually $R'_1C' \ll R_2C$ and $R_1C \ll R'_2C'$; thus the collector voltage approaches V_{CC} while V'_{BE} is still negative; this is a case similar to the one assumed at the instant of starting, except that the roles of the transistors and other corresponding components are interchanged. Thus a similar analysis can be performed for the discharging phase of C, after which a state identical with the beginning state is achieved. The waveforms with their details are shown in a series of diagrams in Fig. 8-17.

The small positive peaks in V_{BE} at T_0, T_2, etc., and in V'_{BE} at T_1, etc., which decay off with a short time constant, are caused by changes in base currents. Immediately after switching, base currents are taken through R_1 and R_2, or R'_1 and R'_2, respectively. When C or C' is discharging, the current through R_1 or R'_1 decays off. Since the form of these peaks does not enter into the timing analysis, we do not consider them in detail.

The collector–base leakage currents have been neglected, because, in fact, their only effect is to lower the asymptotic value of V_{BE} and V'_{BE} from V_{CC} to $V_{CC} - R_2I_{CO}$ or $V_{CC} - R'_2I_{CO}$, respectively.

The pulse widths are obtained from the exponential parts of V_{BE} and V'_{BE}:

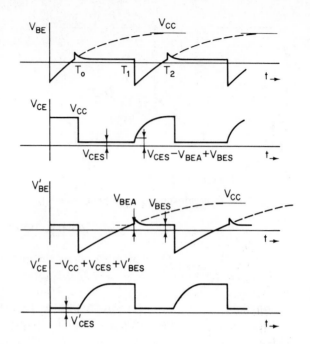

Fig. 8-17. Waveforms of the collector-coupled astable multivibrator.

For $T_0 < t \leq T_1$:

$$V'_{BE} = V_{CC} - (2V_{CC} - V_{CES} - V'_{BES})e^{-(t-T_0)/\tau} \qquad [8\text{-}6]$$

where $\tau = R_2 C$.

For $T_1 < t \leq T_2$:

$$V_{BE} = V_{CC} - (2V_{CC} - V'_{CES} - V_{BES})e^{-(t-T_1)/\tau'} \qquad [8\text{-}7]$$

where $\tau' = R'_2 C'$. Substituting $t = T_1$ in Eq. [8-6] and setting $V'_{BE} = V'_{BEA}$, we obtain

$$T_1 - T_0 = R_2 C \ln \frac{2V_{CC} - V_{CES} - V'_{BES}}{V_{CC} - V'_{BEA}}$$

$$\simeq R_2 C \ln 2 \qquad [8\text{-}8]$$

Substituting $t = T_2$ in Eq. [8-7] and with $V_{BE} = V_{BEA}$, we obtain

$$T_2 - T_1 = R'_2 C' \ln \frac{2V_{CC} - V'_{CES} - V_{BES}}{V_{CC} - V_{BEA}}$$

$$\simeq R'_2 C' \ln 2 \qquad [8\text{-}9]$$

The length of one complete period is thus

$$T_2 - T_0 \simeq \ln 2(R_2 C + R_2' C') \qquad \text{[8-10]}$$

Example 8-7. Typical component values of the collector-coupled astable multivibrator of Fig. 8-16 are

$$R_1 = R_1' = 1 \text{ k}\Omega$$

$$R_2 = R_2' = 10 \text{ k}\Omega$$

$$C = C' = 5 \text{ nF}$$

If the saturation voltages V_{CES} and V_{BES} are assumed to be 0.3 and 0.6 V, respectively, we get the following output frequencies:

$$f = 14.1 \text{ kHz} \qquad \text{for } V_{CC} = 10 \text{ V}$$

$$f = 13.7 \text{ kHz} \qquad \text{for } V_{CC} = 5 \text{ V}$$

The emitter–base junctions of Q and Q' must tolerate reverse voltages of the order of V_{CC} without danger for breakdown. Unless this is the case, extra diodes should be connected in series with the base leads.

Comment. The collector-coupled multivibrator has the disadvantage that there is a possibility for it being latched to a state in which both of the transistors are saturated and the voltages over C and C' are $V_{CES} - V_{BES}'$ and $V_{CES}' - V_{BES}$, respectively. This sometimes happens when supply voltages are turned on. This hazard can be circumvented by making the circuit as unsymmetrical as possible (e.g., by an extra resistor connected to one base from a fixed potential). If a safe starting of the multivibrator is positively necessary, the collector-coupled multivibrator might not be the right solution to the problem.

If sharp-rising edges are needed on the collector output waveforms, they can be obtained by a circuit such as that of Fig. 8-18; when a transistor collector current is cut off, the charging of C or C' occurs through the resistors R_1 and R_1', respectively, but the output voltages are isolated by the diodes and may rise promptly to V_{CC} due to R_3 and R_3'.

Example 8-8. The modified version of the collector-coupled astable multivibrator of Fig. 8-18 can be realized with the following components:

$$R_1 = R_1' = R_3 = R_3' = 2.2 \text{ k}\Omega$$

$$R_2 = R_2' = 22 \text{ k}\Omega$$

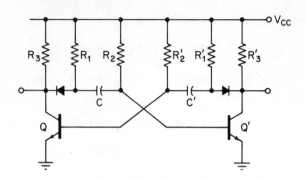

Fig. 8-18. Modified version of the collector-coupled astable multivibrator, with fast rise in collector voltage waveforms.

$$C = 157\,\mathrm{pF}$$

$$C' = 314\,\mathrm{pF}$$

$$V_{CC} = 12\,\mathrm{V}$$

The values are chosen so that the output frequency is 150 kHz and the ratio of the partial cycles is 1 : 2. The diode voltage and the transistor parameters are assumed to be

$$V_D = 0.5\,\mathrm{V}$$

$$V_{CES} = 0.4\,\mathrm{V}$$

$$V_{BES} = 0.7\,\mathrm{V}$$

$$V_{BEA} = 0.3\,\mathrm{V}$$

$$I_{CBO} = 5\,\mu\mathrm{A}$$

Astable Multivibrator as Voltage-to-Frequency Converter. The frequency of a collector-coupled astable multivibrator can be made controllable by an external voltage. Instead of using resistors R_2 and R'_2 as in Fig. 8-16 for the charging of C and C' during respective cutoff states of Q_1 and Q'_1, the capacitors can be charged through constant-current generators formed by the transistors Q_2 and Q'_2 of Fig. 8-19. The waveforms are in general similar to those represented earlier, but the exponential parts of $V_{BE}(t)$ and $V'_{BE}(t)$ are replaced by linear portions

$$T_0 \le t < T_1\!:$$

$$V_{BE}(t) = -V_{CC} + V'_{CES} + V_{BES} + \frac{V_2 - V_1 - V_{BE2}}{R_E C}(t - T_0) \quad [8\text{-}11]$$

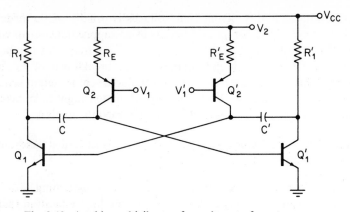

Fig. 8-19. Astable multivibrator for voltage-to-frequency conversion.

$T_1 \leq t < T_2:$

$$V'_{BE}(t) = -V_{CC} + V'_{CES} + V_{BES} + \frac{V_2 - V'_1 - V'_{BE2}}{R'_E C'}(t - T_1) \quad [8\text{-}12]$$

where V_{BE2} and V'_{BE2} are the base–emitter voltages of Q_2 and Q'_2, respectively, and

$$V_{BE}(T_1) = V_{BEA}$$

$$V'_{BE}(T_2) = V'_{BEA}$$

In the same way as with the collector-coupled astable multivibrator, we get

$$T_1 - T_0 = R_E C \frac{V_{CC} + V_{BEA} - V_{BES} - V'_{CES}}{V_2 - V_1 - V_{BE2}} \quad [8\text{-}13]$$

$$T_2 - T_1 = R'_E C' \frac{V_{CC} + V'_{BEA} - V'_{BES} - V_{CES}}{V_2 - V'_1 - V'_{BE2}} \quad [8\text{-}14]$$

Selecting or adjusting

$$V_1 + V_{BE2} = V'_1 + V'_{BE2} = V'' \quad [8\text{-}15]$$

we finally obtain the oscillating frequency,

$$f = \frac{1}{T_2 - T_0} = \frac{1}{k}(V_2 - V'') \quad [8\text{-}16]$$

where

$$k = R_E C(V_{CC} + V_{BEA} - V_{BES} - V'_{CES})$$
$$+ R'_E C'(V_{CC} + V'_{BEA} - V'_{BES} - V_{CES}) \quad [8\text{-}17]$$

The temperature stability of junction voltages is usually of the order of a few millivolts per degree centigrade, and by using supply and control voltages of a few tens of volts, the temperature stability of the frequency can be made better than 10^{-4} per °C. Notice that to avoid breakdown of V_{BE} of the transistors in the cutoff states, extra diodes should be connected in series with the base leads of Q_1 and Q'_1. Voltages over the diodes ought to be taken into account in V_{BEA}, V_{BES}, V'_{BEA}, and V'_{BES} of the previous analysis.

Comment. The circuits for Q_2 and Q'_2 should be designed so that these transistors are never saturated.

Current-Mode Astable Multivibrator. Current switching, also called current-mode switching, provides fast response and high repetition frequency for an astable circuit, because transistors are not saturated. Also, there is no danger for latching in a stable state as the case was with a collector coupled multivibrator.

In the circuit shown in Fig. 8-20 the capacitor C acts as a temporary short circuit during switching, but immediately before and after switching of the current from one transistor to another, the sum of the emitter currents remains constant. The sum of the emitter currents, however, is not constant at all times, as will be seen.

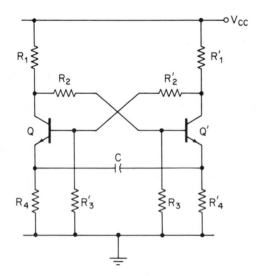

Fig. 8-20. Current-mode astable multivibrator.

We start again with a well-defined state, with Q conducting and Q' cut off. There are direct implications of these assumptions:

(1) $V'_C = V_{CC}$, whence

$$(2)\ V_B = \frac{R'_3}{R'_1 + R'_2 + R'_3} V_{CC}$$

(3) $V_E = V_B - V_{BE}$ (with V_{BE} taken constant)

These are values that are temporarily constant in time. It is assumed further that $V'_E > V'_B$ (because Q' is cut off) and V'_E is falling toward zero with a time constant $R'_4 C$, as a result of the discharge of C through R'_4. The discharge time constant is usually so large that V'_E changes almost linearly with time. Since

$$V'_E(t) = V'_E(0)e^{-t/\tau'} \qquad \text{where } \tau' = R'_4 C \qquad\qquad [8\text{-}18]$$

the emitter current of Q is

$$I_E(t) = \frac{V_E}{R_4} + \frac{V'_E(0)}{R'_4} e^{-t/\tau'} \qquad\qquad [8\text{-}19]$$

Thus

$$V_C(t) = \frac{R_2 + R_3}{R_1 + R_2 + R_3} V_{CC} - I_E(t)[R_1 \| (R_2 + R_3)] \qquad\qquad [8\text{-}20]$$

and this implies that

$$V'_B(t) = \frac{R_3}{R_2 + R_3} V_C(t) \qquad\qquad [8\text{-}21]$$

From the point of view of Q', it is the difference $V'_B(t) - V'_E(t) = V'_{BE}(t)$,

$$V'_{BE}(t) = \frac{R_3}{R_1 + R_2 + R_3} V_{CC} - \frac{R_1 R_3}{R_1 + R_2 + R_3} \left(\frac{V_E}{R_4} + \frac{V'_E(0)}{R'_4} e^{-t/\tau'} \right)$$
$$- V'_E(0)e^{-t/\tau'} \qquad\qquad [8\text{-}22]$$

which determines the break in of the emitter current of Q'. When $V'_{BE}(t) \geq V_{BEA}$, which occurs, say, at $t = T_0$, Q' starts conducting, and through a regenerative action Q cuts off, switching the emitter current $I_E(T_0)$ to Q'. There will be a sudden rise in V'_E of magnitude $I_E(T_0)R'_4$. Since the voltage over C cannot change instantaneously, there is a corresponding step in V_E, after which the roles of the transistors have been interchanged and the analysis may be repeated as before. These calculations are easiest to perform numerically. In Fig. 8-21 we have shown qualitatively the most important waveforms.

Example 8-9. A possible set of component values of the current-mode astable multivibrator of Fig. 8-20 is

$$R_1 = R'_1 = 100\ \Omega$$

$$R_2 = R_2' = 1 \text{ k}\Omega$$

$$R_3 = R_3' = 4.7 \text{ k}\Omega$$

$$R_4 = R_4' = 1.5 \text{ k}\Omega$$

$$C = 1 \ \mu\text{F}$$

$$V_{CC} = 12 \text{ V}$$

If the base currents are neglected and the base–emitter junctions are

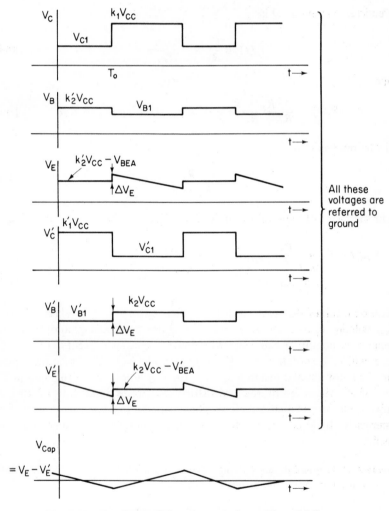

Fig. 8-21. Waveforms of the current-mode astable multivibrator.

assumed to be idealized diodes with $V_{BE} = 0.7\,\text{V}$, we get the output frequency $f = 1.6\,\text{MHz}$.

Collector-Coupled Monostable Multivibrator. The collector-coupled monostable multivibrator (Fig. 8-22) may be thought to be composed of one half of an astable multivibrator and an inverter. There is a rest state to which the circuit always returns, with Q saturated and Q' in the cutoff state.

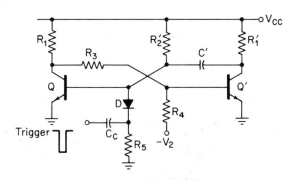

Fig. 8-22. Monostable multivibrator.

This is due to the fact that the current through R_2' alone is sufficient to saturate Q, and (R_3, R_4) form an input circuit for Q' with the same specifications as in an inverter. Thus the collector current of Q' is zero and the right side of C' is charged to the voltage V_{CC}.

When a short negative pulse is applied to the trigger input, the base charge of Q is at first pumped off by the current through C_c and D. Then the collector current of Q is cut off and the voltage at the collector of Q rises to a level classified as a logical 1. Consequently, Q' is saturated by this control signal, and the cutoff state of Q is secured by the negative-going voltage step at the base of Q, coupled through C' from the collector of Q'. The magnitude of this step is $V_{CC} - V_{CES}'$.

In the intermediate state, when Q is cut off and Q' saturated, the only dynamic phenomenon is the discharging of C', or the rise of V_{BE} toward the final value V_{CC}. When $V_{BE} \geq V_{BEA}$, Q starts to conduct, and through a regenerative switching action the current of Q' is rapidly cut off and Q saturated.

A transient phenomenon, the charging of the right side of C' toward the potential V_{CC}, continues with a time constant $R_1'C'$.

The waveforms are shown in Fig. 8-23. The length of the pulse is

$$T = R_2'C' \ln \frac{2V_{CC} - V_{BES} - V_{CES}'}{V_{CC} - V_{BEA}} \simeq R_2'C' \ln 2 \qquad [8\text{-}23]$$

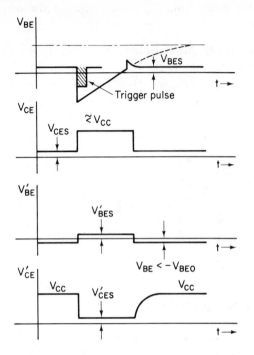

Fig. 8-23. Waveforms of the collector-coupled monostable multivibrator.

It can be seen that the diode D has a reverse voltage as long as the triggering signal is "covered" by the negative area of V_{BE}. As long as the triggering signal is confined in these limits, the shape and magnitude of the triggering signal have no effect on the waveforms.

As a result of the loading effect of R_3 on Q', V_{CE} in the cutoff state remains somewhat smaller than V_{CC}.

Example 8-10. The monostable multivibrator of Fig. 8-22 can be realized with component values

$$R_1 = R_1' = 1 \text{ k}\Omega$$

$$R_2' = 18 \text{ k}\Omega$$

$$R_3 = 8.2 \text{ k}\Omega$$

$$R_4 = 47 \text{ k}\Omega$$

$$V_{CC} = V_2 = 10 \text{ V}$$

$$C' = 10 \text{ nF}$$

The transistor parameters are assumed to be

$$V_{CES} = 0.4 \text{ V}$$

$$V_{BES} = 0.8 \text{ V}$$

$$V_{BEA} = 0.8 \text{ V}$$

$$\beta = h_{FE} = 20$$

Further we assume that there are $3\text{k}\Omega$ loading resistors connected to ground from the collectors. The length of the pulse is $T = 100$ μsec.

8.3. Timing with Operational Amplifiers

To improve the constancy of pulse widths or periods of waveforms in timing, the first task is to increase the accuracy with which voltage discrimination limits are defined. In the previous sections, voltage discrimination limits were usually defined by that value of a *B-E* voltage of transistors at which the transistor became active and drew a certain base current. If a timing accuracy equal to or better than, say, about $1: 10^3$ over a rather wide range of ambient temperatures is needed, the voltage-level detection must be made by special circuits developed for the comparison of voltages.

Operational amplifiers, as is well known, are used in analog computers for the generation of accurately controlled waveforms, whereby the transfer characteristics of an amplifier system are mainly determined by the feedback circuits.

An ideal operational amplifier has two symmetrical input terminals, called *inverting* and *noninverting input,* and an *output* terminal. The output voltage, referred to ground potential, is directly proportional to the voltage difference between the noninverting and the inverting inputs. In an ideal operational amplifier the output voltage ought to be independent of the mean of the input potentials, called *common-mode voltage;* this feature is called *common-mode rejection.* In practical operational amplifiers, a common-mode voltage has a slight effect upon the output voltage. This effect is expressed in terms of a *common-mode voltage attenuation,* which is usually of the order of 100 to 120 dB. The input circuit of the operational amplifier is often adequately described assuming an input impedance Z_{in} (usually resistive) between the input terminals. If the signal voltages are in the linear ranges of operation of the amplifier, Z_{in} may be assumed constant. However, in the nonlinear mode of operation, the input stage and subsequent stages are controlled far beyond the limits of linear operation. In the following circuits the noninverting input is denoted by $+$ and the inverting input by $-$.

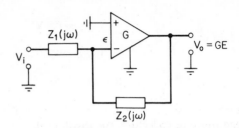

$$V_0(j\omega) = -\frac{Z_2(j\omega)}{Z_1(j\omega)} V_i(j\omega)$$

Figure 8-24

If $Z_1(j\omega)$ and $Z_2(j\omega)$ in Fig. 8-24 are general impedances and the open-loop gain G of the operational amplifier is much larger than 1, the input offset voltage ϵ (and the offset current) may be assumed negligible. Therefore, the input–output voltage-transfer relation is

$$V_o(j\omega) = -\frac{Z_2(j\omega)}{Z_1(j\omega)} V_i(j\omega) \qquad [8\text{-}24]$$

In linear analog-computing circuits, it is postulated that operational amplifiers, by the effect of negative feedback, are always sustained in the active state, in which they can be described by a constant or an almost-constant open-loop gain G. Actually the input–output voltage-transfer relation in operational amplifiers is of the type depicted in Fig. 8-25, where the ϵ scale has been expanded to exaggerate the proportional portion of this relation.

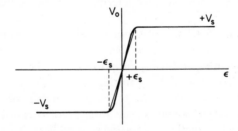

Fig. 8-25. Open-loop input–output voltage-transfer relation of an operational amplifier.

The saturation limits are usually symmetrical with respect to zero, and the limits of ϵ ($= \pm\epsilon_s$) for which saturation begins are approximately defined by V_s and the average open-loop gain G_{av}, over the proportionality region,

$$\epsilon_s \simeq \frac{V_s}{G_{av}} \qquad [8\text{-}25]$$

For example, with $V_s = 10$ V and $G_{av} = 10^4$, $\epsilon_s = 1$ mV.

Circuits Exhibiting Hysteresis Phenomena. As operational amplifiers with negative feedback can be used to implement basic linear amplifier types, so most of the well-known basic waveform generators and multistable circuits may be implemented by the use of *positive feedback* in connection with operational amplifiers. It is further essential that the amplifier is *saturable* [i.e., the output voltage (and current) cannot exceed certain limits]. We shall be mostly concerned with symmetrical saturation limits; this type of amplifier has the input–output voltage relation shown in Fig. 8-25. Later we shall also discuss some unsymmetrical circuits.

For the purpose of graphical analysis, we shall define the *inverse transfer relation* $\epsilon(V_o)$ by

$$\epsilon = g(V_o) \qquad [8\text{-}26]$$

where $g(V_o)$ is a single-valued function given experimentally.

The following circuits, whose operation is based on positive feedback and saturable amplifiers, exhibit characteristics similar to those encountered in systems called *relaxation circuits*.

When a positive feedback is applied to an ideal *linear amplifier* as shown in Fig. 8-26, the closed-loop gain will at first increase with the increasing feedback transfer ratio H until the gain approaches infinity with $H = 1/G$:

$$\frac{V_o}{V_i} = \frac{G}{1 - GH} \qquad [8\text{-}27]$$

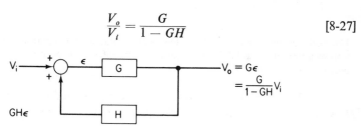

Fig. 8-26. Amplifier with positive feedback.

A similar feedback in a *saturable amplifier* with a linear gain portion will increase the overall gain of the linear portion only, leaving the saturation limits unaltered; with $H = 1/G$, the overall relation $V_o(V_i)$ becomes a step-like function. See Figs. 8-27 to 8-29.

When the feedback transfer ratio H is made $>1/G$, there exists no stable analytical solution for the relation $V_o(V_i)$ in linear amplifiers; with a saturable amplifier, however, $V_o(V_i)$ will become *double-valued* in a certain range of V_i (i.e., there will be a *hysteresis* between V_o and V_i, as will be seen).

Let the open-loop gain, denoted by G in Fig. 8-26, be replaced by the nonlinear output–input relation defined by

$$\epsilon = g(V_o) \qquad [8\text{-}28]$$

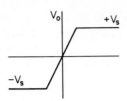

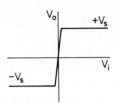

Fig. 8-27. Open-loop transfer relation of an idealized saturable amplifier.

Fig. 8-28. Overall transfer relation of a saturable amplifier with positive feedback H, $0 \leq H < 1/G$.

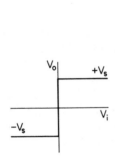

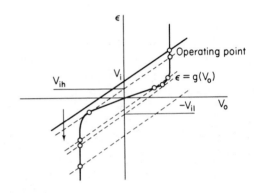

Fig. 8-29. Overall transfer relation of a saturable amplifier with positive feedback $H = 1/G$.

Fig. 8-30. Solution for operating points in a saturable amplifier with positive feedback.

If the feedback attenuation is always represented by a constant factor H, we have a second simultaneous equation which is linear,

$$\epsilon = V_i + HV_o \qquad [8\text{-}29]$$

Notice that in saturable amplifiers with positive feedback, ϵ *usually is not small but of the same order of magnitude as V_i and V_o.*

When the graphs representing Eqs. [8-28] and [8-29] are drawn to $\epsilon - V_o$ coordinates as in Fig. 8-30, the intersection of these graphs gives the operation point V_o corresponding to V_i; here V_i is the ordinate of the intersection of the straight line with ϵ axis, and the slope of the straight line is always the same, namely H.

Five cases are now distinguished. With V_i above a certain limit V_{ih}, or V_i below a certain other value $-V_{il}$, there is only one solution for the operating point. With $V_i = V_{ih}$ or $V_i = V_{il}$, the straight line is tangent to $\epsilon = g(V_o)$ either in the third or in the first quadrant, respectively, and has another intersection in the opposite quandrant. If V_i lies between V_{ih} and

$-V_{il}$, there are three intersections (operating points), of which the one in the middle can be shown to be unstable. The true operating point lies either in the first or the third quadrant, depending on *whether V_i has come to the region of hysteresis down from the top or up from the bottom, respectively.*

Let us redraw the relation $V_o(V_i)$ in Fig. 8-31. For the graphical construction it should be noted that the solid curves, called *branches*, are exact replicas of certain portions of $\epsilon = g(V_o)$, only displaced in new positions with respect to Fig. 8-30.

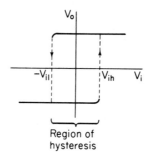

Fig. 8-31. Hysteresis in the input–output voltage relation of a saturable amplifier with positive feedback.

Region of hysteresis

Referring to the two values of V_o for $-V_{il} \leq V_i \leq V_{ih}$, it will be said that the system is in one of two *states* if the operating point is on the upper or lower branch of the graph. This circuit thus has *memory action;* if V_i is in the middle region, the present state unambiguously indicates from which side V_i last entered this region. The circuit can be made to switch its state by pulling V_i over the respective value V_{ih} or $-V_{il}$. At the instant of passing that value, the operating point will suddenly jump or *relax* to the opposite branch, along one of the dashed lines in Fig. 8-31.

What exactly happens during the relaxation, which starts, say, at $t = 0$, can be described by a nonlinear dynamic theory if the dynamic properties of the amplifier are known. Because of the reactive phenomena, Eqs. [8-28] and [8-29] ought to be replaced by a pair of differential equations

$$\epsilon(t) = g\left(V_o(t), \frac{dV_o}{dt}, \frac{d^2V_o}{dt^2}, \ldots\right) \qquad [8\text{-}30]$$

$$\epsilon(t) = V_i(t) + HV_o(t) \qquad [8\text{-}31]$$

If the operating point moves along either branch, and changes in $V_i(t)$ are slow, we may neglect the time derivatives in Eq. [8-30], and thus the operating point is obtained from static equations. On the other hand, when $V_i(t)$ crosses the point V_{ih} or $-V_{il}$, changes in $V_o(t)$ occur fast, and time derivatives must be taken into account. After the elimination of $\epsilon(t)$, Eqs. [8-30] and [8-31] then comprise a single nonlinear differential equation in $V_o(t)$, with $V_i(t)$ as its forcing function. Naturally this equation has solutions, although they might be difficult to find in closed form. The general solution for $V_i(t)$, with stated boundary conditions $V_o(0-)$ and $V_o(\infty)$, has been depicted in Fig. 8-32. The relaxation is thus a transient phenomenon. The settling down time of V_o depends on particular amplifier types. Admittedly, this time is not very short for contemporary operational amplifiers; it may well be of the order of microseconds. In circuits using *differential line receivers* discussed later, the switching time may be shorter than, say, 50 nsec. The transient illustrated in Fig. 8-32 sets a limit for the switching speed of operational amplifiers used in digital technology.

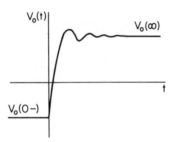

Fig. 8-32. Switching transient during a relaxation.

Schmitt Triggers. Circuits with well-defined hysteresis whose operation is based on positive feedback in saturable amplifiers are called *Schmitt triggers*, by analogy with an early vacuum-tube-circuit version developed for the same purpose. There are numerous variants of Schmitt triggers implemented with solid-state circuits. A couple of rather general forms of them using operational amplifiers will be described in the next sections. Schmitt triggers are used mainly as voltage-level detectors.

SCHMITT TRIGGER I

A circuit whose operation is similar to the general principle described in the previous section is shown in Fig. 8-33, in which the positive feedback

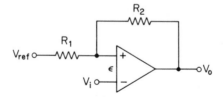

Fig. 8-33. Schmitt trigger I.

is a *serial feedback*. Assume at first that $V_{ref} = 0$. The input voltage V_i is connected to the inverting input of the operational amplifier. For simplicity, assume the following properties for the operational amplifier:

Input impedance $Z_{in} = \infty$

Output impedance $Z_{\text{out}} = 0$
High saturation limit $V_{oh} = +V_s$
Low saturation limit $V_{ol} = -V_s$
Gain on the linear portion G (constant for $-V_s \leq V_o \leq +V_s$)

In this case it will suffice to change the sign of V_i in Eq. [8-29] to find $V_o(V_i)$. The feedback transfer ratio is

$$H = \frac{R_1}{R_1 + R_2} \qquad [8\text{-}32]$$

The input–output voltage-transfer relation $V_o(V_i)$ for $V_{\text{ref}} = 0$ has been shown in Fig. 8-34. The limits of hysteresis are obtained from Eq.

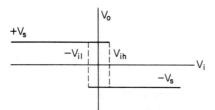

Fig. 8-34. $V_o(V_i)$ of Schmitt trigger I.

[8-29]. At the upper edge of the hysteresis the linear gain relation is still valid,

$$\epsilon_s = \frac{V_s}{G} = -V_{ih} + \frac{R_1}{R_1 + R_2} V_s \qquad [8\text{-}33]$$

and so

$$V_{ih} = V_s\left(\frac{R_1}{R_1 + R_2} - \frac{1}{G}\right) \qquad [8\text{-}34]$$

By symmetry, $V_{il} = V_{ih}$. As long as V_{ih} is positive, $V_o(V_i)$ will display hysteretic behavior. When $G = (R_1 + R_2)/R_1$, the width of the hysteresis is zero, and when H is further decreased, the relation $V_o(V_i)$ becomes single-valued.

The limits of hysteresis V_{ih} and $-V_{il}$ can be shifted up or down by varying V_{ref}; denoting the changes in V_{ih} and V_{ref} by ΔV_{ih} and ΔV_{ref}, respectively, we have

$$\Delta V_{ih} = \frac{R_2}{R_1 + R_2} \Delta V_{\text{ref}} \qquad [8\text{-}35]$$

SCHMITT TRIGGER II

The necessary positive feedback can be implemented with *parallel feedback*, too, by connecting a resistor R_2 from the output to the noninverting input; an input resistor R_1 is connected to the same junction. This circuit is obtained from the circuit of Fig. 8-35 by interchanging V_i and V_{ref}. Assume first that the inverting input is grounded, $V_{ref} = 0$. Referring to Fig. 8-35,

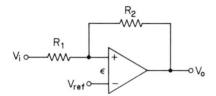

Fig. 8-35. Schmitt trigger II.

with an operational amplifier having $Z_{in} = \infty$, $Z_{out} = 0$, and the open-loop transfer relation described by $\epsilon = g(V_o)$, we have

$$\epsilon = g(V_o) \qquad [8\text{-}36]$$

$$\epsilon = \frac{R_2}{R_1 + R_2} V_i + \frac{R_1}{R_1 + R_2} V_o \qquad [8\text{-}37]$$

From the form of the latter equation we can see that the general method represented in the first section applies when slightly modified; here ϵ is not the exact error voltage of Fig. 8-26.

Comment. Before proceeding with the analysis, it ought to be noted that when Z_{in} is finite and resistive, we can transform the input circuit according to Thévenin's theorem; we can use the circuit of Fig. 8-35 and Eqs. [8-36] and [8-37], replacing V_i by $Z_{in}V_i/(R_1 + Z_{in})$ and R_1 by $R_1 \| Z_{in}$. Then Eq. [8-37] in fact reads

$$\epsilon = \alpha V_i + \beta V_o \qquad [8\text{-}38]$$

with recalculated α and β. Bearing this in mind, we proceed from the idealized model, Eqs. [8-36] and [8-37]. (Also, with nonlinear input circuits, the input current may often be neglected.)

The width of the hysteresis may be calculated approximately assuming that the linear portion of the transfer function extends up to the saturation limits $V_o = \pm V_s$; at the edge is $\epsilon = V_s/G$, and

$$\frac{V_s}{G} = \frac{R_2}{R_1 + R_2} V_i + \frac{R_1}{R_1 + R_2} V_s \qquad [8\text{-}39]$$

giving the negative limit of the hysteresis. Then we have

$$V_i = -V_{il} = \frac{R_1 + R_2}{R_2}\left(\frac{R_1 V_s}{R_1 + R_2} - \frac{V_s}{G}\right) \qquad [4\text{-}40]$$

Because of assumed symmetry, the width of the hysteresis is

$$\Delta V_i = \frac{2R_1 V_s}{R_2}\left(1 - \frac{R_1 + R_2}{GR_1}\right) \qquad [8\text{-}41]$$

In order to shrink the hysteresis to zero, we must have

$$R_2 = (G - 1)R_1 \qquad [8\text{-}42]$$

Astable Multivibrator. A passive RC timing circuit and a level detector can be combined to form an astable multivibrator in the way shown in Fig. 8-36. This circuit is, in fact, a Schmitt trigger in which the input voltage has

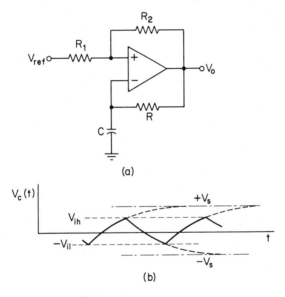

(a)

(b)

Fig. 8-36. (a) Astable multivibrator; (b) waveform of the voltage over the capacitor C.

been replaced by an automatically changing waveform obtained from a dynamic RC circuit. The oscillatory operation of this circuit can be explained in the following way: For simplicity, let the reference voltage V_{ref} initially be zero. As a result of a positive feedback path (R_2, R_1), the circuit acts as a Schmitt trigger whose input voltage has been provided automatically by the voltage $V_c(t)$ over the capacitor. At first, assume that $\epsilon > 0$, whence $V_o = +V_s$. The waveform of $V_c(t)$ which follows is shown in Fig. 8-36(b); when $V_c(t)$ is rising toward a final value $+V_s$ and achieves the upper limit of hysteresis V_{ih}, V_o jumps to a value $-V_s$. After that, $V_c(t)$ starts falling toward $-V_s$ until it achieves the value $-V_{il}$. Now V_o makes another jump to the value $+V_s$, and this cycle is repeated. In Fig. 8-36(b), the rising and falling waveforms are symmetrical and so the output is a square wave with equal positive and negative pulse widths. The relative length of the positive versus the negative pulse can be adjusted, for example, by changing the reference voltage V_{ref}. Thereby, V_{ih} and $-V_{il}$ are changed. The pulse widths T_1 (with rising V_c) and T_2 (with falling V_c) are obtained in the usual way,

$$T_1 = RC \ln \frac{V_s + V_{il}}{V_s - V_{ih}} \qquad [8\text{-}43]$$

$$T_2 = RC \ln \frac{V_s + V_{ih}}{V_s - V_{il}} \qquad [8\text{-}44]$$

This circuit has no stable solutions for $V_c(t)$, and so there is no danger for latching to a stable state.

Example 8-11. If the component values of the astable multivibrator of Fig. 8-36 are

$$R_1 = R_2 = R = 10 \text{ k}\Omega$$

$$C = 10 \text{ nF}$$

$$V_{\text{ref}} = 0 \text{ V}$$

and the saturation levels of the operational amplifier are $+10$ and -10 V, the output frequency is 5 kHz.

MONOSTABLE MULTIVIBRATOR I

We can make the circuit described in the previous section *monostable* by adjusting either V_{ih} or $-V_{il}$ so that the asymptotic value of $V_c(t)$ (either rising or falling branch) remains between the limits of hysteresis. Let us assume that $-V_{il}$ is lowered so that V_o and V_c have a stable-state solution equal to $-V_s$.

When a triggering signal (e.g., a short positive pulse) is applied to the noninverting input, say through a small coupling capacitor, V_o can be made to jump to the value $+V_s$. Then $V_c(t)$ will start rising toward $+V_s$.

Once started, $V_c(t)$ will rise toward $+V_s$ until the level V_{ih} is achieved. Then V_o makes a jump back to $-V_s$ and $V_c(t)$ will approach, without returning, its rest value $-V_s$. For details of waveforms, see Fig. 8-37.

The amplitude of the triggering pulse must exceed $V_s + V_{il}$ to reach the upper level of hysteresis. If a new pulse were triggered after $t_0 + T$, the triggering amplitude, $V_c(t) + V_{il}$, could be correspondingly smaller. This circuit has a drawback: Since the length of an output pulse (V_o) is equal to the time for $V_c(t)$ to rise from the level where it is up to the level V_{ih}, new pulses are always shorter than the first one. If the settling-down time of $V_c(t)$ before a

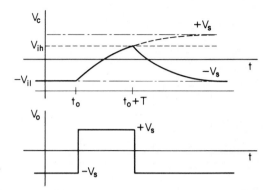

Fig. 8-37. Waveforms of monostable multivibrator I.

new pulse is larger than about $5RC$, this difference will be less than 1 percent,
however. The need for a settling-down time can be avoided by making the
discharging of C faster (e.g., by shunting R with a diode, with its anode side
toward C).

MONOSTABLE MULTIVIBRATOR II

A more practical monostable multivibrator than the former one is
shown in Fig. 8-38. The triggering input is here isolated from the rest of the
circuit and does not pick up any interference due to the switching action of
the circuit. Here the triggering signal is negative and its amplitude must ex-
ceed $|V_{\text{ref}}|$. Normally the inverting input will be kept at a zero level by R_t.

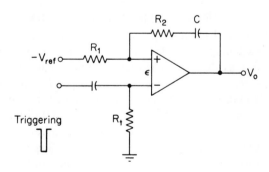

Fig. 8-38. Monostable multivibrator II.

In the rest state, the input offset voltage ϵ is $= -V_{\text{ref}}$, and $V_o = -V_s$.
After a short triggering pulse, V_o is made to jump from $-V_s$ to $+V_s$ (Fig.
8-39) and then ϵ makes an attenuated jump,

$$\Delta\epsilon = \frac{R_1}{R_1 + R_2} 2V_s \qquad\qquad [8\text{-}45]$$

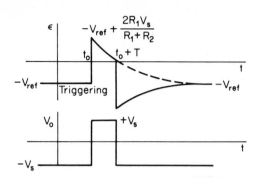

Fig. 8-39. Waveforms of monostable multivibrator II.

since the voltage over C cannot change instantaneously. The value of ϵ after triggering at time $t = t_0$ is

$$\epsilon(t_0+) = -V_{ref} + \frac{2R_1 V_s}{R_1 + R_2} \qquad [8\text{-}46]$$

Now C begins to discharge through R_2 and R_1, and $\epsilon(t)$ starts falling toward $-V_{ref}$, with a time constant $(R_1 + R_2)C$. When $\epsilon(t)$ crosses the zero level, V_o jumps to $-V_s$, and by positive feedback, ϵ makes a respective negative jump equal to $-\Delta\epsilon$, corresponding to a value given by Eq. [8-45]. After this, $\epsilon(t)$ starts rising toward the asymptotic value $-V_{ref}$ with a time constant $(R_1 + R_1)C$. The pulse width is now

$$T = (R_1 + R_2)C \ln \frac{2R_1 V_s}{(R_1 + R_2)V_{ref}} \qquad [8\text{-}47]$$

Differential Line Receivers in Multivibrators. Operational amplifiers intended for analog applications have, as has been mentioned, rather long switching times. Much faster operation, with a switching speed almost as fast as that of other digital circuits, can be achieved using *differential line receivers*, which have been developed for the reception of differential digital voltages from transmission lines and which produce normalized digital signals referred to ground at their outputs. Admittedly, the stability of discrimination levels of line receivers is not comparable to that of operational amplifiers but is sufficient for most purposes. Since the output saturation limits are unsymmetrical, output waveforms are directly compatible with logic circuits. Some examples for which the previous analysis can be applied with minor modifications (replacing $-V_s$ by the LOW voltage) are shown in Figs. 8-40 to 8-42. Notice that the output inverter belonging to the line receiver inverts the phase of signals.

Example 8-12. Typical component values of the astable multivibrator of Fig. 8-40 (using Fairchild line receiver 9620) are

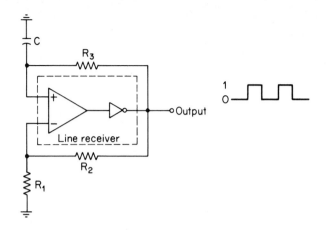

Fig. 8-40. Astable multivibrator.

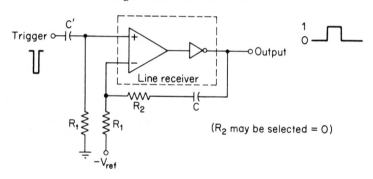

Fig. 8-41. Monostable multivibrator.

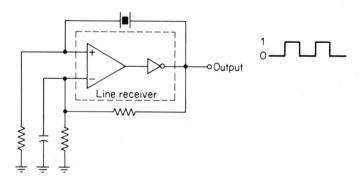

Fig. 8-42. Crystal oscillator.

$$R_1 = 1.6 \, \text{k}\Omega$$

$$R_2 = 2.7 \, \text{k}\Omega$$

The length of the period is

$$T = 1.3 R_3 C$$

Example 8-13. When using Fairchild line receiver 9620 a possible set of component values for the monostable multivibrator given in Fig. 8-41 is

$$R_t = 1.2 \, \text{k}\Omega$$

$$R_1 = 1 \, \text{k}\Omega$$

$$R_2 = 0$$

$$C' = 0.1 \, \mu\text{F}$$

With these values the length of the output pulse is

$$T(\text{nsec}) = 50 + 3.5 \times 10^3 C_2 (\text{farads})$$

8.4. Linear Time-Base Circuits

For graphical display, and also in analog measurements in which a time interval is made directly proportional to a given voltage, there is a need to produce linearly rising or falling waveforms. Such waveforms could be used for timing, too, although exponential waveforms serve as well for the definition of fixed pulse widths or periods. In the following, circuits producing linear waveforms (called *ramp voltages*) which have a closely controlled slope and which are intended for timing purposes are called *time-base circuits*.

The design of time-base circuits (Fig. 8-43) can be divided into two phases:

1. Generation of a linear ramp voltage.
2. Detection of the level of voltage and subsequent resetting of the voltage.

The ramp voltage may be triggered by an external control signal, or the system may be freely oscillating so that a new ramp begins immediately after the previous has been reset.

A modification of the time-base circuit described above is shown in Fig. 8-44, where the level detection and the resetting are separately controlled operations. There may be several level detectors connected to a single ramp

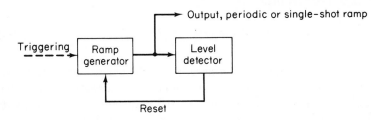

Fig. 8-43. Time-base circuit.

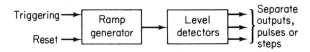

Fig. 8-44. Timing system.

and each one provides an output with a different delay with respect to the triggering. The outputs can be used for further triggering in timing circuits.

Generation of a Linear Ramp. A voltage that rises or falls linearly with time can be obtained by integrating a constant voltage or current. Where simpler circuits are accurate enough, a single-stage transistor current generator such as the one shown in Fig. 8-45 can be used to charge a capacitor.

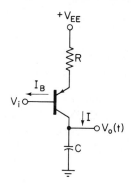

Fig. 8-45. Transistor linear ramp generator.

The charging current in the circuit of Fig. 8-45 is

$$I = \frac{V_{EE} - V_i - V_{BE}}{R} - I_B \qquad [8\text{-}48]$$

Usually a high-gain transistor can be selected for this circuit and $V_{EE} - V_i \gg V_{EB}$ (sometimes $V_{EE} - V_i$ is several tens of volts), whence

$$I \simeq \frac{V_{EE} - V_i}{R} \qquad [8\text{-}49]$$

The voltage over C is then

$$V_o(t) = V_o(t_0) + \frac{I}{C}(t - t_0) \qquad [8\text{-}50]$$

Using constant-current properties of field-effect transistors, these components could also be used as constant-current generators (see Fig. 8-46). Since the value of current depends on the parameters of the MOSFET, this circuit is not recommended for accurate timing circuits.

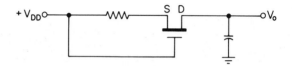

Fig. 8-46. MOSFET linear ramp generator.

There are applications where a linearity achievable with the previous circuits (of the order of 1 percent) is not sufficient. A linearity of, say, 0.1 percent or even better can be obtained using operational amplifiers. A usual integrator can be applied for this purpose (Fig. 8-47).

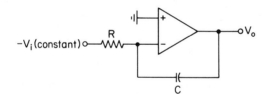

Fig. 8-47. Integrator-type ramp generator.

The previous circuit has the disadvantage that neither terminal of the capacitor C can be grounded. When C has to be discharged, an electronic switch ought to be connected between the output and the inverting input of the amplifier, which is not a good practice. For this reason, a *bootstrap circuit* might be preferred.

Bootstrapped Ramp Circuit. Of the various circuits using the bootstrap principle, the most general one is shown here. The operational amplifier in Fig. 8-48 is connected as a unity follower; we can assume that its output voltage is equal to the input voltage to a high degree of accuracy.

The capacitor C is charged through the resistor R. When the voltage V_c over C is rising, the current through R is kept constant by sampling V_c and lifting the potential of the upper terminal of R by the same amount. This action is called *bootstrapping*. Instead of a battery voltage V, a Zener diode can be used. In this case a bias current is necessary (Fig. 8-49). V and R should

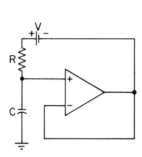

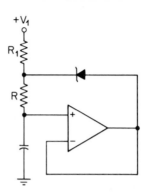

Fig. 8-48. Principle of a boot-strap ramp generator.

Fig. 8-49. Practical form of a bootstrap ramp generator.

be rather high to guarantee an approximately constant bias current through the Zener diode.

Level Detection. There are two forms of level-detection circuits— those with and without hysteresis.

1. The output V_o of circuits without hysteresis is always a single-valued function of the input signal V_i. High-gain saturable differential amplifiers may be used to produce a sharp step-like relation $V_o(V_i)$.
2. To prevent multiple switching of the output voltage due to noise voltage when the comparison is due, it is usually advantageous to have hysteresis in the relation $V_o(V_i)$. Schmitt triggers have this property. When the input voltage is rising, a transition to the

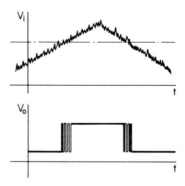

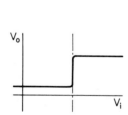

Fig. 8-50. Level-detection rela-tions in a high-gain saturable amplifier.

Fig. 8-51. Bouncing of V_o due to noise.

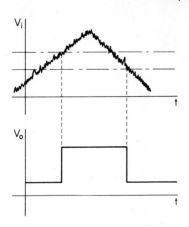

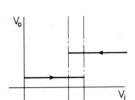

Fig. 8-52. Detection levels in a hysteretic amplifier.

Fig. 8-53. Elimination of bouncing by hysteresis.

upper value of V_o happens at a higher input voltage limit than when the input voltage is falling. This does not affect the level-detection accuracy, however.

Resetting Circuits. With voltage time bases using integrating capacitors, the accumulated charge must be pumped away in a short time, which means that a high discharge current must be drawn. The maximum tolerable current is usually limited. The discharging could be made with current switches; however, separate information should be provided to indicate when the voltage has fallen to a reference (starting) value.

Shunt Transistor Switch. In simpler circuits, a saturating transistor switch may be used to shunt the capacitor (see Fig. 8-54). According to the

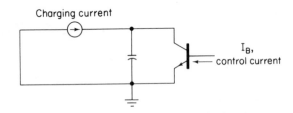

Fig. 8-54. Resetting with a switching transistor.

Ebers–Moll equations, the $I_C(V_{CE})$ relation of a saturating transistor is nonlinear. With $V_{CE} \gg kT/q$, the transistor is almost equivalent to a current generator, with a value $h_{FE}I_B$. When V_{CE} approaches ηV_T, h_{FE} becomes very small and the final approach to the value V_{CES} given by Eq. [8-51] is rather slow:

$$V_{CES} = \eta V_T \ln \frac{I_{EO}(\alpha_N I_E + I_C)}{I_{CO}(\alpha_I I_C + I_E)} \qquad [8\text{-}51]$$

Therefore, a transistor switch is useful only when a new ramp does not follow immediately.

Pulse Integrating Discharging Circuit. The following discharging method is used in some pulse-frequency-modulation circuits. After the ramp has achieved the discrimination level, the output voltage is dropped down by a fixed amount, subtracting a constant charge from the integrating capacitor. A rectangular (or otherwise fixed) current pulse corresponding to a fixed charge is applied to the summing junction. Figure 8-55 shows the general arrangement.

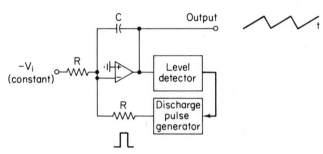

Figure 8-55

Discharging Circuit Using Nonlinear Feedback. The feedback circuit of Fig. 8-56 has the property that when $V_i < V_c$, the diode conducts and V_c

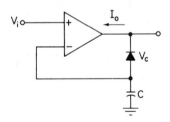

Fig. 8-56. Rapid discharging circuit using nonlinear feedback.

follows V_i with an accuracy determined by the input voltage offset, which is a few microvolts for good operational amplifiers. The rate of fall of $V_c(t)$ is limited by the maximum deliverable output current of the amplifier,

$$\max \left| \frac{dV_c}{dt} \right| = \frac{1}{C} \max |I_0| \qquad [8\text{-}52]$$

When $V_i > V_c$, the diode will be reverse biased, and only a small leakage current (e.g., about 10 nA with silicon epitaxial diodes) flows through it. If the leakage current were constant and equal to I_s, the rate of fall would be

$$\frac{dV_c}{dt} = -\frac{I_s}{C} \qquad [8\text{-}53]$$

which is usually negligible; with $C = 1\ \mu\text{F}$ and $I_s = 10\ \text{nA}$, the rate is only $10^{-2}\ \text{V/sec}$.

Example 8-14. A precision time-base system with adjustable slope and amplitude is shown in Fig. 8-57. This system is made of operational blocks.

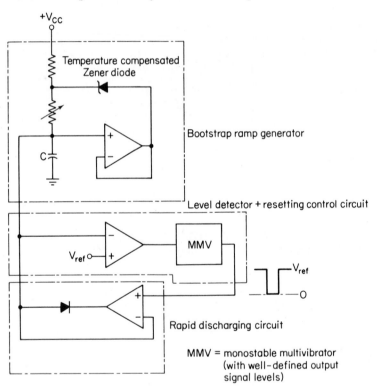

Fig. 8-57. Precision time-base circuit.

PROBLEMS

8-1   Figure P8-1a represents a pulse stretcher. Sketch the detailed forms of the voltages V_a, V_b, and V_c as a function of time, when input V_i is given by Fig. P8-1b and the input–output voltage transfer relation

of the gates is given by Fig. 7-7. The output impedance of the inverter
is 4 kΩ when the output is HIGH and 50 Ω when the output is LOW.

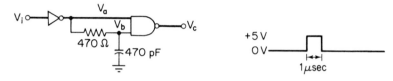

Figure P8-1a   **Figure P8-1b**

8-2   Draw the extender input voltage of gate G_1 and the output waveforms
of both gates of the monostable multivibrator shown in Fig. 8-10
when the gates are defined as in Problem 8-1. $C = 10$ nF, $R = 1$ kΩ.

8-3   Are the noise margins of the monostable multivibrator of Fig. 8-10
smaller than the noise margins of the corresponding logic family, and
for what reasons, if the noise appears
(a) at the input?
(b) in the supply line?

8-4   Design an astable collector-coupled multivibrator similar to the one
shown in Fig. 8-16 that produces a symmetrical square wave with a
frequency 100 kHz, using a supply voltage $V_{CC} = 5$ V. For the collector resistors, take $R_1 = R_1' = 1$ kΩ. For the transistor parameters,
use the following values:

$$V_{BES} = 0.6 \text{ V} \qquad V_{CES} = 0.2 \text{ V}$$

$$h_{FE} = 30 \qquad I_{co} = 0$$

(*Hint:* For R_2, select a value which is approximately such that the base
current of a saturated transistor is at least twice the value needed for
saturation.)

8-5   Calculate the output frequency of the astable multivibrator of Fig.
8-19 for the following values:

$$V_1 = V_1' = 5 \text{ V} \qquad V_2 = V_{CC} = 10 \text{ V}$$

$$R_1 = R_1' = 1 \text{ k}\Omega \qquad R_E = R_E' = 22 \text{ k}\Omega$$

$$C = C' = 5 \text{ nF}$$

The current gains of the transistors are assumed high and the saturation voltages $V_{CES} = 0.2$ V. The base–emitter voltages of transistors
in the conducting state are assumed to be $V_{BE} = 0.6$ V.

8-6   Analyze the detailed waveforms V_{BE}, V'_{BE}, V_{CE}, and V'_{CE} of the monostable multivibrator shown in Fig. 8-22 for the following component values:

$$R_1 = R'_1 = 1 \text{ k}\Omega \qquad R'_2 = 22 \text{ k}\Omega$$

$$R_3 = 18 \text{ k}\Omega \qquad R_4 = 100 \text{ k}\Omega$$

$$C' = 10 \text{ nF}$$

The voltages are as follows:

$$V_{CC} = V_2 = 12 \text{ V}$$

$$V_{CES} = 0.3 \text{ V} \qquad V_{BES} = 0.6 \text{ V}$$

The triggering circuit is assumed to give a correct triggering.

8-7   Plot the output voltage V_o of the circuit shown in Fig. P8-7 versus input voltage V_i. The line receiver is saturated when its output voltage reaches the values $+5$ V or 0 V and its gain is assumed to be very high.

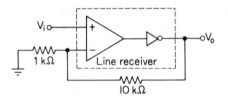

Figure P8-7

8-8   Find the voltage waveforms of the astable multivibrator shown in Fig. 8-36 and plot them in a suitable time and amplitude scale. The component values are as follows:

$$R_1 = R_2 = 10 \text{ k}\Omega \qquad C = 1 \text{ }\mu\text{F}$$

$$R = 50 \text{ k}\Omega \qquad V_{\text{ref}} = 0$$

The gain of the operational amplifier is very high and its saturation levels are $+10$ V and -10 V.

8-9   (a) Calculate the output frequency of the precision time generator shown in Fig. 8-57. The output current of the operational amplifiers is limited to 100 mA and the output pulse width of the MMV is 100 μsec. The Zener diode voltage is 5 V and the variable resistor is 5 kΩ. The value of C is 1 μF and V_{ref} is (1) 2 V and (2) 5 V.

   (b) Plot the voltage over C versus time in both cases.

REFERENCES

Basic waveform generators can be found in

> Millman and Taub: *Pulse, Digital, and Switching Waveforms* [95].

The implementation of timing circuits with the aid of logic gates can be found in

> Burlingame, "MDTL multivibrator circuits," *Motorola Application Note AN*-409 (Nov. 1967).

The use of operational amplifiers in timing circuits has been explained in

> Widlar, R. J., "The operation and use of a fast integrated circuit comparator," *Fairchild Application Bulletin APP*-116 (Feb. 1966) (reprinted from *END*, May 1965.); Integrated Circuits Applications Engineering Staff of Motorola, "Monostable multivibrator design using an I/C operational amplifier," *Motorola Application Note AN*-258 (May 1969).

chapter 9

Memories

This chapter is devoted to a review of the mass memories needed in a computer system. Since this book deals primarily with fundamental digital electronic circuits, we do not intend to describe other than electronic digital equipments in detail. Memories, however, play so important a role in central processors of digital computers that we have decided to present the principles of their operation in this book.

A preliminary discussion of memories was given in Chapter 1. All digital memories consist of storage cells or equivalent items of memorizing media, each of which usually stores a binary unit of information (bit). In Sections 9.2 and 9.3 we discuss general principles of active (electronic) memories. Sections 9.4 and 9.5 describe ferrite-core memories, which are the most important main memories at this writing. Read-only memories are described in Section 9.6. In Section 9.7 the most usual backing or auxiliary memories, the magnetic surface memories, are reviewed.

9.1. Classification of Memories

Classification of Memories According to Access. The most usual type of memory is the *direct-access* or *random-access memory* (*RAM*), in which information can be written to and read from storage locations identified by unique *addresses*. *Access time* is the time needed to transfer information from an addressed location to output circuits. In random-access memories, this time is constant. Such memories are used as mainframe memories in computers.

Various mass memories are used as backing storage in computer systems. Mass memories are much slower than mainframe memories but are capable of storing large amounts of information. These memories are

361

usually of the *sequential-access type*, which means that in order to specify a location in which information is stored, a large number of address locations must be scanned in succession, and the location at the nth step of scanning (where n is specified by the control signals) is selected for writing or reading. Access is sequential in magnetic drum, disk, and tape memories, in which storage locations are scanned mechanically.

A rather special type of memory is the *content-addressable* or *associative memory* (*CAM*), in which locations are selected according to the type of information stored in them (e.g., in reading, the memory system may transmit all words that have a certain bit combination at specified bit positions).

Read/Write and Read-Only Memories. In general-purpose computers, the running of different programs implies that a variable set of instructions has to be loaded into the memory for each problem, and working space for the constants and intermediate results must also be provided during computing. These features call for alterable memories. Information may sometimes also be permanently stored [e.g., by the predetermined structure of components or their interconnections ("hard-wired memory")]. The system is then called *read-only memory* (ROM) and *in principle it is simply a special combinational circuit, because an input signal combination (address, entry) defines a unique output signal combination.*[1] Read-only memories are used in control operations of digital systems, for the storage of data in tabulated form, and so on. They are particularly suitable to large-scale integration, due to the regularity of wiring and the general organization of addressing. In special-purpose computers, such as those intended for process control with fixed computational programs, large portions of alterable program memories can be replaced by more reliable ROM's.

Destructive and Nondestructive Reading. The reading of a memory may be *destructive* or *nondestructive*. In the first case, information stored in a selected position is destroyed during reading (i.e., zeros will replace the previous contents). Sometimes the restoring of information may not be necessary, but usually the contents are written back while the same location is selected. In nondestructive memories the binary contents are in no way interfered with by the reading.

Active Memories. The name *active memory* has been reserved for alterable memories in which the basic storage element is an active bistable electronic circuit, such as a flip-flop. Active memories are used for temporary

---

[1] Actually, the truth table of a combinational circuit is usually implemented by simplified logic circuits, whereas in ROM's the number of variables is so large and the bits so cheap that truth tables can be implemented directly, without prior simplification of Boolean expressions.

storage of information, and a set of registers, of which a particular one can be selected, might be called an active memory.

To provide working space for programs and for the fast execution of short subroutines, active memories are often used as *scratchpad memories*. Special components such as magnetic thin-film arrays are also used as scratchpad memories and the readout procedure may then be destructive.

9.2. Addressing and Address Selection

Logical Selection of Addresses in Active Memories. A characteristic feature of most memories is that the stored information is *addressable*. An identifiable address may be assigned to every bit in an array, in which case the memory is said to be *bit-oriented* (*bit-organized*). This kind of organization is common in some integrated arrays fabricated as bit-addressable building blocks.

In digital computing, the data are usually organized as words with a fixed length, and so one identifiable address may refer to one word, whereby all bits of the word attain the same address. If this is the case, we speak of *word-oriented* (*word-organized*) memories. The address code of a bit or a word is usually given as a set of parallel binary signals. When information is stored (written to) or read from an addressed location (bit or word) of a memory, it is said that the addressed location is *selected*. Selection is similar in bit and word-oriented memories; in the latter case, the same selection signals are used to control the selection of all bits of a word. Thus we may consider the selection of bits only. For each bit separately, a control line may be assigned, and this line is selected by a decoder. A logic 1 occurs on one and only one of these lines at a time. This principle is called *linear selection*. If the number of addresses is large, it is more advantageous to make available the inputs of an addressed bit by an AND gate or its equivalent, whereby these gates are driven by *coordinate lines*. An example of two-dimensional *coordinate selection* is shown in Fig. 9-1; it is also possible to implement the selection by more than two coordinates. In two-dimensional selection one X line and one Y line has to be selected.

Reading from a Logically Organized Active Memory. The reading of a binary memory element M_{ij} in an array consisting of words indexed by a single index i and of bit positions indexed by j is a logical operation. For this purpose, let us assume that the reading is done by words, and a word is selected if a *word line*, identified by a variable W_i, attains a logical value 1 while all other W_k, $k \neq i$, are logical zeros. The binary value of the jth bit in a selected word, denoted by an output signal B_j, is

$$B_j = \sum_{i=1}^{m} W_i M_{ij} \qquad (j = 1 \text{ to } n) \qquad [9\text{-}1]$$

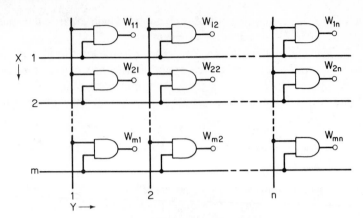

Fig. 9-1. Coordinate selection.

where m is the number of words. Since m is usually a large number, the implementation of a logical sum may be an awkward task if done by the usual logic gates. If a Wired AND is used for the logical sum, the number of interconnections is greatly reduced (Figs. 9-2 and 9-3).

In some methods using bistable memory elements other than logical flip-flops, the implementation of logical sums is replaced by a linear summing (superposition) of signal voltages delivered by the selected M_{ij} elements. In this procedure, an *interrogation signal* is transmitted into a selected word line. By the physical nature of the system, an electrical response telling of the

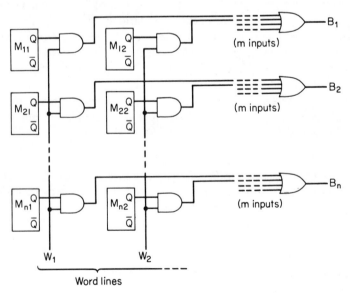

Fig. 9-2. Logical implementation of Eq. [9-1].

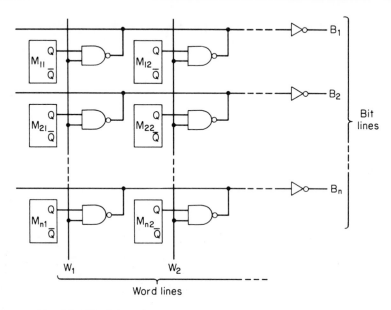

Fig. 9-3. Wired AND implementation of the logical sum in Eq. [9-1].

state of the binary storage element is relayed to the corresponding bit line, from which it is transformed into a standardized logic voltage by a *sense amplifier*. In ferrite-core memories, ferromagnetic material simultaneously constitutes the bistable element and a transformer core by which the interrogation signal induces a response to the bit line. Thus no extra AND or NAND selector gates are needed in magnetic memories.

Writing into a Logically Organized Active Memory. If the bistable element is a direct *R-S* flip-flop, the *S* or *R* input of it must be activated depending on whether 1 or 0, respectively, has to be written in it. These inputs are provided with AND gates that receive information about address selection and the logical value to be stored. Thus the gates have two or more inputs, depending on whether a linear or a coordinate selection method, respectively, has been used. The same address lines are utilized in reading and in writing, and if separate read and sense bit lines are used, the logic structure

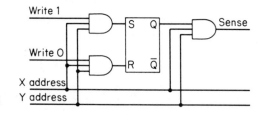

Figure 9-4

of an active memory cell is shown in Fig. 9-4. Each cell needs one or two word lines, two write lines (write 1 or 0), and a bit sense line: this totals four or more lines, plus the supply voltage lines.

9.3. Special Active Memory Arrays

TTL Active Memory. As an example of practical active bipolar transistor memories we consider the operation of a TTL memory cell and its interconnections in an array. By proper voltage-level-adjusting techniques, the electronic configuration of a memory cell can be made simple. Admittedly, this memory cannot then be described by standard logic gates.

The basic memory cell is depicted in Fig. 9-5. The two multiemitter

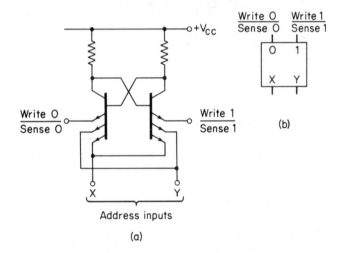

Fig. 9-5. TTL memory element.

transistors can be connected as a flip-flop by cross-coupling the respective collectors and bases. All cells are connected in parallel to common Write/Sense lines, and X and Y inputs are driven by coordinate lines. Two of the emitters on both sides are used for address selection and they are controlled by standardized logic voltages varying between 0 and 3.5 V. The third emitters attain voltages that are confined to narrower limits: With the aid of associated circuits, they are allowed to vary between a typical V_{BE} voltage of about 0.75 V, and 1.5 V, which is about twice a V_{BE}. Thus the emitter current of the conducting transistor flows through the third emitters if the address inputs are driven HIGH. Since the collector circuits are not loaded, this arrangement will suffice to guarantee bistability.

The address line inputs are normally sustained at the LOW voltage.

When selected, both X and Y inputs are switched HIGH, and the corresponding junctions become reverse biased. The third emitters are then used for writing as well as for reading. For *writing*, one of these emitters is driven LOW (actually to a voltage which is of the order of V_{BE} of an active transistor), whereas the other is sustained at about twice that value. The second transistor will then be turned off by this voltage difference. If the Write 1 voltage was higher, a state 1 will be set to the flip-flop; if Write 0 was higher, a state 0 will be set. For the *reading* of this flip-flop both address inputs are made HIGH. Now the third emitters are used to provide sense signals indicating the state of the flip-flop. Since the input impedance of sense amplifiers is rather low and a voltage of about 1.5 V is normally sustained on the lines, the potentials of both of the third emitters will rise up to approximately this level. The emitter current of the conducting transistor will now be switched to the corresponding sense amplifier input and the state of the flip-flop will be detected.

A 16-bit memory matrix that is bit-oriented is shown in Fig. 9-6.

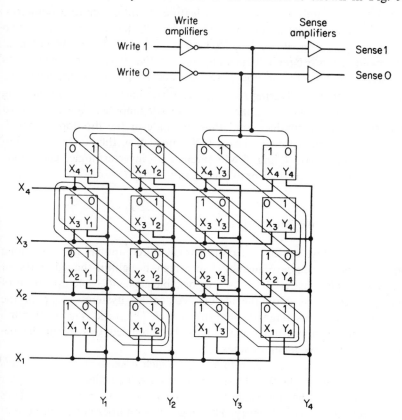

Fig. 9-6. Sixteen-bit TTL memory matrix.

MOSFET Active Memory. In order to reduce the number of lines needed, the write lines in a bipolar memory could be simultaneously used as bit sense lines. A special switching circuit was necessary to isolate the write drivers from the sensing circuits. Figure 9-7 shows a similar MOSFET

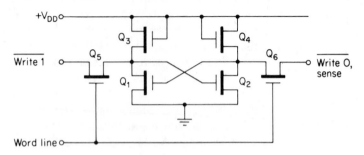

Fig. 9-7. MOSFET memory element.

memory implementation in which this feature is still easier to accomplish. Because of the bilateral nature of MOSFET's, Q_5 and Q_6 are used in two ways: During writing, $\overline{\text{Write 1}}$ or $\overline{\text{Write 0}}$ is sustained at a LOW potential by a low-impedance circuit, depending on whether it is 1 or 0 which has to be written. Thus the flip-flop is set as usual in inverter logic, when the word line is activated. During reading, the $\overline{\text{Write}}$ lines are left to a floating potential and $\overline{\text{Write 0}}$ line is connected to a write amplifier. When the word line is activated, Q_6 conducts and thus the state of the flip-flop is sensed. However, since the sense line has a high impedance, the state of the flip-flop is not changed. Using AND circuits of inverter logic, coordinate selection of words is also accomplished easily.

9.4.　Ferrite-Core Memories

The cheapest working memory for digital computers has for a long time been the ferrite-core memory, in which doughnut-shaped cores of saturable magnetic material are used as bistable elements. The operation of memory cores is based on the almost rectangular hysteresis loop of the magnetic flux ϕ versus magnetic field strength H. The static H–ϕ relation has many characteristic features, which must be discussed to explain the behavior of ferrite-core memory stacks.

If H makes an excursion to a value which exceeds the *coercive field strength* H_c, the magnetic flux density B attains a value called *remanence flux density* $\pm B_r$, which has the same sign as the value of H applied last. Thus the core acts as a binary memory. The magnetization does not occur at once over the volume but starts at the inner surface of the core, where H is largest.

The magnetization moves outward over the core volume with a velocity v which depends on H, being approximately

$$v = k(H - H_c) \qquad k = \text{constant} \qquad [9\text{-}2]$$

Thus the time needed for the complete magnetic switching of the core is inversely proportional to $H - H_c$, and of course inversely proportional to the radial thickness of the doughnut. A typical switching time is 0.5 μsec.

A core may also be partially switched, and several successive applications of H not exceeding H_c may drive the flux remanence to a point that is substantially smaller than the value after initial magnetization. Let this fact be illustrated graphically by Fig. 9-8, where the point denoted by B'_r is the remanence to which B converges after many partial switchings.

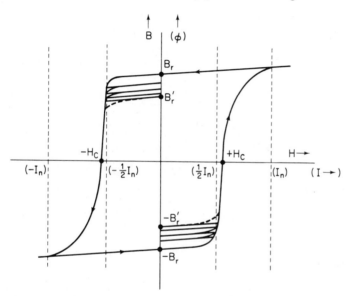

Fig. 9-8. Hysteresis loop of a ferrite core.

The total magnetic flux ϕ in a core is an integral of $B(r)$ over a cross section extending from the inner to the outer radius of the core. Thus, ϕ versus magnetizing current I through the core has almost the same shape as B versus H. A certain current I_n piercing the core is always able to drive the magnetization to $+B_r$, whereas $-I_n$ will drive it to $-B_r$, independent of the previous state of magnetization. On the other hand, a value $\pm I_n/2$ is never able to change the direction of magnetization but may only slightly lower the remanence flux density. Repeated applications of $\pm I_n/2$ drive B_r or $-B_r$ to points B'_r or $-B'_r$, respectively. The positive remanence flux values $B'_r \leq$

$B \le B_r$ are called logic 1, and the corresponding negative values of flux are logic 0's.

The switching of a core can be observed by an induction voltage generated in a secondary winding. Assume that the initial flux is $-\phi_r$. An open-circuit voltage $v(t)$ on the secondary is

$$v(t) = -\frac{d\phi}{dt} \qquad [9\text{-}3]$$

If changes in $I(t)$ are so slow that the time for I to change from 0 to I_n is always longer than the shortest switching time given by Eq. [9-2], we may assume that ϕ follows I according to

$$\phi(t) = f[I(t)] \qquad [9\text{-}4]$$

Thus $v(t)$ has a waveform that is derived from the static $\phi(I)$ relation graphically as shown in Fig. 9-9(a). On the other hand, if changes in $I(t)$ are faster

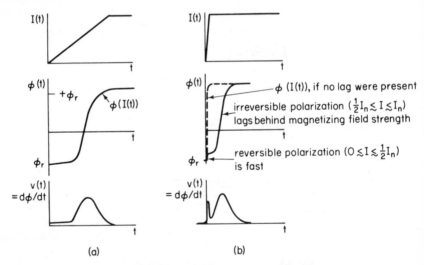

Fig. 9-9. Waveforms of a switching ferrite core.

than the shortest switching time, $\phi(t)$ lags behind $I(t)$, and this results in the waveform for the observed induction voltage depicted in Fig. 9-9(b). The observed waveform in Fig. 9-9(b) has the following explanation: Various parts of the hysteresis curve are due to different physical phenomena. The sections of this curve which have a small slope are due to *reversible magnetic polarization*, which is usually described in terms of magnetic permeability dB/dH. Thus induction voltages due to changes in B on the small-slope sections are fast and follow I with practically no delay. On the other hand, the steep sections of the H–B curve are due to permanent (irreversible) changes in the magnetization in which volumes called ferromagnetic domains change their magnetization. It is this effect which is slower, and $d\phi/dt$ on the steep sides lags behind dI/dt.

The short peak in the observed induction voltage of Fig. 9-9(b) is due to flux changes in the range of reversible polarization. Notice that $d\phi/dt$ may attain large values even if $d\phi$ is small. Because of this, it is advisable to limit the rise times of magnetizing current pulses to values for which reasonable switching speed is still achieved but the initial voltage peak remains low.

Assume first that the state of magnetization is in the range $-B'_r$ to $-B_r$ and a positive pulse of magnetizing current is applied. If the magnitude of this current is $I_n/2$, it will cause a reversible magnetization, and only the short peak depicted in Fig. 9-10 results. A magnetizing current I_n will reverse

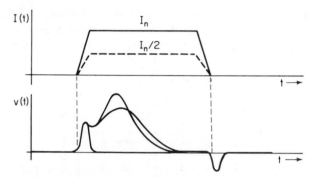

Fig. 9-10. Waveforms from magnetic cores during reading.

the magnetic state, and a large induction voltage results. If the initial magnetic state was "disturbed" (driven to $-B'_r$ by several successive current pulses of the magnitude $I_n/2$), a somewhat lower induction voltage results.

If the initial state was $+B'_r \leq B \leq +B_r$, a current pulse $+I_n$ would not change the direction of magnetization but would only drive B to the "undisturbed" value. The induction voltage is of the same form (short initial peak) as with $I = I_n/2$ when B was negative.

As we shall see later, cores in a memory array are usually selected by a coincident method, which means that magnetization currents of the magnitude $I_n/2$ may be applied to a core even if it is not selected. Thus the magnitude of the remanence flux density may usually be anything between $|B'_r|$ and $|B_r|$, depending on the history of computing operations.

9.5. Organization of Ferrite-Core Memories

The reading of ferrite cores is usually destructive, because the simplest way of indicating the magnetic state of an individual core is to apply a magnetizing current $I < -I_n$ through the core. Cores in state 1 (with $B'_r \leq B \leq B_r$) are switched to the opposite state, delivering a large and long induction voltage pulse; cores already in state 0 ($-B'_r \leq B \leq -B_r$) undergo only a reversible change of flux, and the induction voltage is smaller and short. If the state was 1, it must be restored after reading.

An initial writing of information is done by first resetting all cores to state 0 and then setting the 1's.

A coincident current method is usual in ferrite cores: A current $I_n/2$ cannot flip the state, whereas $I = 2 \times I_n/2 = I_n$ can. Thus if there are two wires piercing the core and a current $I_n/2$ is defined as a logical 1, the core will switch if it was previously set to the opposite state and if both current lines are simultaneously activated, corresponding to an AND operation.

Linear Select Word-Organized Ferrite-Core Memory. Memory core arrays are usually word-organized. A simple word-organized memory core array is shown in Fig. 9-11. The selection of a word line is accomplished

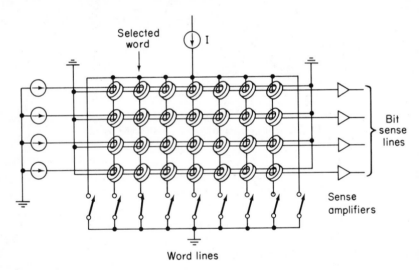

Fig. 9-11. Linear select ferrite-core memory.

by switching the lower end of a word line to ground by a logically controlled electronic switch which is able to conduct currents in both directions. The upper ends of the word lines are tied together and provided with a common current generator. In reading, this current is $-I_n$ and it is diverted to the selected word line only. During writing, the value of the current is $+I_n/2$. Two bit lines (a sense line and a write line) are threaded through each core, one pair for each bit position. The bit sense lines are connected to sense amplifiers which give a logical 1 output if any core along the corresponding line switches from state 1 to state 0. The bit-write lines are provided with current generators, which give a magnetizing current $+I_n/2$ to lines corresponding to bit positions in a selected word to which a 1 has to be written.

Three-Dimensional Word-Organized (Coincident Current) Ferrite-Core Memory. With a large number of words, the two-dimensional memory array may become rather expensive, because individual address decoders and bidirectional switches are needed for each word line separately. The usual organization of ferrite-core memories in computers is the coincident current memory, in which four lines pass through every core. Two of these are the *address selection wires* (called X and Y wires); one is a *bit sense line*, and one line, called the *inhibit wire*, is used in writing. A *memory stack* is a three-dimensional array in which one two-dimensional array, called the *memory plane*, is reserved for each bit position and a word has the same (X, Y) address in every memory plane. The selection of a word is implemented by X and Y wires which are threaded through all memory planes in the way shown in Fig. 9-12, so when, for example, a current $-I_n/2$ is applied to a pair of

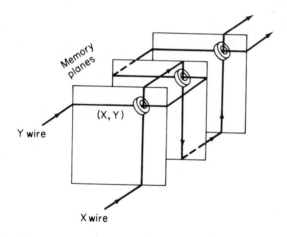

Fig. 9-12. Threading of coordinate wires in a ferrite-core memory stack.

X and Y lines, only cores existing at the corresponding address of each memory plane obtain a total current $-I_n$.

A bit sense line and an inhibit wire are threaded through all cores in every memory plane.

A word is selected in the same way during reading and writing (i.e., one end of an X wire and one end of a Y wire are switched to ground by electronic switches). The opposite ends of the address wires are tied together and driven with common current generators, one for the X wires, the other for the Y wires. The value of each current generator is $-I_n/2$ during reading and $-I_n/2$ during writing. Let us denote the current of the X wire of a particular core by I_x and the corresponding current of a Y wire by I_y. The inhibit wire carries a current I_z. A core is said to be *selected* if both X and Y wires carry a current $\pm I_n/2$. A core is called *half-selected* if only one of the coordinate wires carries a current $\pm I_n/2$ while the other current is zero. Half-selected cores lie along the selected X and Y wires (see Fig. 9-12 for clarity), excluding the intersection. All other cores are called *not selected*.

During reading, the currents I_x, I_y, and I_z are as seen from Table 9-1, in which the sum of these currents is denoted by $\sum I$. As will be seen, only the selected cores which are in state 1 are switched over and deliver due signals to the sense wires of memory planes.

During writing, the currents I_x, I_y, and I_z are defined as in Table 9-2. It is assumed that the core before writing is always in state 0 (e.g., after a reading operation). We see that $\sum I = I_n$ only in case a core is selected and if

Table 9-1. Currents of Ferrite Cores during Reading

Wire	Not Selected	Half-selected		Selected
I_x	0	0	$-I_n/2$	$-I_n/2$
I_y	0	$-I_n/2$	0	$-I_n/2$
I_z	0	0	0	0
$\sum I$	0	$-I_n/2$	$-I_n/2$	$-I_n$

Table 9-2. Currents of Ferrite Cores during Writing

Wire	Write "0"			Write "1"		
	Not Selected	Half-selected	Selected	Not Selected	Half-selected	Selected
I_x	0	$I_n/2$	0	$I_n/2$	0	$I_n/2$	0	$I_n/2$
I_y	0	0	$I_n/2$	$I_n/2$	0	0	$I_n/2$	$I_n/2$
I_z	$-I_n/2$	$-I_n/2$	$-I_n/2$	$-I_n/2$	0	0	0	0
$\sum I$	$-I_n/2$	0	0	$I_n/2$	0	$I_n/2$	$I_n/2$	I_n

$I_z = 0$. Thus to write a 1 to a specified bit position, the inhibit current generator of that memory plane is disabled, whereas for the writing of a 0, the inhibit current must be $-I_n/2$.

Let us now return to the noise problem mentioned before. Since every half-selected core in a memory plane causes a short peak due to the reversible magnetization, and there is a large number of half-selected cores along the X and Y lines, which all contribute to the sense line signal, the superposition of these peaks is much higher than the true signal (Fig. 9-13).

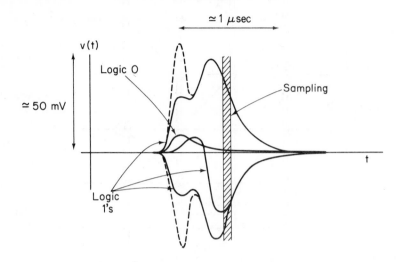

Fig. 9-13. Some typical sense signals from a coincident current memory. (The δ peaks may be much larger, as sketched by the dashed lines.)

To reduce this noise, the cores in a memory plane are tilted in alternate directions and wires are threaded in such a way that the polarity of one half of the induction voltages is opposite to the other half. Thus the noise is largely compensated. Figure 9-14 shows the direction of ferrite cores in a memory plane and the way in which different wires are threaded through the plane.

Notice that every second X wire carries a current that has been defined positive in the upward direction and that in the rest of the X wires the positive direction is downward. The positive directions of Y currents alternate also.

However, since the noise peaks are not identical but depend, for example, on the states of cores that can be randomly distributed (as well as switching history), and since cores are nonidentical, a residual peak voltage remains. This type of noise is often called a δ *peak*. When compensated, the δ peaks in a large array may even exceed the true signal from the selected core. To circumvent this difficulty, use is made of the fact that the true signal is always delayed with respect to the noise voltage and can be *sampled* at a moment when the noise has decayed off. Therefore, sense amplifiers are activated by a sampling (strobe) signal which is triggered at the leading edge of reading currents I_x and I_y and subsequently delayed.

The δ peak noise sets a practical limit for memory planes at about

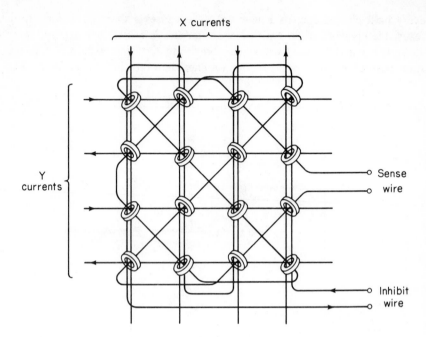

Fig. 9-14. Threading of wires in a memory plane.

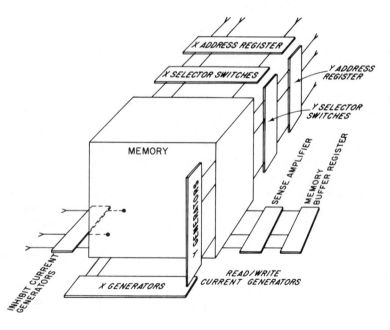

Fig. 9-15. Ferrite-core memory system.

64 by 64 = 4096 words. For a higher number of words, separate sense wires should be provided for sections of this or of smaller size.

The complete three-dimensional configuration of a ferrite-core memory stack with associated electronic circuits is illustrated in Fig. 9-15, and a photograph of a memory plane is shown in Fig. 9-16.

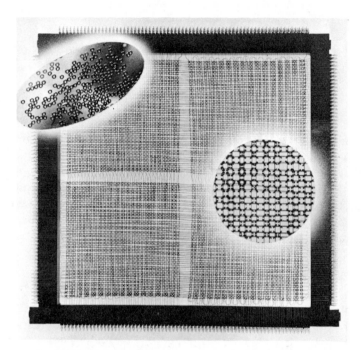

Fig. 9-16. Ferrite-core memory plane. This plane is from a so-called 2½-dimensional memory, in which writing is accomplished by a linear selection and reading by the coordinate selection. Accordingly, the sense wire is threaded along coordinate wires. (*Courtesy of International Business Machines Corp.*)

Selector Switches, Current Generators, and Sense Amplifiers for the Coordinate Wires. Bidirectional switches which can be controlled by logic voltages and which are able to conduct currents of the order of 0.5 A in both directions can be implemented in several ways (e.g., by a symmetrical switching transistor). Rectifying bridges can also be used. There are also several ways to implement the two-way current generators. One of the simplest solutions is a pulse transformer in which a separate primary winding is provided for the read and write controls (Fig. 9-17). Two-way electronic current generators are also available.

The most difficult problem is usually presented by the sense amplifiers. Output voltages of sense wires are of the order of 50 mV delivered by low-

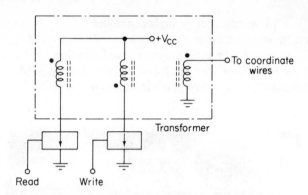

Fig. 9-17. Read/Write current generator using transformer.

impedance circuits. These signals are usually transmitted through twisted pairs of wires to sense amplifiers that have symmetrical inputs. Notice, however, that the polarity of output pulses may be either positive on negative, depending on the compensation method used. Accordingly, the sense amplifier must be rectifying. There are special integrated twin operational amplifiers with built-in rectifying circuits; a sampling gate for strobe pulses is also incorporated in the circuit. Figure 9-18 shows a symbolic representation and

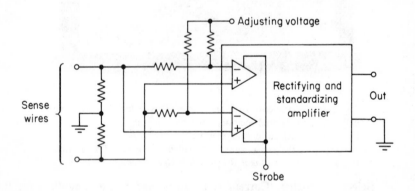

Fig. 9-18. Sense amplifier.

the symmetric input balancing circuit of a sense amplifier. The output is single-ended and may be used to drive logic circuits.

Selection of Coordinate Wires by the Matrix Method. In the selection of coordinate wires, when closing the circuit through one of the wires and the current generator, a selector switch can be placed at both ends of the wire. With M switches at one end and N switches at the other, one of $M \times N$ wires can then be selected if the wires are driven in a matrix as shown in Fig.

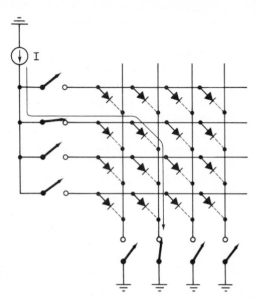

Fig. 9-19. Selection of one coordinate wire by the matrix method. Dashed lines represent wires to be selected.

9-19. This will result in substantial savings in decoders. Notice that some electronic switches must be located at a HIGH potential. Notice also that in order to avoid short-circuiting of adjacent lines in the matrix, a diode must be placed in series with each wire to be selected.

Example 9-1. A 4K memory (4096 words) organized as a two-dimensional array would require 4096 decoders for the selection of word lines. A coincident current memory of the same size needs 64 X decoders and 64 Y decoders, 128 in all. If both ends of the wires are selected, only 8 decoders are needed at each end of the X and Y wires, a total of $4 \times 8 = 32$ decoders.

9.6. Read-Only Memories

The binary information in read-only memories (ROM's) is usually represented by the presence or the absence of a signal coupling between word and bit lines. This is so in all the following cases, which comprise the most usual ROM principles: resistor matrix, capacitor matrix, diode matrix, bipolar transistor matrix, MOSFET matrix, and inductive coupling.

Resistor Matrix. The resistor matrix is, in fact, a set of special threshold logic RTL OR gates, connected in parallel to the word lines in such a way that there is a resistor from a word line to the bit line if the corresponding stored information bit is 1; there is no connection if this bit is 0. In large

RTL arrays, there is a problem of crosstalk between bit lines, and for this reason two provisions are necessary:

1. The circuits driving word lines ought to have a very small output impedance, at least when the signal is 0. This is usually valid if saturating transistors are used as logic switches.
2. The input impedance of the amplifiers to which the logic resistors are connected ought to be as low as possible. Therefore, current amplifier inputs (e.g., common base transistors or parallel feedback operational amplifiers) are preferable.

Figure 9-20 shows the general arrangement of a resistor matrix.

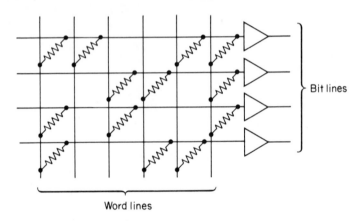

Fig. 9-20. Resistor matrix read-only memory.

Capacitor Matrix. Using photolithographic techniques, it is often easier to produce a capacitor than a resistor. Figure 9-21 shows the geometric form of a capacitor matrix in which capacitor plates are fabricated between all word and bit lines, but a connecting bridge is broken where a bit is 0. The breaking can be made, for example, by digitally controlled automatic devices.

Only ac signals (e.g., short pulses) can be transmitted through the coupling capacitors. In principle, the capacitor matrix works according to same principles as the resistor matrix, with ac impedances in place of the resistors. Bit signals transmitted by the capacitances are superimposed on the bit lines. Impedance restrictions on word line drivers and sense amplifiers are approximately the same as for a resistor matrix ROM.

Diode Matrix. An improved method of switching matrices is based on DTL OR logic gates arranged in a matrix as shown in Fig. 9-22. This circuit is electrically superior to both of the previous ones because the diode is a

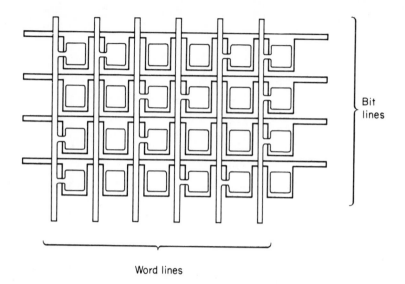

Bit
lines

Word lines

Fig. 9-21. Capacitive read-only memory.

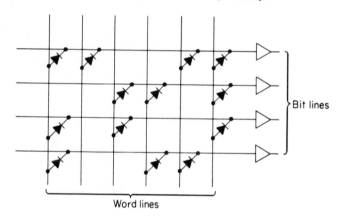

Bit lines

Word lines

Fig. 9-22. Diode matrix read-only memory.

unidirectional signal coupler that isolates crosstalk from adjacent bit lines. Therefore, the sense amplifiers can have a higher input impedance than in resistor or capacitor matrices, and there are no restrictions on the word line drivers. These advantages are not as important, however, as the higher cost of diodes. Diode matrices are advantageous if the number of bits is high. Also, AND gates can be used (see Problem 9-3).

Bipolar Transistor Matrix. Better immunity to crosstalk between lines is provided by a transistor matrix. When made of bipolar transistors, this may take a form such as is shown in Fig. 9-23. (Sometimes there are

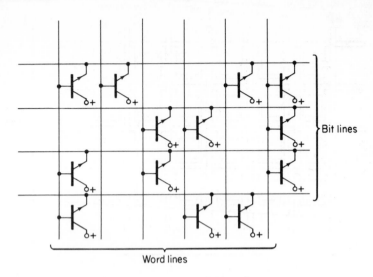

Fig. 9-23. Bipolar transistor read-only memory.

external resistors in series with the bases. These are not shown in Fig. 9-23.) Transistors are usually fabricated at every crossing, but in the final customization, the emitter leads of transistors corresponding to stored 0's are cut. Inactive transistors are not shown in the figure. Bipolar transistor matrices are used, especially in LSI arrays, where thousands of bits are stored on one chip.

MOSFET Matrix. Since the fabrication of MOSFET's by photolithographic techniques is a rather inexpensive process, ROM's with good electrical characteristics can be made, replacing the passive coupling components by MOS's, which are unilateral devices. This principle allows for a very high number of stored words because the coupling impedances are switched on only during the interrogation of a particular word, and the impedances associated with only one word at a time are controlled. A stored "1" is represented by a normal MOSFET, which conducts during the word interrogation, whereas a stored "0" is represented by a MOSFET in which the gate insulator is so thick that an interrogate voltage does not cause any enhancement conduction in the FET. The stored binary information is defined by the fabrication of the oxide layer only (which is a cheap method to be controlled accurately), whereas the rest of the FET structure is always identical for all bits, independent of the stored information, and thus allows automatic fabrication of a uniform, regular, and cheap component array. Figure 9-24 depicts a MOSFET ROM in which only the active MOSFET's are shown.

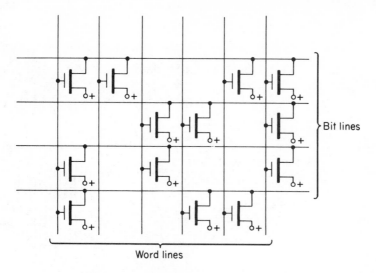

Fig. 9-24. MOSFET read-only memory.

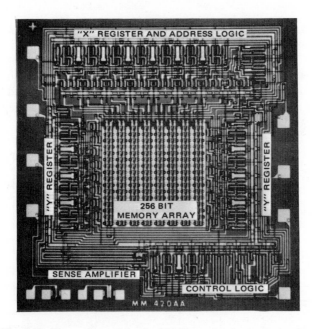

Fig. 9-25. Photomicrograph of a 256-bit read-only memory using MOS *P*-channel enhancement-mode FET's. (*Courtesy of National Semiconductor Corp.*)

383

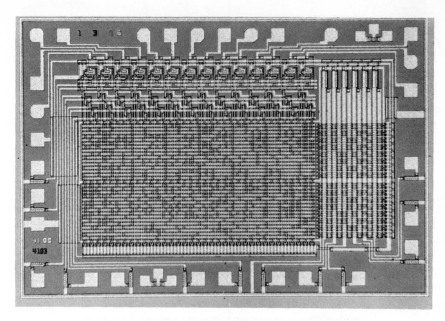

Fig. 9-26. MOS 2240-bit read-only memory for character generation. (*Courtesy of Texas Instruments, Inc.*)

Inductive Memory. Signal coupling between word and bit lines may also be inductive. To illustrate the principle of reading, the dashed horizontal lines in Fig. 9-27 corresponding to bit lines in other memories represent the

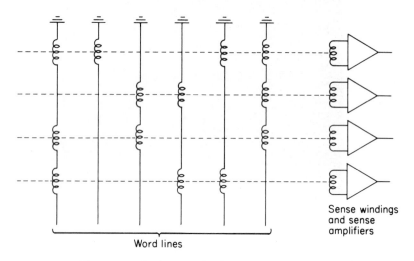

Fig. 9-27. Inductive coupling in a read-only memory.

inductive coupling between a word line and the secondary sense windings. Inductive coupling can be implemented by special electroplated transformers. A more widespread practical arrangement, however, is that shown in Fig. 9-28, in which a stored bit 1 (inductive coupling) is implemented

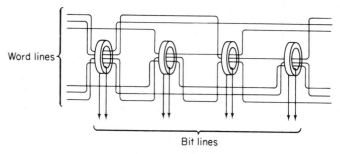

Fig. 9-28. Braid memory.

by threading the word line through a sense core, while a stored 0 is represented by a missing coupling (i.e., by a word line passing the sense core). There are usually several windings on the secondary side of each sense transformer, to provide a high-level signal to sense amplifiers.

This system is called *braid memory*, and the threading of wires is usually done automatically (e.g., by a device called a *Jacquard loom*). Word line conductors may also be metalized on plastic cards, in the form of a ladder. One of the legs of the ladder is cut by punching. Thus the remaining conductor forms a current path through or by the cores (Fig. 9-29). A stack of such cards comprises the braids. To provide easy threading of conductors, ferrite rods or tubes are often used instead of ring cores.

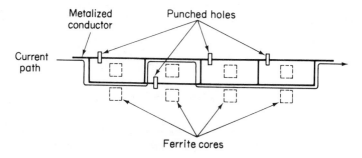

Fig. 9-29. Printed conductors for a braid memory.

Magnetic-Core Microprogram Memory. Popular in special-purpose applications such as military computers, a variant of braid memory using switching magnetic cores is used where higher-speed operation is needed. With this principle, not only can a word be read from an addressed location but complete micro-operations sequences of machine instructions can be implemented with this memory. Here we describe this more complete mode of use. The operation has two major cycles: During the first one, a microprogram corresponding to a particular machine instruction is loaded into a ferrite-core memory matrix. The second cycle consists of reading this information from the matrix by rows, whereby a sequence of parallel binary words is obtained from the read amplifiers. The number of sequential steps during the reading cycle is the same as the number of rows (or maybe shorter). In the basic form of this device, branching conditions are not taken into account during the reading; however, the main branching decision was already made, with the selection of a particular machine instruction during loading.

Figure 9-30 gives an idea of the magnetic-core microprogram memory.

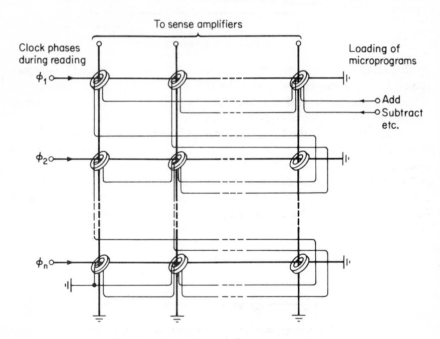

Fig. 9-30. Magnetic-core microprogram memory.

Each column in the matrix has a common sense wire and a sense amplifier. Different microprograms can be loaded (preset) to the cores by braids, each one of which corresponds to an arithmetic, logic, or other machine operation in the computer (terminals on the right). Only one of these braids is selected at a time. After loading, the cores are read by the application of read currents on the rows, starting from the top (terminals on the left). At the clock phases ϕ_1 through ϕ_n, control signals corresponding to the preset information are obtained from the sense amplifiers (terminals at the top).

9.7.   Magnetic Surface Memories

There are discrete storage elements for all bits in active and ferrite core memories, and the words are completely selected by electronic circuits. Such methods are fast, especially since the access to all storage locations is direct. On the other hand, this method becomes rather expensive per bit because of the large number of address decoders needed.

Substantial savings at the cost of access time can be obtained if consecutive bits can be stored on a surface that is scanned by mechanical motion; the information may be recorded by magnetization on a magnetic surface, optically, etc. Magnetic recording provides high speed in reading as well as writing of information and is the only method considered here. A row of bits that are scanned mechanically by magnetic reading and writing heads is called a *track*. The number of bits per track is limited by bit density and by the geometry of the medium (i.e., by the length of the track, which may be circular, linear, or spiral).

In this section we shall discuss three types of memories used extensively as backing storages in computers: magnetic drum memories, magnetic disk memories, and magnetic tape memories. The magnetic surface in drum and disk memories is in continuous motion, whereas the tape is moved only on request.

Some desk calculators also use magnetic cards with linear tracks. Such cards are not standardized. Magnetic ink records do not belong to this category because it is the presence or absense of printed magnetic marks (characters) that is sensed in the latter.

Magnetic Recording. The problems of recording and reading are approximately similar in all the previously mentioned memories and will be discussed first. A magnetic head with a narrow gap between the poles (typically 10 μ) is used for writing and reading of bits in much the same way as in tape recorders. Sometimes a dual gap is used for immediate readout after writing, to check the recorded information (Fig. 9-31).

In drum and disk memories, a small clearance (typically 10 to 20 μ) is used to separate the magnetic head from the surface. The magnetic film is usually made of ferromagnetic oxides, but in drum memories Ni–Co alloys are also used. Typical film thickness is 5 to 15 μ.

The width of the magnetic track is usually much larger than the gap length, typically 1 mm. The density of bits along a track depends on many factors: the length of the gap, gap-to-surface distance, film thickness and material, and speed. On magnetic tapes, for example, where the head is in contact with the surface, standard values for bit densities (on one track) are 200, 556, and 800 bits per inch (bpi). On drums and disks, the density is lower, as a result of the clearance between the head and the surface.

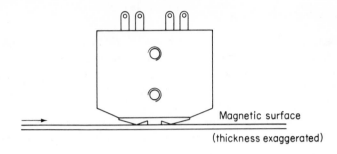

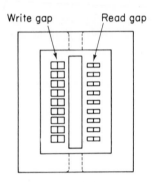

Write gap   Read gap

Fig. 9-31. Dual gap magnetic head (for nine tracks).

The bit values are indicated by the direction of saturation magnetization, and the writing of these values is thus made by positive or negative magnetization currents through the magnetic head. However, there are several ways in which flux may be put in correspondence to binary information.

In the following, a lengthwise division along a track, reserved for one bit, is called a *bit cell;* there are several methods of recording a bit on it. Some of these methods need an *external clock signal* for timing reference; others are *self-clocked.* Because during operation, written information may be changed several times, both the 0's and 1's on the track must be indicated by positive or negative magnetization. (In other words, 0's cannot be represented by unmagnetized areas.) The differences in these methods lie in the way in which saturation flux is changed between bit cells. Externally clocked recording methods are the *RZ,* or the *return to zero,* method; *NRZ,* or the *non-return to zero,* method; and *NRZI,* or the *non-return to zero inverted,* method. Self-clocked recording methods are *phase modulation,* or the *Manchester* method; and the *frequency modulation* method. In the following we shall discuss each method separately.

In the *RZ method,* every bit is magnetized separately by a positive or a negative current; the surface between bits remains unmagnetized. During reading, the induction voltage in the reading coil is proportional to the time derivative of flux, and it is possible to deduce the binary value of every bit from the polarities.

In order to strive for a higher packing density of information, the *NRZ method* is often used. The magnetization remains constant between similar consecutive bits. The maximum packing density is about twice as much as in the RZ method, but the reading head gives an output voltage only when consecutive bits are different. Thus the last positive or negative sensing signal must be memorized by the reading circuits to reproduce intervening similar bits, and the reading circuits must be synchronized with the bits (e.g., using a separate clock track on the medium on which synchronizing pulses are permanently recorded).

In the *NRZI method* the magnetization is reversed every time a 1 is to be written, but its direction is kept constant for all 0's. With this method, which is more common than NRZ, a positive or negative sense pulse indicates a 1 during reading. A separate clock track for synchronizing pulses is also needed in this method.

Waveforms for magnetizing current $i(t)$ (and magnetization) as well as for the output voltage $v(t)$ of reading heads are shown in a series of diagrams (Fig. 9-32) for these three methods.

Of the three methods, RZ has no much practical importance. The other two, the NRZ and the NRZI methods, are mainly used in low-cost, low-capacity memories. In reading, strobing pulses derived from the clock

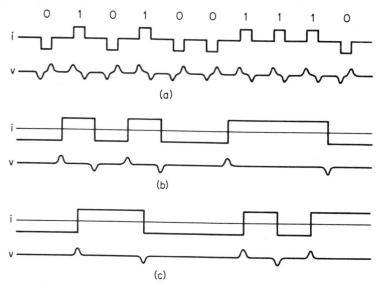

Fig. 9-32. Recording methods: (a) RZ; (b) NRZ; (c) NRZI.

track signal may be used to increase noise rejection. The strobing signal must then be delayed with respect to the bit cell boundary, by half the bit cell length, and, if the speed of the surface is varying, this delay should be altered accordingly to compensate for speed variations. At high bit density, this method for sampling cannot be used, because the clock and data easily get out of phase. It is therefore advantageous to have a method in which the reading signal itself contains the timing reference. Such a reference is provided by the other two recording methods, which are used in high-quality magnetic surface memories.

Besides the difficulty of clocking, there is another drawback in the NRZ and NRZI methods: A large bandwidth is needed in the read amplifiers. Notice that it must be possible to record any binary information on a track, and arbitrarily long sequences of 0's and 1's require that the gain of the read amplifiers must be constant from dc to the highest operational frequency. If we can be sure that the flux is changed at least, say, every two bit cells, we could use an ac-coupled read amplifier for better stability. This has been accomplished in the following methods.

In the *phase modulation* or *Manchester method*, flux direction *always* changes at the center of the bit cell boundary, and the direction of the change tells whether the bit is 0 or 1. (Let us assume that negative-going flux indicates a 0 and positive-going flux a 1.) There may or may not be a flux change at the bit cell boundary, according to need. In this way there are frequent changes in the output signal, and the binary value of a bit cell can be deduced by sampling the output signal at the center of the bit cell. Figure 9-33(a) shows the waveforms for this method. The method is widely used in high-quality devices.

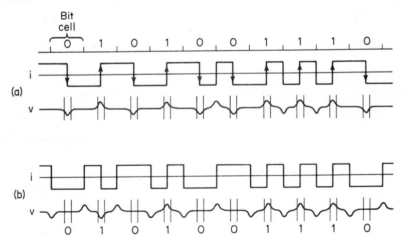

Fig. 9-33. Recording methods: (a) phase modulation; (b) frequency modulation.

In the *frequency modulation method* there is a flux reversal at the middle of the bit cell for a 1 but no change for a 0. The flux is always reversed at the bit cell boundary. The method is illustrated in Fig. 9-33(b).

There are several more-complicated methods, for example, the *QHC* and *Potter methods* (see, e.g., [213]), which are used in very large memories.

Address Selection. Words can be arrayed in two ways on magnetic surfaces, *in series* and *in parallel*. In the serial method, which is more common with drums and disks, the consecutive bits of one word are recorded serially on one track. To specify a word it is thus necessary to indicate a track and the relative position of the word on it, measured from a reference point. In parallel arrays, the bits of the same word are recorded on separate tracks. Parallel–serial recording is used in magnetic tape devices in which, say, nine bits (of which some are check marks) are simultaneously recorded on one row (i.e., in parallel on all nine tracks). Several consecutive rows, however, are assigned for one word.

On drums and disks, the surfaces are divided into circular tracks, and every track is further divided into an integral number of lengths called *sectors*. For to specify a bit position on the perimetry, there are usually two tracks with permanently recorded synchronizing marks on them. These tracks are provided with reading heads only. One track contains a single mark or pulse, called a *synchronizing pulse*. On the other track there is a *clock pulse* for every bit position. The borders of the sectors are not marked on the surface. There is a separate binary *bit counter* in the memory system which counts clock pulses modulo N, where N is the number of pulses on the clock track. The counting is started from the synchronizing pulse with contents zero in the counter. If there are 2^w bits in a sector, the w last bits of the bit counter will be zeros at the beginning of every word, and this fact is used to identify the borders of the sectors. The remaining bits of the bit counter comprise the *sector counter*, which holds the *sector address*. To select a sector, a given sector address is compared with the counted sector address; if all bits coincide, we have an indication that we are for the present scanning a selected word. This indication may then be used to control reading or writing circuits.

Thus, to select a complete address of a word, it is first necessary to select a track and then to wait for the selected sector. The reading or writing of bits is started at the border of a selected sector, transferring the bits in series and in synchronism with the clock pulses.

In memories with fixed magnetic heads, tracks are selected by static decoders. Disk memories have special provisions for track selection.

Drum Memories. Of the three memories discussed in this section, drum memories are considered the most reliable. The magnetic coating of drum memories has been deposited on a cylinder that is rotating with high constant speed around its axis. The surface may be divided into a number of tracks, ranging from a few to a thousand, each provided with a magnetic head. Notice that the heads may be displaced to arbitrary positions along the track, which is a common arrangement for their mounting at a rather high track density.

Drums vary in size and speed of operation. Typical diameters are from 2 in. to a couple of feet, and lengths range from $\frac{1}{4}$ in. to a few feet. Rotating speeds usually vary from 500 to 10,000 rpm, being 3000 rpm typically. There

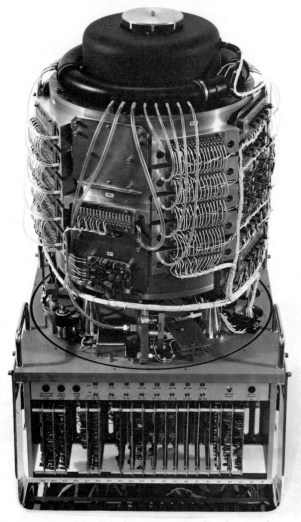

Fig. 9-34. Magnetic drum memory. (Plastic tubes are for pressurized air, and electronic parts are housed at the bottom. (*Courtesy of Digital Development Corp.*)

are 15 to 30 tracks per inch and the bit density in the channel may be 50 to 100 bpi. Surface coatings consist usually of Fe_2O_3 mixed with other oxides, or Ni–Co alloy. Figure 9-34 shows a typical small drum memory.

In the 1950s, drums were generally used as main memories. They are still needed as backup storages in medium and large-sized computers. Average access times are typically 10 msec, and large drum memories may contain 10^8 bits.

Disk Memories. Disks coated with magnetic material are also extensively used as backing storages. There may be as many as 50 disks on one axis driven together, which makes 100 magnetic surfaces in one device. The tracks consist of concentric circles, typically 100 of them on one side. The disks are usually removable. This makes it possible for every user to keep his own disk files, and disks can thus be regarded as one form of input–output media.

Some disk memories have only one pair of magnetic heads, one for all top surfaces and another for the bottom sides (Fig. 9-35). This pair of heads has been mounted on a movable arm which can reach any surface and any track on a surface. Sometimes there is one arm for each disk. Moreover, there

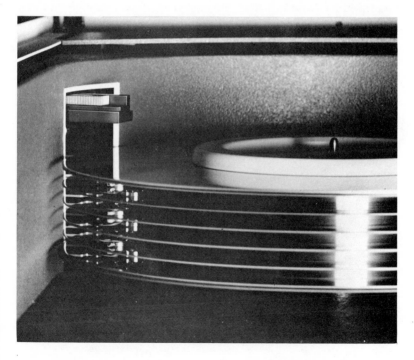

Fig. 9-35. Near view of a magnetic disc memory, showing moving-arms mechanism. (*Courtesy of International Business Machines Corp.*)

are disk memories with fixed arms, so one magnetic head is permanently provided for each track; in this case, disks are not removable.

Especially in disk memories with movable arms, a special form is given to the magnetic heads so that the head and the disk are separated by an air cushion. The clearance between the head and the surface thus remains constant, although the arms are not sturdy and the disks sometimes become slightly warped.

To search a word from a disk, three steps are necessary: searching the surface, moving the arm to a desired track, and waiting for the desired sector.

Typical access times in disk memories are from 100 msec to seconds, and one disk may contain 10^6 to 10^7 bits. Disk memories can be used as backing storages in almost any type of computers.

Tape Memories. The largest files are usually implemented by magnetic tape devices, which record seven or nine tracks on a $\frac{1}{2}$-in.-wide magnetic tape. Nine-track devices are more common. The tape is driven only on request, and special drive mechanisms are needed to match the feeding of tape from the reels with the actual movements of tape over the magnetic heads. This is accomplished by an arrangement depicted in Fig. 9-36. For high acceleration

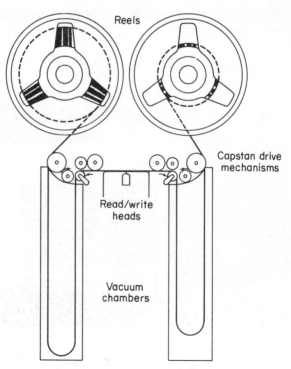

Fig. 9-36. Moving mechanism for magnetic tape.

on starting and stopping, and to advance the tape with constant speed, the moving of tape is accomplished by a capstan drive and made independent by a reel drive. Tape reels are rotated by a servomechanism that maintains a proper length of slack tape in vacuum chambers near the magnetic heads (called a *take-up system*). There are also moving mechanisms with one capstan only which are able to move the tape in both directions.

The stopping accuracy of tape is some millimeters, and a single row cannot be selected. For this reason, information is written on tape in *blocks* which contain a large number of consecutive characters. The blocks are separated by 0.5 to 0.75 in. A block is always written or read in one operation, and at the beginning of each block, an identification code is recorded.

A parity check for each character as well as for all bits along every track (longitudinal check) is usually performed.

The access time of tape memory depends on the length of the tape. Typical speeds are 20 to 150 in./sec. Tape memories are used as files with medium- and large-sized computers.

9.8. Trends in Memory Technology

The types of memories discussed in this chapter are representatives of devices encountered in most computers, and it was our purpose to describe only the most widespread technologies. It must be admitted, however, that there are some principles not reviewed here (e.g., magnetic thin film, which has found use in scratchpad memories of some large commercial computers). It seems, however, that the active memories invade these areas, too. There are also a lot of other memory principles which are in the development stage and which might become commercial in a few years. To mention a few, we have optic memories, especially based on holography; cryogenic memories; ferroelectric memories; Ovshinsky diode memories, which are able to store information in a currentless state; and local charge distributions in FET structures. On the other hand, some technologies reviewed in textbooks during recent years [e.g., dynamic (delay line) memories, including ultrasonic quartz delay lines, multiaperture magnetic cores, and tunnel diode memories] are currently used primarily in special applications and are not reviewed here.

A good picture of the state of the art and future trends of memories can be obtained from the series of articles in *Electronics* listed in the Bibliography.

PROBLEMS

9-1 Discuss the relative merits and shortcomings of active and ferrite-core memories.

9-2   Using illustrations, point out how the directions of induction voltages generated by half-selected cores on the sense wire in a ferrite-core memory plane are compensated.

9-3   Figure P9-3 represents a ROM including an address decoder with a two-bit address (formed of signals A and B and their negations). Obtain the Boolean functions $X_1(A, B)$ to $X_4(A, B)$ using positive logic. (H = logical 1.)

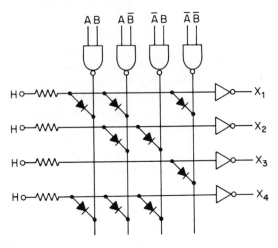

Figure P9-3

9-4   Consider an 8×8 resistor ROM matrix. The amplifiers driving the word lines have an output impedance of 7 Ω; the input impedance of sense amplifiers is also 7 Ω. The matrix resistors are 4900 Ω each. What is the relative crosstalk between two bit lines (one carrying a logical 1, the other a logical 0 signal) for the worst case in which there are 63 resistors in the matrix?

9-5   Design a ROM, including word line decoders, which transforms any 2421 code to a 8421 code, as shown in Table 1-3. Use diode matrix ROM and DTL NAND gates. (*Hint:* The two 2421 code representations in rows 2 to 7 in Table 1-3 must yield the same 8421 code output. See also Problem 9-3.)

REFERENCES

An economic and fast mass memory has been sought for many years. The number of articles published on memories, especially active ones, is very large. Active memories are described in applications reports of semiconductor companies; for example,

Motorola, *Developments in Large Scale Integration* [233].

Memories based on moving magnetic surfaces are described in

Hawkins, *Circuit Design of Digital Computers* [57].

For additional reference lists, look at the series of 34 special articles published in *Electronics* beginning October 28, 1968. These articles are listed in the Bibliography. Microprogram memories are described in

Husson, *Microprogramming: Principles and Practices* [236].

chapter ten

Transmission of
Digital Signals

10.1. General

Lumped-parameter models for electrical components and circuits are approximately realistic only insofar as changes in signal voltages or currents occur in a time that is much longer than the time for electromagnetic fields or electric currents to travel from one point to another in the system. The phase velocity of electromagnetic fields in vacuum (i.e., also in a space confined between metal conductors, such as waveguides or coaxial lines with vacuum isolation) is the same as the velocity of light c, which is approximately 3×10^8 m/sec. In a similar space filled with dielectric, with relative dielectric constant ϵ_r and relative magnetic permeability μ_r as compared to the corresponding constants of vacuum, the phase velocity of electromagnetic fields is $c/\sqrt{\epsilon_r \mu_r}$. For example, in plastics, the phase velocity of fields is of the order of $c_p \simeq 2.5 \times 10^8$ m/sec. The propagation times of signals in typical digital systems are thus of the order of 10^{-10} to 10^{-9} sec, and lumped-parameter models for circuits are applicable if switching times are, say, longer than of the order of 10^{-8} to 10^{-7} sec. With contemporary logic elements, quite easily achievable switching times are of the order of 10^{-9} to 10^{-8} sec, and this is why we have to consider signal transmission phenomena according to *distributed parameter models*. Not so seldom, digital signals are also transmitted on lines over large distances, ranging up to miles, without intermediate amplification.

In practice, circuits located in different parts of the system can never have a common reference potential, termed a *ground*. Although we could consider a metal plate or a metalized sheet extending over some square feet as a reasonably good signal ground over this area, we must recall that between every two points of any conductor there is a certain amount of inductance, and every geometrical body has a certain capacitance with respect to its surroundings, both of which will manifest themselves when signals are fast enough.

Phenomena that we do not meet with lumped-parameter models of circuits are the *reflections* of signals from receiving ends of transmission lines; this effect is similar to the reflection of electromagnetic waves at surfaces between two different media. There will be, in general, an infinite series of reflections accompanying every signal in the lines unless they are correctly terminated.

The purpose of the discussion of this chapter is to show in which way the amplitudes of reflections depend on line parameters and how optimal attenuation of reflections can be implemented, in order that they should decay off as soon as possible.

10.2. Transmission Lines

Telegraph Equations. Transmission lines are the simplest form of *distributed parameter systems.* They are usually described with the following simplified model, in which ohmic phenomena are not taken into account. Every conductor has a certain amount of inductance along its length arising from the interaction of current elements through magnetic fields, and we shall present this inductance as a series connection of similar differential inductances δL. But every piece of conductor has also a certain capacitance with respect to its surroundings, and in the suggested model this capacitance is represented by differential capacitances δC shunting the inductances to a common line (i.e., earth or equivalent piece of conductor having a practically zero impedance between any two separate points). Besides this asymmetrical model represented in Fig. 10-1, we might deal with a symmetrical model which describes the behavior of a line formed of two symmetrical conductors. This model is shown in Fig. 10-2. Only the former case is discussed, since the symmetrical model can be easily reduced to the asymmetrical one by a transformation.

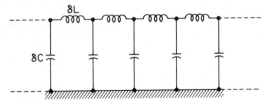

Fig. 10-1. Asymmetrical delay-line model.

In the infinite network shown in Fig. 10-3, we have the following relations for the voltages and currents at two neighboring nodes located around the place x:

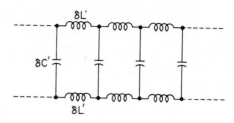

Fig. 10-2. Symmetrical delay-line model.

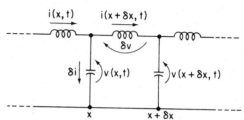

Fig. 10-3. Definition of voltage and current differentials.

$$\delta i = i(x + \delta x, t) - i(x, t) = \frac{\partial i(x, t)}{\partial x} \delta x = -(C_0 \, \delta x)\frac{\partial v(x, t)}{\partial t} \qquad [10\text{-}1]$$

$$\delta v = v(x + \delta x, t) - v(x, t) = \frac{\partial v(x, t)}{\partial x} \delta x = -(L_0 \, \delta x)\frac{\partial i(x, t)}{\partial t} \qquad [10\text{-}2]$$

where C_0 and L_0 are the capacitance and the inductance per unit length of line, respectively.

Dividing by δx we get the partial differential equations

$$\frac{\partial i}{\partial x} = -C_0 \frac{\partial v}{\partial t} \qquad [10\text{-}3]$$

$$\frac{\partial v}{\partial x} = -L_0 \frac{\partial i}{\partial t} \qquad [10\text{-}4]$$

After operating both sides of the first equation by $-L_0(\partial/\partial t)$ and the latter by $\partial/\partial x$, and noticing that $\partial^2/\partial t \, \partial x = \partial^2/\partial x \, \partial t$, a comparison of the equations yields the first of the *telegraph equations*,

$$\frac{\partial^2 v}{\partial x^2} = L_0 C_0 \frac{\partial^2 v}{\partial t^2} \qquad [10\text{-}5]$$

Similar operations by $\partial/\partial x$ and $-C_0 \, \partial/\partial t$, respectively, give us the second telegraph equation,

$$\frac{\partial^2 i}{\partial x^2} = L_0 C_0 \frac{\partial^2 i}{\partial t^2} \qquad [10\text{-}6]$$

and the variables $v(x, t)$ and $i(x, t)$ have thus been separated. Equations [10-5] and [10-6] are *wave equations;* their solutions are one-dimensional waves traveling with the *phase velocity* $c_p = (L_0 C_0)^{-1/2}$, as will be shown: Denote

$$\boxed{L_0 C_0 = \frac{1}{c_p^2}} \qquad \text{[10-7]}$$

$$\frac{\partial^2 v}{\partial x^2} = \frac{1}{c_p^2} \frac{\partial^2 v}{\partial t^2} \qquad \text{[10-8]}$$

It can be shown by direct substitution that

$$v(x, t) = v_1\left(t - \frac{x}{c_p}\right) + v_2\left(t + \frac{x}{c_p}\right) \qquad \text{[10-9]}$$

is a solution of Eq. [10-5] or [10-8], $v_1(\cdot)$ and $v_2(\cdot)$ being arbitrary functions of their respective arguments, and thus independent of each other. Similarly,

$$i(x, t) = i_1\left(t - \frac{x}{c_p}\right) + i_2\left(t + \frac{x}{c_p}\right) \qquad \text{[10-10]}$$

is a solution of Eq. [10-6], $i_1(\cdot)$ and $i_2(\cdot)$ being arbitrary functions. However, $v_1, v_2, i_1,$ and i_2 cannot all be independent of each other. Substituting $i(x, t)$ from Eq. [10-10] and $v(x, t)$ from Eq. [10-9] into Eq. [10-3] (or Eq. [10-4]), and denoting

$$\xi = t - \frac{x}{c_p} \qquad \text{[10-11]}$$

$$\eta = t + \frac{x}{c_p} \qquad \text{[10-12]}$$

we obtain

$$\frac{di_1}{d\xi} \frac{\partial \xi}{\partial x} + \frac{di_2}{d\eta} \frac{\partial \eta}{\partial x} = -C_0 \left(\frac{\partial v_1}{\partial t} + \frac{\partial v_2}{\partial t}\right) \qquad \text{[10-13]}$$

Now, since

$$\frac{\partial i_1}{\partial t} = \frac{di_1}{d\xi} \frac{\partial \xi}{\partial t} = \frac{di_1}{d\xi} \qquad \text{[10-14]}$$

and

$$\frac{\partial i_2}{\partial t} = \frac{di_2}{d\eta} \frac{\partial \eta}{\partial t} = \frac{di_2}{d\eta} \qquad \text{[10-15]}$$

a substitution yields an equation

$$-\frac{1}{c_p}\frac{\partial i_1(\xi)}{\partial t} + \frac{1}{c_p}\frac{\partial i_2(\eta)}{\partial t} = -C_0\left[\frac{\partial v_1(\xi)}{\partial t} + \frac{\partial v_2(\eta)}{\partial t}\right] \qquad [10\text{-}16]$$

The reader should notice that ξ and η are independent variables, because the simultaneous choice of arbitrary x and t generally gives different values for them. Equation [10-16] can hold only if separately

$$\frac{\partial i_1}{\partial t} = c_p C_0 \frac{\partial v_1}{\partial t} = \frac{1}{Z_0}\frac{\partial v_1}{\partial t} \qquad [10\text{-}17]$$

and

$$\frac{\partial i_2}{\partial t} = -c_p C_0 \frac{\partial v_2}{\partial t} = -\frac{1}{Z_0}\frac{\partial v_2}{\partial t} \qquad [10\text{-}18]$$

Hereby we have also defined the parameter

$$\boxed{Z_0 = (c_p C_0)^{-1} = \sqrt{\frac{L_0}{C_0}}} \qquad [10\text{-}19]$$

which has the dimension of resistance and is called the *characteristic impedance of the line*, or simply *impedance*. If i and v are sinusoidal and can be expressed in symbolic (complex) form, we would have from Eqs. [10-17] and [10-18]

$$j\omega I_1(x, \omega) = \frac{1}{Z_0} j\omega V_1(x, \omega) \qquad [10\text{-}20]$$

$$j\omega I_2(x, \omega) = -\frac{1}{Z_0} j\omega V_2(x, \omega) \qquad [10\text{-}21]$$

and, for any ω,

$$\frac{V_1(x, \omega)}{I_1(x, \omega)} = \frac{V_2(x, \omega)}{-I_2(x, \omega)} = Z_0 \qquad [10\text{-}22]$$

so Z_0 is real and independent of ω. (The minus sign has been written to $-I_2$, because the value of reflected current has been defined positive when flowing to the right, but reflected current signals are propagated to the left.)

Note that Z_0 is the ratio of symbolic voltages and currents in the same wave, and similarly it is the ratio of time derivatives of voltages and currents in the forward or backward waves, respectively. Let us now integrate Eq. [10-17] with respect to time:

$$\int_0^t \frac{\partial i_1(x, t)}{\partial t} \, dt = \frac{1}{Z_0} \int_0^t \frac{\partial v_1(x, t)}{\partial t} \, dt \qquad [10\text{-}23]$$

$$i_1(x, t) - i_1(x, 0) = \frac{1}{Z_0}[v_1(x, t) - v_1(x, 0)] \qquad [10\text{-}24]$$

Similarly, we obtain, from Eq. [10-18],

$$i_2(x, t) - i_2(x, 0) = -\frac{1}{Z_0}[v_2(x, t) - v_2(x, 0)] \qquad [10\text{-}25]$$

Thus, if the line originally was voltageless and currentless,

$$v_1(x, 0) = Z_0 i_1(x, 0) \qquad [10\text{-}26]$$

and we can write

$$v_1(x, t) = Z_0 i_1(x, t) \qquad [10\text{-}27]$$

with analog expressions for v_2 and i_2. Equation [10-27] thus holds for *initially discharged lines, and for ac voltages and currents*.

Termination of Delay Lines. Reflections. At a certain place $x = x_0$ of an infinitely long line, let us observe an imaginary section shown in Fig. 10-4. Normally there is a connection between *a-b* and *a'-b'*. Let there be a

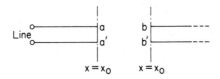

Line

x = x₀   x = x₀

Fig. 10-4. Imaginary section of a delay line.

primary voltage wave

$$v(x, t) = v_1\left(t - \frac{x}{c_p}\right) \qquad [10\text{-}28]$$

traveling in the line to the right, and no other waves present. If the line at $x = x_0$ before the incidence of the wave was voltageless and currentless, the current wave corresponding to $v(x, t)$ is, according to [10-27],

$$i(x, t) = Z_0^{-1} v(x, t) \qquad [10\text{-}29]$$

Now, if the portion of the line to the right of the section is removed and replaced by a resistor $R = Z_0$ between *a-a'*, a voltage $v(x_0, t)$ over this resistor would cause a current

$$i(x_0, t) = Z_0^{-1}v(x_0, t) \qquad [10\text{-}30]$$

in it. Comparing Eqs. [10-29] and [10-30], we will be assured that from the point of view of voltage and current observations performed to the left of $a\text{-}a'$, it is not possible to conclude whether there is a resistor equal to $R = Z_0$ or an infinitely long line connected to $a\text{-}a'$. This is called the *correct termination of the line*, and it should be noted especially that there is no reason at hand that would call for or allow other waves to be present.

Now, if the value of R were different from Z_0, we seem to have a contradiction. On one hand, a voltage $v(x_0, t)$ would cause a current

$$i(x_0, t) = R^{-1}v(x_0, t) \qquad [10\text{-}31]$$

in the resistor. But on the basis of line equations, on the other hand, we ought to have for the current wave

$$i(x_0, t) = Z_0^{-1}v(x_0, t) \qquad [10\text{-}32]$$

which is in contradiction with Eq. [10-31]. But we notice that Eq. [10-32] need not hold if we assume the simultaneous presence of forward and backward waves v_1 and v_2, respectively,

$$v(x, t) = v_1\left(t - \frac{x}{c_p}\right) + v_2\left(t + \frac{x}{c_p}\right) \qquad [10\text{-}33]$$

$$i(x, t) = \frac{1}{Z_0}v_1\left(t - \frac{x}{c_p}\right) - \frac{1}{Z_0}v_2\left(t + \frac{x}{c_p}\right) \qquad [10\text{-}34]$$

Thus $v(x, t)$ and $i(x, t)$ satisfy all line equations, and Eq. [10-31] can still be valid at $x = x_0$ at any time t. Stated mathematically, Eqs. [10-33] and [10-34] constitute the solutions of Eqs. [10-5] and [10-6] with imposed boundary conditions at $x = x_0$ and $x = -\infty$.

Now we solve Eqs. [10-33] and [10-34] for $x = x_0$, together with Eq. [10-31]. Thus we get the following for the ratio of the variables: From

$$v_1\left(t - \frac{x_0}{c_p}\right) + v_2\left(t + \frac{x_0}{c_p}\right) = \frac{R}{Z_0}\left[v_1\left(t - \frac{x_0}{c_p}\right) - v_2\left(t + \frac{x_0}{c_p}\right)\right] \qquad [10\text{-}35]$$

we have

$$\boxed{\frac{v_2(t + x_0/c_p)}{v_1(t - x_0/c_p)} = \frac{R/Z_0 - 1}{R/Z_0 + 1} = \rho} \qquad [10\text{-}36]$$

which is called the *reflection coefficient*. Thus

$$v(x_0, t) = v_1\left(t - \frac{x_0}{c_p}\right) + \rho v_1\left(t - \frac{x_0}{c_p}\right) \tag{10-37}$$

$$i(x_0, t) = \frac{1}{Z_0}\left[v_1\left(t - \frac{x_0}{c_p}\right) - \rho v_1\left(t - \frac{x_0}{c_p}\right)\right]$$

$$= i_1\left(t - \frac{x_0}{c_p}\right) - \rho i_1\left(t - \frac{x_0}{c_p}\right) \tag{10-38}$$

To have equations *for all x*, we have to make use of the fact that v_1 is a wave traveling forward ($+x$ direction) and v_2 a wave traveling backward ($-x$ direction) with the velocity c_p; similarly for i_1 and i_2. That is, we want $v(x, t)$ to be of the form

$$v(x, t) = v_1\left(t - \frac{x}{c_p}\right) + \rho v_1\left(t - D + \frac{x}{c_p}\right) \tag{10-39}$$

where the latter wave travels to the left and contains an undetermined parameter D. This is determined by requiring that Eqs. [10-37] and [10-39] are identical for $x = x_0$, which gives us the relation

$$-D + \frac{x_0}{c_p} = -\frac{x_0}{c_p} \tag{10-40}$$

or

$$D = \frac{2x_0}{c_p} = 2T \tag{10-41}$$

where T is a transit time from point $x = 0$ to $x = x_0$.

Example 10-1. The primary wave $v_1(t - x/c_p)$ is a short pulse launched at $x = 0$ at $t = 0$ (Fig. 10-5). (Note that v_1 is always nonzero when its argument

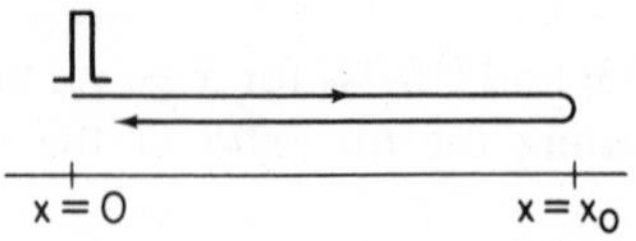

Fig. 10-5. Illustration for Example 10-1.

is zero.) At $t = T$, the pulse will reach an open-ended termination at $x = x_0$ and then the argument is also zero,

$$T - \frac{x_0}{c_p} = \frac{x_0}{c_p} - \frac{x_0}{c_p} = 0 \tag{10-42}$$

The reflection $v_2(t - 2T + x)$ will be back at $x = 0$ when its argument becomes zero at $x = 0$,

$$t - 2T + 0 = 0 \quad \text{or} \quad t = 2T \qquad [10\text{-}43]$$

The signal has thus traveled the way $0 \longrightarrow x_0 \longrightarrow 0$ in the time $2T$.

Multiple Reflections. In practice the length of any transmission line is finite. A usual arrangement is shown in Fig. 10-6, where the generator

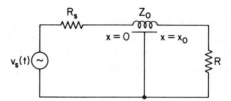

Fig. 10-6. Finite transmission line.

has an internal impedance R_s. The line is represented by its symbol. The left end is marked by $x = 0$ and the right end by $x = x_0$.

Along the length of the line, the solutions for v_1 and v_2 of Eq. [10-9] must be valid,

$$v(x, t) = v_1\left(t - \frac{x}{c_p}\right) + v_2\left(t + \frac{x}{c_p}\right) \qquad [10\text{-}9] \text{ or } [10\text{-}44]$$

In Eq. [10-44], v_1 is the complete wave traveling to the right and v_2 is the complete wave traveling to the left. However, in the first place there exists a primary wave $v_1^{(1)}(t - x/c_p)$ transmitted into the line at $x = 0$. This wave will first reach the termination and a fraction $\rho = (R/Z_0 - 1)/(R/Z_0 + 1)$ of it will be reflected back. The reflected wave is, in fact, a new primary wave traveling toward the left terminating resistor R_s. [According to the superposition theorem, which holds for delay lines also, we need not take the voltage generator into account a second time, since it already gave rise to $v_1^{(1)}(t - x/c_p)$.] Denote the left-end reflection coefficient by

$$\rho' = \frac{R_s/Z_0 - 1}{R_s/Z_0 + 1} \qquad [10\text{-}45]$$

whence a fraction ρ' of the backward wave will now be reflected forward. This, again, when reaching the termination will cause a new reflection, and so on *ad infinitum*.

Denote a one-way propagation delay in the line by T. Rearranging all the forward reflections and all the backward reflections into separate series, we obtain *for the complete forward wave*

$$v_1\left(t - \frac{x}{c_p}\right) = v_1^{(1)}\left(t - \frac{x}{c_p}\right) + \rho\rho' v_1^{(1)}\left(t - 2T - \frac{x}{c_p}\right) + \cdots$$

$$+ (\rho\rho')^k v_1^{(1)}\left(t - 2kT - \frac{x}{c_p}\right) + \cdots \qquad [10\text{-}46]$$

and *for the complete backward wave*

$$v_2\left(t + \frac{x}{c_p}\right) = \rho v_1^{(1)}\left(t - 2T + \frac{x}{c_p}\right) + \rho(\rho\rho')v_1^{(1)}\left(t - 4T + \frac{x}{c_p}\right) + \cdots$$

$$+ \rho(\rho\rho')^{k-1}v_1^{(1)}\left(t - 2kT + \frac{x}{c_p}\right) + \cdots \qquad [10\text{-}47]$$

These are just the expressions which, when added, give us the total voltage at any place along the line and at any time. We have only to specify $v_1^{(1)}(t - x/c_p)$, the primary wave sent into the line. A relation between the input voltage and input current (for an initially discharged line) is

$$v_1^{(1)}\left(t - \frac{x}{c_p}\right) = Z_0 i^{(1)}\left(t - \frac{x}{c_p}\right) \qquad [10\text{-}48]$$

where $i^{(1)}(t - x/c_p)$ is the primary forward current wave.

In other words, as seen from the generator end of the line, the line appears as an impedance Z_0 in the circuit. (This is no longer true when reflections are added to the primary wave.) We can thus draw *the sending-end equivalent circuit for the primary wave*, as in Fig. 10-7, so

$$v_1^{(1)}(t) = \frac{Z_0}{R_s + Z_0}v_s(t) \qquad [10\text{-}49]$$

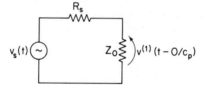

Fig. 10-7 Sending-end equivalent circuit.

We might also be interested in obtaining a lumped-parameter equivalent circuit for the receiving end. (It must be emphasized that for traveling signals there are no equivalent circuits but the delay-line model, consisting of a very large number of sections. The simpler models are valid only for primary waves at a certain place.) Compare Figs. 10-8(a) and 10-8(b). If the primary wave in Fig. 10-8(a) is $v_1^{(1)}(t - x/c_p)$, it will cause a voltage over

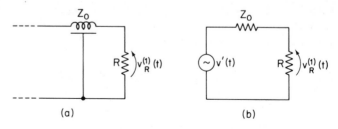

(a)   (b)

Fig. 10-8. Derivation of the receiving-end equivalent circuit.

R, which is

$$v_R^{(1)}(t) = (1 + \rho)v_1^{(1)}(t - T) = \frac{2R}{Z_0 + R}v_1^{(1)}(t - T) \qquad [10\text{-}50]$$

On the other side, $v_R^{(1)}$, according to Fig. 10-8(a), is

$$v_R^{(1)}(t) = \frac{R}{Z_0 + R}v'(t) \qquad [10\text{-}51]$$

Thus Fig. 10-8(b) is the equivalent of Fig. 10-8(a), if we set

$$v'(t) = 2v_1^{(1)}(t - T) \qquad [10\text{-}52]$$

The receiving-end equivalent circuit for the primary wave is now as shown in Fig. 10-9.

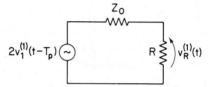

Fig. 10-9. Receiving-end equivalent circuit.

Example 10-2. Let us discuss the line shown in Fig. 10-10, with $v_s(t) = Vu(t)$, where V is constant and $u(t)$ is the unit step function. For the reflection coefficients we have $\rho = -\frac{1}{2}$ and $\rho' = +\frac{1}{2}$. Thus

$$v_1^{(1)}(t - T) = \frac{V}{4}u(t)$$

$$v_R^{(1)}(t) = \frac{1}{2}v_1^{(1)}(t - T) = \frac{V}{8}u(t - T)$$

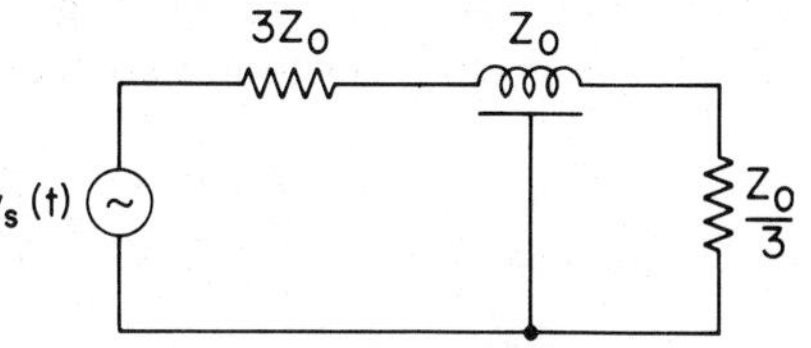

Fig. 10-10. Illustration for Example 10-2.

Voltages at the sending end, $x = 0$, are, from Eqs. [10-46] and [10-47] (see Fig. 10-11),

$$v_1(t - 0) = \frac{V}{4}\left[u(t) - \frac{1}{4}u(t - 2T) + \frac{1}{16}u(t - 4T) - \cdots\right]$$

$$v_2(t + 0) = \frac{V}{4}\left[-\frac{1}{2}u(t - 2T) + \frac{1}{8}u(t - 4T) - \cdots\right] \qquad [10\text{-}53]$$

$$v(0, t) = v_1(t - 0) + v_2(t + 0)$$

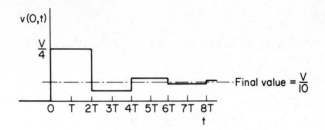

Fig. 10-11. Waveform of $v(0, t)$ for Example 10-2.

Voltages at the receiving end, $x = L$, are

$$v_1(t - T) = \frac{V}{4}\left[u(t - T) - \frac{1}{4}u(t - 3T) + \cdots\right]$$

$$v_2(t - T) = \frac{V}{4}\left[-\frac{1}{2}u(t - T) + \frac{1}{8}u(t - 3T) - \cdots\right]$$

[10-54]

We can see that the sum of these is similar to $v(0, t)$ but delayed with respect to it by an amount T.

From Eqs. [10-46] and [10-47] we see how to avoid multiple reflections; either ρ or ρ' must be zero. Thus it is usually sufficient to have either R_s or R equal to Z_0.

Reactive Terminations. For the sake of brevity we discuss only the cases in which the termination is a pure reactance (i.e., capacitance or inductance). Generalization is a straightforward task.

There are no restrictions in the foregoing derivations, which would prevent the termination R from being a general impedance as long as the line was initially discharged. Denoting a general termination by a Laplace operator $Z(s)$, the general reflection coefficient becomes

$$\rho(s) = \frac{V_2(x_0, s)}{V_1(x_0, s)} = \frac{Z(s) - Z_0}{Z(s) + Z_0}$$

[10-55]

where $V_1(x_0, s)$ and $V_2(x_0, s)$ are the Laplace transforms of the forward and backward voltages, respectively, at $x = x_0$ (termination).

REFLECTION OF A STEP VOLTAGE FROM A CAPACITIVE TERMINATION

Denote

$$Z(s) = \frac{1}{sC}$$

[10-56]

$$\rho(s) = \frac{1 - Z_0 Cs}{1 + Z_0 Cs} = \frac{1 - \tau s}{1 + \tau s}$$

$$= \frac{1}{1 + \tau s} - \frac{\tau s}{1 + \tau s} \qquad \text{[10-57]}$$

where $\tau = Z_0 C$. The first term of the last expression in Eq. [10-57] is the transfer function of a first-order low-pass circuit and the second term is the negative of the transfer function of a first-order high-pass circuit.

When the input is a step function, which for simplicity is assumed to arrive at $x = x_0$ at time $t = 0$, we have

$$v_1(x_0, t) = Eu(t)$$

$$V_1(x_0, s) = \frac{E}{s} \qquad \text{[10-58]}$$

We know from the foregoing the form of $v_2(x_0, t)$; see Fig. 10-12. The first term of Eq. [10-57], when multiplied by $V_1(x_0, s)$, corresponds to a time function represented by the uppermost curve in Fig. 10-12. The second term

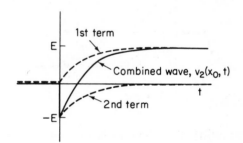

Fig. 10-12. Response of a transmission line with capacitive termination for a step voltage.

corresponds to the lowermost curve in Fig. 10-12. The sum of these partial waves is the total voltage at the termination, which is represented by the solid curve in Fig. 10-12.

When the reflection $v_2(x_0, t)$ is superimposed onto $v_1(x_0, t)$, we obtain the total voltage at C, as in Fig. 10-13.

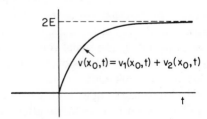

Fig. 10-13. Resultant waveform at a capacitive termination.

REFLECTION OF A STEP VOLTAGE FROM THE INDUCTIVE TERMINATION

Analogously with the previous case we have

$$Z(s) = sL \qquad [10\text{-}59]$$

$$\rho(s) = \frac{sL - Z_0}{sL + Z_0} = \frac{\tau s - 1}{1 + \tau s}$$

$$= \frac{\tau s}{1 + \tau s} - \frac{1}{1 + \tau s} \qquad [10\text{-}60]$$

where $\tau = L/Z_0$.

This is obviously the negative of the transfer function for the capacitive case, as long as the time constants are equal (i.e., $Z_0 C = L/Z_0$). In Fig. 10-14, the reflection $v_2(x_0, t)$, as well as its superposition with $v_1(x_0, t)$, are shown.

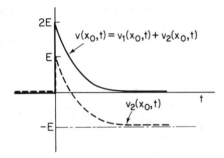

Fig. 10-14. Waveforms for an inductive termination.

10.3. Reflections from Nonlinear Terminations

By their nature, electronic logic circuits are devices with highly nonlinear terminal voltage–current relations. The interconnecting wiring of such circuits should always be regarded as a set of transmission lines with certain characteristic impedances. It is therefore to be expected that step voltages with short rise times, such as digital signals, immediately after switching do not achieve their final levels since the generator voltages are divided between the internal source impedances of the driving stages and the characteristic impedances of the lines, the latter usually being rather low, of the order of 50 to 100 Ω. The signal voltages therefore do not rise to the final level determined by dc conditions of the circuit until reflections from the receiving ends of the lines have arrived. Because of nonlinearities there is usually a severe mismatching at both ends of every line, and therefore the final voltage level is not achieved until after an appreciable number of back-

and-forth reflections. This effect does not play any role in very short intercon-nections, in which propagation delays are short compared to the rise times of signals. On the other hand, commercial logic circuits (e.g., TTL) with rise times as short as 5 nsec are in widespread use nowadays, and the back-and-forth propagation delay in a line should be shorter than this in order that the line could be classified as short. About 2 ft of line is the dividing line between short and long. Accordingly, reflection phenomena are not usually severe on printed circuit boards, where lines are short, whereas they may present a problem in long wired interconnections. For the transmission of signals over longer distances, linear circuits called line drivers and line receivers must be used to facilitate proper matching of line impedances.

Stray capacitances and inductances may also be associated with the terminals, and their effect ought to be considered separately. In what follows it is assumed that transmission lines are terminated with circuits that can be described by nonlinear static voltage–current relations.

Graphical Solution for Reflections in Transmission Lines with Non-linear Terminations. We shall restrict ourselves to the case in which primary signals are step functions, although the discussion can easily be generalized for other signals. Let us recall from the linear theory of transmission lines that an arbitrary selection for integration constants in Eqs. [10-9] and [10-10] was possible, and a constant voltage v_0 and current i_0 can therefore be applied on the line according to the dc conditions of the circuit. A discon-tinuous jump in a circuit voltage and current is a primary signal that travels in the line and is assumed to appear at the other end of the line a time T later. In the case of linear terminations, a certain relative fraction of this step is reflected back. In the nonlinear case, partial forward and backward waves must also be present, and they will also be shown to have a step-like form. Thus a complete wave in the transmission line is a staircase function. In the total voltage and current observed at the ends of the line, discontinuities occur at $t = kT$, where $k = 0, 1, 2, \ldots$. A graphical procedure will be presented for the solution for the relative amplitudes of partial waves. Since partial waves are not directly observable, expressions for the total voltage and current amplitudes are derived. These expression relate voltages (and currents) before and after every step, observed at both ends of the line.

The characteristic impedance of the line is Z_0 and the nonlinear volt-age–current relation at the termination is denoted by

$$v = f(i) \qquad\qquad [10\text{-}61]$$

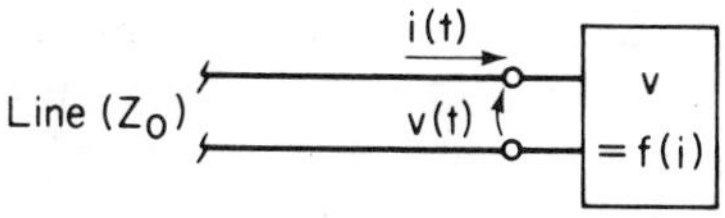

Fig. 10-15. Nonlinear termina-tion.

(see Fig. 10-15). In conformity with the linear case, let us denote all waves traveling to the right by a subscript 1 and those traveling to the left by 2. The total voltage at the termination is

$$v(t) = v_0 + v_1(t) + v_2(t) \tag{10-62}$$

where $v_1(t)$ is the forward wave and $v_2(t)$ is the reflection. The current in the termination is

$$i(t) = i_0 + i_1(t) + i_2(t) = i_0 + i_1(t) - \frac{v_2(t)}{Z_0} \tag{10-63}$$

where $i_1(t)$ is the forward current wave and $i_2(t)$ the reflection, and the last expression follows from Eq. [10-25]. A substitution for $v_2(t)$ from Eq. [10-62] and a rearrangement of terms yields

$$v(t) - v_0 - v_1(t) = -Z_0[i(t) - i_0 - i_1(t)] \tag{10-64}$$

In i–v coordinates, this is an equation for a straight line that passes through the point $(i_0 + i_1, v_0 + v_1)$ and has a slope $-Z_0$. Denoting the voltage and current values of the total incident wave by

$$v_i(t) = v_0 + v_1(t)$$

$$i_i(t) = i_0 + i_1(t)$$

we have a pair of equations

$$v(t) = f[(i(t)]$$
$$v(t) - v_i(t) = -Z_0[i(t) - i_i(t)] \tag{10-65}$$

in which time t occurs as a parameter and from which $v(t)$ and $i(t)$ can be solved as a function of $v_i(t)$ and $i_i(t)$ using graphical methods.

For secondary reflections from the driving end, the previously obtained $v(t)$ and $i(t)$ appear in the role of a new incident total voltage and current, respectively, whereafter the new total voltage and current are obtained from Eq. [10-65], and so on. In the following we consider a case in which a line is driven by an output circuit and terminated by a gate input.

Since the output circuits of logic gates usually act as current sinks for the inputs of logic gates, and the input as well as output voltage–current relations are usually also given in terms of the sink current, we shall define the positive directions of voltages and currents as indicated in Fig. 10-16 (deleting the time variable). For brevity of notation, if there is no danger of confusion, we shall identify the voltage and current variables by a letter A

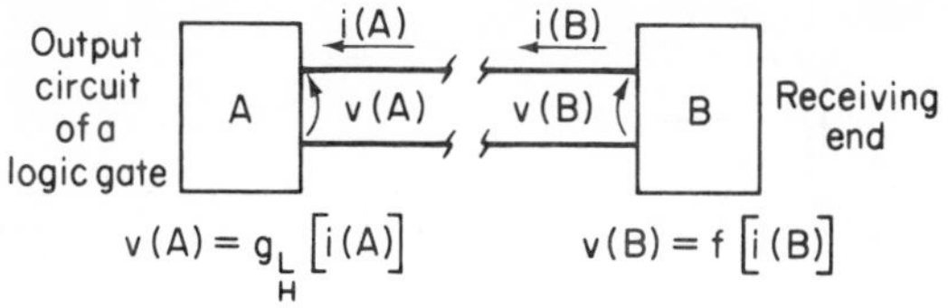

Fig. 10-16. Definition of currents and voltages for the graphical solution.

at the sending end and by B at the receiving end, respectively, as shown in Fig. 10-16. The output voltage–current relations are thus denoted

$$v(A) = g_{\substack{L \\ H}}[i(A)] \qquad [10\text{-}66]$$

where the double subscripts L and H stand for the LOW and the HIGH state, respectively. As before, the receiving end is described by a unique relation

$$v(B) = f[i(B)] \qquad [10\text{-}67]$$

The steady-state values of voltage and current v_0 and i_0, respectively, are obtained by neglecting the line, whence we obviously must have

$$v_0 = g_{\substack{L \\ H}}(i_0) = f(i_0) \qquad [10\text{-}68]$$

The switching between the states LOW and HIGH occurs in the i–v space as depicted in Figs. 10-17 through 10-19. Taking into account the new conventions for the directions of voltages and currents as illustrated in Fig. 10-16, the total voltage and current amplitudes, $v(A)$ and $i(A)$, in the line immediately after switching are obtained from Eqs. [10-69] and [10-70], which are the equivalent of Eqs. [10-24], [10-25], [10-31], and [10-65]:

$$
\begin{array}{|c|}
\hline
\\
v(A) = g_{\substack{L \\ H}}[i(A)] \\
\\
v(A) - v_i(A) = -Z_0[i(A) - i_i(A)] \\
\\
\hline
\end{array}
\qquad [10\text{-}69]
$$

$$
\begin{array}{|c|}
\hline
\\
v(B) = f[i(B)] \\
\\
v(B) - v_i(B) = Z_0[i(B) - i_i(B)] \\
\\
\hline
\end{array}
\qquad [10\text{-}70]
$$

Here the subscript i is used to denote total voltage and current values of incident waves (which, for the moment, are assumed to be known). The first pair of equations, Eqs. [10-69], defines a solution which constitutes

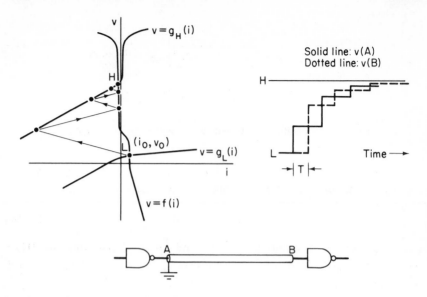

Fig. 10-17. Graphical solution for voltage waveforms at both ends of a transmission line, interconnecting two TTL gates. (Switching from LOW state to HIGH state.)

the new voltage and current amplitudes at the driving end. These values are determined graphically as the intersection of a straight line which goes through the point $(i_r(A), v_r(A))$ and has a slope $-Z_0$, and of the voltage–current graph of the output circuit. Equations [10-70] define the solution for the voltage and current amplitudes at the receiving end, and the new values for voltages and currents are obtained as an intersection of a straight line which goes through the point $(i_i(B), v_i(B))$, and of the characteristic voltage–current relation of the receiving end. Since new amplitudes obtained at one end of the line at the same time serve as amplitudes for a new wave which is incident at the opposite end, the graphical solution depicted in Figs. 10-17 and 10-18 is used to obtain the successive amplitudes of the staircase wave. Let us examine a case in which the output circuit is switching from the LOW state to the HIGH state. In the first place, i_0 and v_0 obtained from Eq. [10-68] are the "incident" variables, and Eq. [10-69] is used to determine the wave transmitted into the line. A straight line drawn with a slope $-Z_0$ intersects the HIGH-state voltage–current relation in Fig. 10-17 and gives the initial values for voltage and current. From this point, a line with a slope $+Z_0$ is drawn and the next intersection with the voltage–current relation of the receiving end gives the new voltage and current at this end. The drawing of lines is now continued, using the slopes $-Z_0$ and $+Z_0$ alternately. Thus a kind of "trajectory" is obtained which converges toward steady-state values.

A similar switching transient for the HIGH–LOW transition is analyzed in Fig. 10-18. Notice that the initial current drawn by the output circuit is very large, owing to the low dynamic impedance of the line. The negative step is reflected from the receiving end with the same polarity, and therefore the total voltage becomes negative for a time $2T$ at the receiving as well as at the driving end.

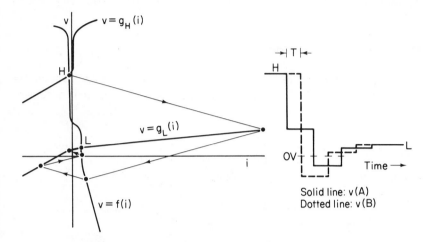

Fig. 10-18. Graphical solution for the switching from HIGH state to LOW state in a TTL system depicted in Fig. 10-17.

The occurrence of negative voltages in the TTL circuits is a short-coming that must be circumvented. The input clamping diodes are one way to alter the input voltage–current relation $v = f(i)$ in a suitable way which also improves the matching of the line. Another way is to utilize the

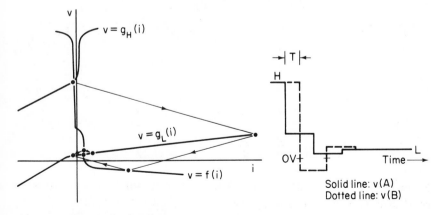

Fig. 10-19. Switching from HIGH state to LOW state in a TTL system with improved input characteristics.

collector-to-substrate *PN* junction of the multiemitter transistor which is always present in TTL circuits. Its properties can be controlled in the diffusion process in such a way that the input characteristics of the multiemitter transistor are altered approximately in the same way as with input clamping diodes. A new analysis of the HIGH–LOW switching transition with improved input characteristics exhibits much lower negative spikes (Fig. 10-19).

10.4. Driving Considerations

Capacitive Effect in Short Lines. Let us examine the circuit model of Fig. 10-20(a), in which a low-impedance transmission line is used to transmit

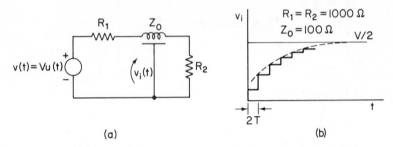

(a)   (b)

Fig. 10-20. (a) Circuit model for the discussion of the capacitive effect; (b) case example of $v_i(t)$.

a generator signal $v(t)$ to the load. Assume that the generator impedance R_1 and the load resistance R_2 are much larger than the characteristic impedance Z_0, and $v(t)$ is a step voltage equal to $Vu(t)$. According to Eqs. [10-46] and [10-47], the input wave $v_i(t)$ in the line is

$$v_i(t) = Vr[u(t) + \rho(\rho' + 1)u(t - 2T) + (\rho\rho')\rho(\rho' + 1)u(t - 4T)$$
$$+ (\rho\rho')^2\rho(\rho' + 1)u(t - 6T) + \cdots] \qquad [10\text{-}71]$$

where

$$r = \frac{Z_0}{R_1 + Z_0}$$

$$\rho = \frac{R_2 - Z_0}{R_2 + Z_0} \qquad \rho' = \frac{R_1 - Z_0}{R_1 + Z_0}$$

From the second term on in the brackets, the step amplitudes constitute a geometric series with a ratio $\rho\rho'$ between successive terms. Since $\rho\rho' < 1$, the series converges. A typical case with $Z_0 = 100 \ \Omega$ and $R_1 = R_2 = 1000 \ \Omega$ is depicted in Fig. 10-20(b).

   The staircase wave shows a salient resemblance to the step response of a first-order low-pass RC circuit; particularly, Fig. 10-21(a) shows the same circuit with a capacitor C in place of the transmission line, and Fig. 10-21(b) is its output waveform. The envelope of the staircase wave in Fig.

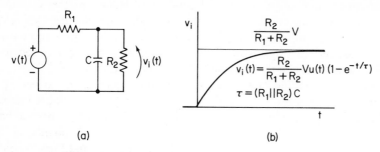

(a)   (b)

Fig. 10-21. (a) Lumped-constant approximation of Fig. 10-20(a); (b) waveform over C.

10-20(b) does not pass through the origin, but a comparison of the ratio of the successive steps (neglecting the first one), obtained from Eq. [10-71] with the time constant τ, yields the result (cf. Problem 10-6)

$$\tau = -\frac{2T}{\ln(\rho\rho')} \qquad [10\text{-}72]$$

Thus we may speak of an effective "time constant" of the transmission-line system, as well as the "capacitance" of the line. This capacitance depends on external components R_1 and R_2, and when $Z_0 \ll R_1, R_2$, it approaches $C_0 l$, where l is the length of the line. From Eq. [10-72] we also recognize that when ρ approaches zero (i.e., when the receiving-end load approaches the correct termination), C also approaches zero, and it is then no longer pertinent to speak of the capacitance of the line.

   Driving of Short Lines. In lines whose delays are short as compared to the rise and fall times of signals, multiple reflections are added to the original signals while their values are changing. Thus reflections are not recognized as separate, because the stairs will be smoothed out. If the driving and loading impedances are high, wired interconnections seem to appear as capacitances according to Eq. [10-72].

   Interconnections in digital systems are usually made by open wiring. A cylindrical wire with a diameter d over a ground plane at a distance h from it, when referred to the axis, has a characteristic impedance

$$Z_0 = \frac{1}{2\pi}\sqrt{\frac{\mu_0}{\epsilon_0}}\ln\frac{4h}{d} \simeq 60\ln\frac{4h}{d}\ \Omega \qquad [10\text{-}73]$$

The values obtained from Eq. [10-73] are usually rather small. For $4h/d = 5$, the impedance is 96.5 Ω, but in order to achieve an impedance of 1000 Ω, $4h/d$ ought to be about 1.8×10^7, which is certainly an inattainable value in practice. We may intuitively generalize this result by stating that practical wired connections in the vicinity of conducting plates or other conductors correspond to transmission lines with a low, although not necessarily constant, characteristic impedance. If reflections in such lines were to be totally eliminated, terminating resistors with low values ought to be used at all inputs. This is obviously impossible because of the decrease in signal amplification and the increase in power dissipation, and nonlinearities of the circuits. In most cases, the impedance of the receiving end should therefore be left as high as possible in order to provide a high fan-out. In normal backpanel wiring and with strip conductors on printed circuit boards, no precautions need usually be taken for correct termination, and there is seldom danger of reflections. However, because of the capacitive effect, circuits with a low output impedance ought to be used for the driving of short lines. Especially when the length of interconnections is in excess of 1 ft, output circuits with resistor pull-up ought to be avoided. Average output impedances of active pull-up bipolar output circuits are of the order of 100 Ω. If several parallel lines must be driven, it is often necessary to use *line drivers*, an example of which (Fairchild type 9621) is shown in Fig. 10-22.

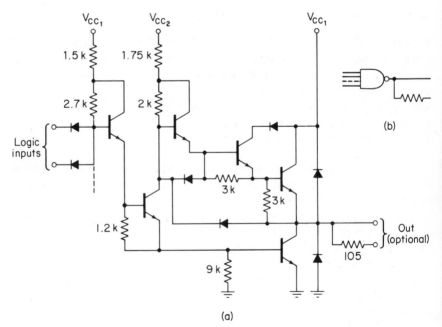

Fig. 10-22. Low-impedance line driver: (a) circuit; (b) symbol.

Transmission Problems in Lines of Medium Length. When propagation delays and switching times of signals are of equal order of magnitude, lines can no longer be regarded as capacitive loads, but termination requirements are still not very stringent. This is due to the fact that the dynamic noise margins of gates are usually much larger than the static ones. Also, in the case of medium-long lines, low output impedances of output stages may be helpful, because multiple reflections can be avoided if at least one end of the transmission line is correctly terminated. Output circuits of logic gates with active pull-up have an output impedance that in all states of circuits is about 100 Ω or less, as mentioned earlier. However, as indicated by the nonlinear analysis of Section 10.3, the voltage at the driving point in the LOW–HIGH switching does not immediately rise to the final level; there is an initial step in the logic voltage caused by the voltage divider effect, also called the *pedestal effect*. It does not usually play any role at the far end of the line, because the incident wave and the reflection add up, as seen from Fig. 10-17. However, since the voltage level of the pedestal at the near end may fall into the forbidden gap of logic voltages, *it is not advisable to connect any logic gates directly to the output as long as the output simultaneously drives a long or medium-long line.*

If all loads cannot be located to the end of the line, it is safest to use special low-impedance line drivers to avoid pedestal effects. Further, correct termination at the receiving ends ought to be made. Figure 10-22 gives an example of common (single-ended) line drivers.

Digital signals can also be transmitted over medium to long distances through coaxial cables. The grounding principles in digital systems are not quite the same as in the transmission of analog signal voltages, because the problem is to avoid disturbances caused by driving currents. Care should be taken in designing the current return path at the driving end in order that it form a closed loop through the line driver. Thus a driving logic gate or a line driver circuit ought to be grounded directly at the outer conductor of the cable, to which the supply line is bypassed by a capacitor also. This arrangement (shown in Fig. 10-23) guarantees that the drain current does

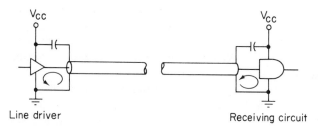

Line driver   Receiving circuit

Fig. 10-23. Elimination of disturbances from ground and supply lines. Current paths are indicated by arrows.

not affect other circuits. At the receiving end, a similar grounding may be made. There is no harm of ground voltage interference between the driving and receiving ends, because a transmitted signal acts between the inner and the outer conductor of the cable, and the current loop at the receiving end forms a closed local loop.

Attenuation and Deformation of Digital Signals in Long Lines. For the transmission of digital signals over long distances, either coaxial cables or symmetrical transmission lines can be used. We have neglected resistive effects in the previous models of all transmission lines. However, a certain series resistance per unit length ought to be added to the model, and a parallel leakage conductance per unit length should also be incorporated. An analysis which is very similar to the analysis with no losses yields the result that the amplitudes of waves propagated in the line are attenuated according to a law

$$V(x) = V(x_0)e^{-\beta(x-x_0)} \qquad [10\text{-}74]$$

where β is called the *linear attenuation coefficient*. If losses other than those mentioned before (e.g., eddy currents and dielectric losses) are also present, the linear attenuation coefficient is a function of frequency, and thus the transfer function of a line depends on its length. Attenuation and deformation of fast signals are usually obtained from empirical relations. In high-quality lines several hundreds of feet long, the bandwidth may be of the order of 100 MHz. If the voltage levels of received signals do not meet standards for logic signals, the signals must be amplified by *line receivers*.

Common-Mode Interference and Balanced Lines. Noise voltages induced between signal lines and the ground potential, or between two separate grounding points, are called *normal-mode interference*. It is obvious that the normal-mode interferences in two closely parallel conductors are equal in amplitude and phase. The difference between such noise voltages is zero, and therefore balanced lines are preferred in the transmission of signals over long distances or in places where the interference level is high. To be completely balanced, a symmetrical line must have equal terminating impedances with respect to ground at both conductors. Twisted pairs of wires are commonly used for the transmission of signals over long distances (up to about 1 mile). Well-balanced telephone lines are examples of such lines used for digital communication.

The feeding of signals into the line must be made by a symmetrical output circuit. Dual line drivers of which each half is a single-ended driver are available as integrated circuit packages. The signal and noise voltages occurring in the lines are described by Fig. 10-24, in which noise voltage generators are indicated by lumped elements. We have, for an ideal differen-

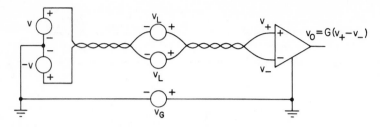

Fig. 10-24. Noise model for a balanced line.

tial amplifier,

$$v_+ = v + v_L - v_G$$

$$v_- = -v + v_L - v_G \qquad\qquad [10\text{-}75]$$

$$v_0 = G(v_+ - v_-) = 2v$$

Thus the line and ground noise are eliminated. The voltage

$$v_{\text{CM}} = \frac{v_+ + v_-}{2} = v_L - v_G \qquad\qquad [10\text{-}76]$$

is called the *common-mode interference.*

With practical differential amplifiers, v_{CM} has a slight effect on v_0. Let us call this signal $v_{0\text{CM}}$. The figure

$$\frac{\text{CMR}}{\text{dB}} = 20 \log_{10} \frac{Gv_{\text{CM}}}{v_{0\text{CM}}} \qquad\qquad [10\text{-}77]$$

is called the *common-mode rejection ratio.* For integrated operational amplifiers, CMR is usually of the order of 100 to 120 dB. For special differential amplifiers used as line receivers, this figure is smaller. Allowable common-mode voltages usually range from volts to some tens of volts.

Drivers and Receivers for Long Transmission Lines. Figure 10-25 shows an arrangement for a long transmission line. Current generators at the driving end are used because an accurate matching of line impedances is necessary. This is made by passive components. For the same reason, the line receivers ought to have a high input impedance. Sometimes several signal-transmission systems use the same transmission line (e.g., telephone lines). A *party-line* arrangement is shown in Fig. 10-26.

If the transmission line is terminated at both ends and the joining circuits have high generator and input impedances, the driving and receiving circuits may be hooked up at arbitrary points on the line. Of course, only

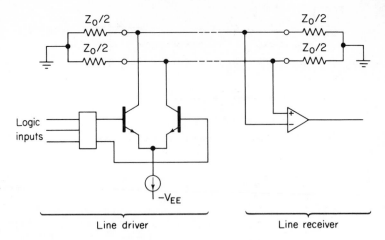

Fig. 10-25. Principle of line driver and receiver for a long transmission line.

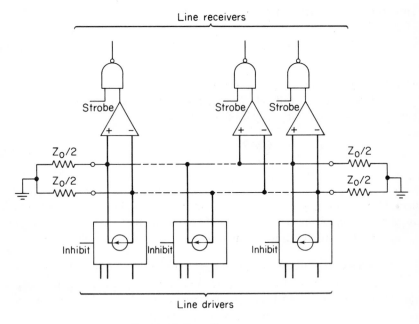

Fig. 10-26. Party-line arrangement.

one party at a time may use the line and therefore the Inhibit and Strobe inputs enable only one driver–receiver pair at a time. This is called *time-division multiplexing*. The multiplexing may be periodic or made on request. Time-division multiplexing is usually more economical than the hiring of separate lines at a distance longer than about 1 mile.

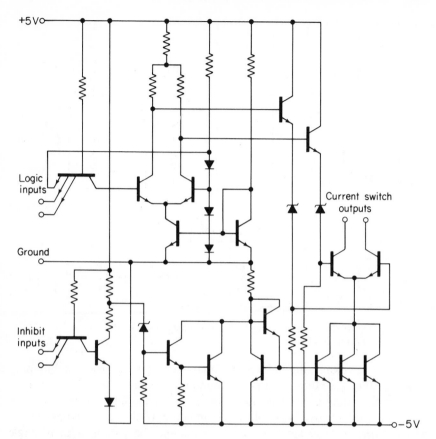

Fig. 10-27. Line driver (Texas Instruments type 75109) for long transmission lines.

Typical circuits for a line driver and line receiver are shown in Figs. 10-27 and 10-28. Digital line receivers must have a much higher switching speed than what can be achieved with conventional operational amplifiers. This can be achieved at the cost of input and output impedances and stability, using fewer cascaded stages.

10.5. Generation of Noise in Signal Lines

There are numerous kinds of noise in large digital systems: external disturbances and noise generated by the system itself. If large enough spurious signals are picked up by the signal lines and added up, they might affect the operation of the system. Noise problems are most difficult with fast circuits, especially with TTL. It is important to know the mechanism of noise generation to take the proper precautions to avoid them.

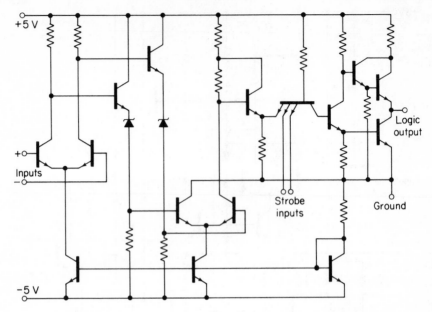

Fig. 10-28. Line receiver (Texas Instruments type 75107) for long transmission lines.

Noise voltages are induced in signal lines by three basic coupling phenomena; through voltage losses in common impedances, and by capacitive or inductive coupling. *Voltage losses in common impedances* of two or more circuits are usually due to coupling in common portions of ground or voltage supply lines. In large systems, all circuits cannot be grounded to a single point, and there is always a certain amount of inductance and resistance between two separate points of every conductor. Figure 10-29 shows how a poor ground return path, which is mainly inductive, may cause a spurious signal. If gate G_1 is of active pull-up type, and if it is loaded by a low-impedance transmission line or otherwise has a significant load, or if the output

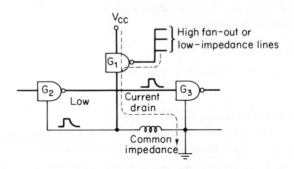

Fig. 10-29. Explanation of the coupling of noise by a common impedance [237].

draws a surge current during the switching transient, a substantial return current is transmitted into the common ground line of G_1 and G_2. A voltage loss is generated in the impedance of the current return path. If G_2 was in the LOW state, the impedance between its output and the ground is low; in a third gate G_3, which is grounded to a separate point, the noise voltage is seen at its input as a signal spike possibly exceeding the LOW noise margin.

Example 10-3. It is possible that G_1 in Fig. 10-29 draws 50 mA during a short period. An inductance as large as 0.1 μH in the ground current return path is not uncommon. (In fact, inductances in long lines may be much larger.) If the output circuit is capable of switching the said current in 5 nsec and a linear rate of rise of current is assumed, the induced noise voltage amplitude is (0.1 μH $\times$ 50 mA)/5 nsec = 1 V, which already may exceed the allowed noise margin.

Small ground line impedances can be achieved by using ground planes (e.g., metal foils on multilayered printed circuit boards).

Another common impedance effect comes from abrupt current drain in supply lines. If output circuits are at their HIGH states, whereby the impedances from the supply line to the outputs are low, noise voltage spikes occurring in the positive supply lines are easily coupled to driven gate inputs. The most usual method for the avoiding of disturbances due to supply-line voltage losses is the bypassing of supply lines to ground at all places where several circuits or high current drains are present. The bypassing capacitors may be electrolytic ones, but in the fastest circuits (e.g., with TTL) electrolytic capacitors may exhibit inductive effects. Thus, for example, ceramic capacitors of a few nF are recommended for bypassing purposes and such capacitors should be placed in the vicinity of every gate.

Overshoot effects due to bypass capacitors: Voltage spikes due to abrupt changes in current drain remain small if supply lines are bypassed by capacitors located with sufficiently short spacings. Let us examine what happens if this is not the case, and circuits are fed through inductive supply lines. According to Fig. 10-30, assume that a supply line with an inductance L is bypassed by a capacitor C and an initial current drain is set by a switching

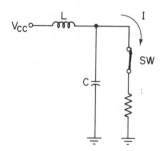

Fig. 10-30. Analytical model for overshoot in bypass circuit.

circuit. When the switch SW is opened, we have a series resonance circuit (with no damping in the model). The energy stored in L due to the initial current I will be added up to the previous energy of C, and the maximum amplitude V of voltage swing will be obtained from Eq. [10-78] when the current of L has become zero:

$$\frac{CV^2}{2} = \frac{CV_{cc}^2}{2} + \frac{LI^2}{2} \qquad [10\text{-}78]$$

$$V = \sqrt{V_{cc}^2 + \frac{L}{C}I^2} \qquad [10\text{-}79]$$

A ringing voltage with an amplitude $V - V_{cc}$ is thus superimposed on V_{cc} at the feeding point and may cause harm to the circuits. If L is large, C must be selected sufficiently large.

Capacitive pickup of noise is significant if the coupling capacitance is of the same order of magnitude as the capacitance of the disturbed circuit with respect to ground and if the resistive part of the impedance at that point is high. Figure 10-31 shows a simplified equivalent circuit in which the noise

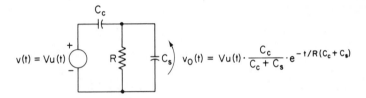

Fig. 10-31. Equivalent circuit for capacitive noise pickup.

voltage is a step $v(t) = Vu(t)$. Analysis shows that the maximum amplitude of noise voltage is attenuated by the capacitive voltage divider, and the noise voltage decays off with a time constant that is R times the sum of all capacitances. Notice that R is the impedance of a signal line with respect to any fixed potential. If active pull-up output circuits are used, this impedance is about 150 Ω in the HIGH state, about 10 Ω in the LOW state, and about 100 Ω during switching. In semiconductor circuits, capacitive pickup of noise usually has to be considered only in the case of long parallel lines. The capacitance per unit length of two parallel cylindrical conductors with a diameter d and a mutual distance D between their axes is

$$C_0 = \frac{\pi\epsilon_0}{\ln(2D/d)} \qquad [10\text{-}80]$$

With $D/d = 5$, C_0 is about 12 pF/m, or about 3.6 pF/ft.

Inductive coupling of noise is most severe in low-impedance circuits, in which small voltages may cause considerable currents. Inductive effects due to the circuits themselves are usually not able to generate high noise voltages, except possibly for noise caused by electromagnetic devices such as solenoids of output devices. However, external disturbances from adjacent power lines or radiofrequency interference in the vicinity of sources of electromagnetic radiation may be significant. Notice, however, that in memory systems we are usually dealing with weak signals and low-impedance circuits, and it is these circuits which are most sensitive to inductive coupling of noise. The effects of inductive coupling can be reduced by using coaxial cables or twisted pairs of signal wires with a short pitch; or by selecting the distance between conductors and the radius of conductors small to reduce the effective "aerial" which is receiving magnetic fields.

Noise Generated by Transmission-Line Reflections. As discussed in Section 10.2, reflections at nonlinear terminations may cause overshoots and ringing in the lines. Digital signals must usually be distributed to several points, and a correct termination of lines is seldom possible. Open wires correspond to transmission lines with rather high impedances, but still not higher than a couple hundreds of ohms. With high-level output circuits, the lines may be approximately matched at the driving ends, which eliminates multiple reflections. However, there is a possibility that reflections from different portions of lines may be piled up.

In principle, either radial or point-to-point wiring, or any combination of them, may be used in the interconnections (Fig. 10-32).

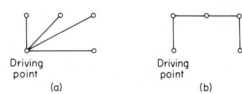

Driving point (a)   Driving point (b)

Fig. 10-32. Types of interconnections: (a) radial; (b) point to point.

Radial lines have the disadvantage that the characteristic impedances of driven lines are coupled in parallel to the driving point, which causes a bad mismatching of impedances. In point-to-point wiring, only one line is loading the driving end, which provides a better impedance matching to the driving stage than the radial interconnections. On the other hand, since most input circuits of logic gates exhibit capacitive effects, a point-to-point wiring with short interconnections may be approximated by a transmission line in which the stray capacitance is added up to the capacitance per unit length of the line. If the average capacitance per input is C' and the average spacing of loads is d, the capacitance may be distributed over the length of the line as shown in Fig. 10-33. The model of Fig. 10-33 is equivalent with a transmission line with a characteristic impedance

$$Z_0' = \sqrt{\frac{L_0}{C_0 + C'/d}} \qquad [10\text{-}81]$$

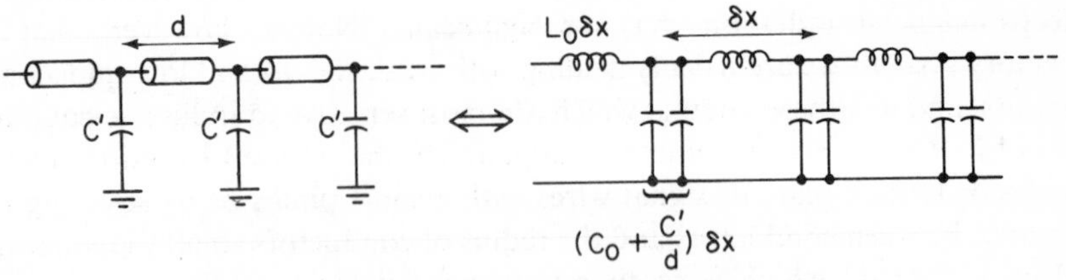

Fig. 10-33. Transmission-line model of a distributed capacitive load.

and phase velocity

$$c_p' = \sqrt{L_0 \left(C_0 + \frac{C'}{d} \right)} \qquad [10\text{-}82]$$

It is seen that although C' is fixed by the gate type, the characteristic impedance can be effected by the selection of the spacing d.

Very often several loading points must be driven by a long initial section of the line. We discuss two such typical cases in more detail. Practical situations are combinations of these.

CLUSTERED LOAD AT THE END OF A LINE

If the spacing between driven loads is short (much shorter than 1 ft, for example) it is immaterial whether these loads are radially or point-to-point connected. If the impedance of a single load is Z_L and this load may be a complex one, a load of n parallel circuits presents a termination Z_L/n to the line. The maximum value for n is determined by the maximum tolerable reflection, which is discussed in Section 10-2.

DISTRIBUTED LOAD AT THE END OF A LINE

In this case, only capacitive loading usually needs to be taken into account. If a point-to-point wiring with average spacing d and a loading capacitance C' at each joint is assumed, the distributed load is equivalent to a transmission line with a characteristic impedance Z_0' according to Eq. [10-81]. At the junction of a line with impedance Z_0 to a line with impedance Z_0', a reflection is defined by the coefficient

$$\rho = \frac{Z'_0 - Z}{Z'_0 + Z} \qquad [10\text{-}83]$$

The maximum tolerable ρ determines the minimum allowable Z'_0. *Since C' is fixed, Eqs.* [10-83] *and* [10-81] *define a limit for the minimum allowable spacing d.*

Unless at least one end of every transmission line can be correctly terminated, the reflection phenomena become very complicated, and for a large system, the maximum amplitudes of reflections should be checked by a computer.

10.6. Long-Distance Data Transmission

With increasing distance from the central processing unit to the input–output devices, wired interconnections become a decisive cost factor. Therefore, at a distance of 500 ft, but more so when signal-transmission paths exceed 1 mile, we must consider other transmission methods. Similar problems have been encountered in the transmission of voice and video signals, and standard telephone networks are extensively used for data transmission. Simultaneously transmitted messages are said to be *multiplexed*, or carried in separate *channels*. For human voice transmission, it is necessary to reproduce a frequency spectrum from about 300 to 3000 Hz, and for radio programs, 50 to 5000 Hz is considered to be sufficient. Using high-frequency carrier waves and signal modulation, a contemporary coaxial telephone cable may carry 11,000 telephone channels spaced about 4000 Hz from each other. Similar methods are used when the modulated carrier waves are transmitted over wireless microwave links.

What is the necessary bandwidth for the transmission of certain digital signals? This is one of the first questions of practical nature when planning transmission of signals to remote locations. The answer obviously depends on the degree of conformity by which digital signals must be reproduced to be recognizable, and first we have to decide *how many bits per second* we want to transmit in one channel. The cost of a channel is relatively small at short distances, but certainly we cannot afford one channel for each signal between interconnected devices. On the other hand, since the speed of operation of most remote terminal devices is relatively slow as compared with central units, parallel logic signals may be first converted into serial form and the bits are transmitted in succession. At the receiving end, serial signals are again transformed into parallel form. In this way, several signals are transmitted using only one pair of wires or an equivalent transmission path; this method is called *time-division multiplexing* (TDM).

Another practical question, closely related to the previous one, is whether binary signals ought to be transformed somehow before transmission because attenuation of signals depends on their frequency and there is certainly an optimal way to utilize channel properties for the least noise and distortion. The three types of modulation are amplitude modulation, frequency modulation, and phase modulation.

Bandwidth Limitation. A usual measure of the speed of pulsed signals is the *rise time t_r*, or the time for a step-like signal to rise from 10 to 90 percent of its final amplitude. The *fall time* is the time needed for a reverse excursion of the signal. There is a semiempirical relation between t_r and the bandwidth BW of the frequency spectrum of this signal: for most pulse signals occurring in practical switching circuits,

$$BW \simeq \frac{0.35}{t_r} \qquad [10\text{-}84]$$

For example, if the rise time of a signal is 1 μsec, the waveform, when decomposed into its frequency components by a Fourier transformation, is equivalent to a frequency spectrum with a bandwidth of 0.35 MHz. Using Eq. [10-84] in reverse order, we may deduce that the shortest rise time of pulsed signals which can be preserved in a telephone channel with BW = 3500 Hz is 100 μsec. Reliability (and possibly also crosstalk between channels due to high-frequency components) determines the ultimate speed of pulsed signals in voice channels.

The speed of data transmission is usually expressed in *bauds*. One baud means 1 bit per second. The lowest speeds used on public telephone lines for data transmission are 100 to 200 bauds, but telephone companies offer lines with typical speeds of 1200, 2400, and 4800 bauds, respectively. For special purposes, transmission links that provide speeds up to 500,000 bauds and even higher have been built. Notice that if necessary, modern transmission methods (e.g. using modulated laser rays) would facilitate a speed of some 10^9 bauds, if necessary.

Modulation. Existing telecommunication lines have been constructed and compensated (balanced) primarily for voice transmission. The use of acoustical carrier waves on these lines is therefore advantageous, although it would perhaps not be so if these lines had been designed for data transmission from the beginning. The attenuation and distortion on existing lines is minimal if the carrier frequency is centered around 1500 Hz, which lies in the middle of the spectrum of speech. *Amplitude modulation* means that a carrier with a constant amplitude is turned on for HIGH signals and interrupted for LOW signals [Fig. 10-34(a)]. Thus the average voltage level is not changed. Although the carrier is usually switched on and off when the phase angle of the carrier is zero, this type of interrupted signal is always

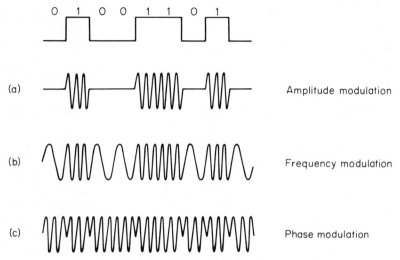

Fig. 10-34. Types of modulation.

distributed over a certain range of frequencies, as shown by an expansion of this signal in its frequency components. What is cumbersome, however, is that amplitude-modulated systems are very susceptible to surge interference (clicks) resulting from switching operations, atmospheric disturbances, etc. Such noise is easily detected as erroneous bits.

Frequency modulation means that the frequency of carrier has a constant value for all transmitted 0's, and is switched to another (usually somewhat higher) value during the 1's [Fig. 10-34(b)]. To avoid the generation of a broad frequency spectrum, the waveform together with its first derivatives must be continuous.

In *phase modulation*, the phase angle of the carrier is suddenly switched 180° out of phase for all 1's in the signal and returned to 0° for all 0's [Fig. 10-34(c)].

Frequency and phase modulation are far less susceptible to noise than amplitude modulation, and they are both in common use. Phase modulation, although a little more expensive than the others, is the best with regard to noise rejection and speed of operation.

Normal telephone channels are designed for *simplex, duplex* (or *full duplex*), and *half-duplex* operation. Simplex operation means that there is equipment for one-way transmission only. Duplex means that the same channel can be used for transmission in both directions simultaneously. Half-duplex means that there is equipment for two-way transmission on the same line but that transmission can occur in only one direction at a time. For most digital applications, the half-duplex mode of operation is what is used.

At both ends of the transmission line there is an electronic set called *modem* (*mo*dulator-*dem*odulator combination), which modulates binary signals for transmission and converts the received message back into binary signals at the receiving end. Sometimes such a system is called an *acoustical coupler*, if voice transmission is used, and it is very commonplace in modern time-sharing computer service where user terminals and large computers form a mutually interwoven data network. The user can dial any computer within the network and get into contact with it.

There are extra problems in addition to those mentioned before with practical data transmission. For example, every voice channel is also used for *signaling*, which is done on a narrow range of frequencies. The rest of the frequency band, usually from 1000 to 2000 Hz, is left for data transmission. Signaling operations are not supposed to interfere with the data channels, but faults occur in all transmission systems. For example, one bit in 10^7 transmitted may be incorrectly detected, for one reason or another. Therefore, error-correcting-and-detecting codes must be used. Messages may also be sent in duplicate and compared. At longer distances, a message is often sent back for comparison while the original information is still buffered at the transmission end. If the two messages do not coincide, new messages are requested and transmitted as many times as is needed to make certain that they have been received correctly. For this mode of operation, the duplex mode of communication is advantageous.

Serial Data Transmission. In principle, the data transmitter and receiver could be synchronized by signaling. In practice, a safer method is to give the start, stop, and synchronizing signals within the data. The information is usually transmitted in bursts of six data bits. A seventh bit follows for parity checking, and this binary word is called a *character*, according to a corresponding teletypewriter code.

On normal telephone lines, the signalless condition is indicated by a HIGH voltage classified as logical 1. When a character is transmitted, a *START* pulse is given by switching the line voltage LOW. After that, data bits are transmitted, followed by the parity bit. One character in which the data bits are labeled in a special way is shown in Fig. 10-35.

Usually there is only a single line between the transmitter and the receiver, and the time intervals between bits must then be known. The

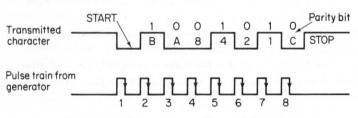

Fig. 10-35. Timing diagram of a serial character.

START signal triggers a pulse-train generator at the receiver, which produces eight pulses for the sampling of bits. Pulse-train generators are discussed in Chapter 5, and the shift register shown in Fig. 10-36 may be used for the buffering of information. Notice that the decoder is not activated until the *START* signal indicates that the whole word has been received.

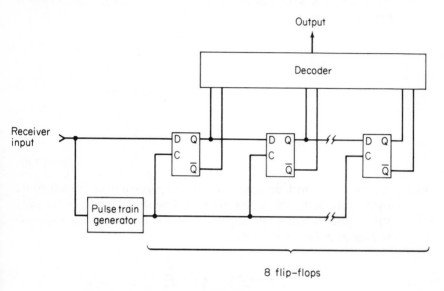

Fig. 10-36. Sequential decoder for serial data transmission.

In asynchronous data transmission, the serial pulse pattern may be initiated at an arbitrary time, but the timing of binary values with respect to the preceding *START* signal must be defined with an accuracy that depends on the number of clock phases transmitted in one row. At a high rate of data transmission it is often preferable to use continuously running synchronized clock oscillators at the transmitter and the receiver. At a pulse rate of 4800 baud, this means that in messages lasting a few seconds, about 10^4 bits are transmitted in a row. The clock oscillators may then be synchronized with each other by a particular binary signal pattern given at the beginning of the transmitted sequence. The same pattern may be used intermittently for resynchronizing, if messages are longer.

PROBLEMS

10-1   A 1-nsec, 1-A current pulse from a current generator (with a very high output impedance) is injected to a 2-m-long cable having a capacitance of 100 pF/m and an inductance of 0.25 μH/m. The cable is terminated by a 75-Ω resistor. Sketch the voltage across the termination as a function of time.

10-2   Figure P10-2 shows a circuit for the generation of short pulses. Find the voltage $v(t)$ when the current $i(t)$ is given by $i(t) = 100 \text{ mA} \cdot u(t)$ [where $u(t)$ is a unit step]. The characteristic impedance of the transmission line is 50 Ω and the inductance is 0.2 μH/m. The length of the line is 1 m.

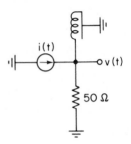

Figure P10-2

10-3   A current step with an amplitude I is propagated from the left to the right in the semi-infinite transmission line shown on the left in Fig. P10-3, and it meets the capacitor at $t = 0$. Derive the capacitor voltage as a function of time.

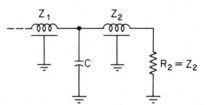

Figure P10-3

10-4   The input–output voltage relation of the inverters of Fig. P10-4 is represented by Fig. P6-4 (lower curve), and the output impedances are 150 Ω. The input impedances are very high. The signal propagation time in the line is 1 μsec and the characteristic impedance is 150 Ω.

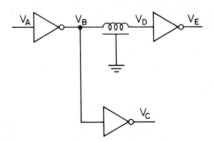

Figure P10-4

(a) Calculate and draw the voltages V_A, V_B, V_C, V_D, and V_E when V_A is suddenly changed from 2.4 to 0.4 V.

(b) Solve the same problem graphically.

10-5   Figure P10-5 shows the output voltage–current characteristics of a line driver for HIGH and LOW states, and the input voltage–current characteristics for a line receiver, respectively.

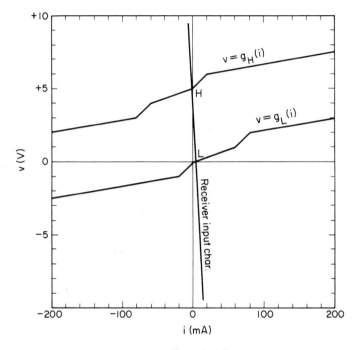

Figure P10-5

(a) Find graphically both transition trajectories for a driver–receiver pair, when the driver is connected to the receiver through a 50-Ω transmission line.

(b) Draw the receiver input voltage corresponding to part (a) when the signal propagation time in the transmission line is 50 nsec.

10-6   Derive Eq. [10-72].

10-7   An initially correctly terminated 50-Ω transmission line has a phase velocity $c_p = 2.5 \times 10^8$ m/sec. What is the maximum amount of evenly distributed capacitive load per unit length that can be added in order that reflections in the line would not exceed 10 percent of the signal?

REFERENCES

The topics of this chapter are discussed in textbooks on pulse techniques. A classical text is

> Millman and Taub, *Pulse, Digital, and Switching Waveforms* [95].

A good discussion on transmission lines with nonlinear terminations, as well as internally generated noise of digital systems, has been presented in

> Saenz and Fulcher, "An approach to logic circuit noise problems in computer design," *Computer Design*, **8**, 84–91 [229].

A few topics discussed in connection with long transmission lines have been described in

> Talley, "Monolithic interfacing in computers," *Texas Instruments Application Report CA* 122,

as well as

> Pippenger and Eljarrat, "Line drivers and receivers SN 55107 series," *Texas Instruments Application Report CA* 130.

Some aspects associated with long-distance data transmission are described in

> Martin, *Design of Real-time Computer Systems* [90].

Also the text

> Morris and Miller (eds.), *Designing with TTL Integrated Circuits* [237]

is a reference for topics discussed in this chapter, as well as in other chapters of this book.

appendix A

Standard Character Codes (ASCII)

For the transmission of information between various computers and peripheral devices, as well as for serial transmission of information over telephone and telegraph lines, a standard code has been adopted. It is called ASCII (American Standard Code for Information Interchange). This code consists of eight bits by which all characters of a standard teletypewriter keyboard, as well as auxiliary signaling marks, can be encoded. The following is a list of character codes of the ASCII code expressed in octal form. (To obtain an octal expression for an eight-bit code, add a 0 to the left end of the code.)

There are also other control signals, encoded in ASCII, not shown in this list.

Character	ASCII (Octal)	Character	ASCII (Octal)
(blank)	(0 0 0)	0	2 6 0
leader/trailer	2 0 0	1	2 6 1
line feed	2 1 2	2	2 6 2
carriage return	2 1 5	3	2 6 3
space	2 4 0	4	2 6 4
!	2 4 1	5	2 6 5
"	2 4 2	6	2 6 6
#	2 4 3	7	2 6 7
$	2 4 4	8	2 7 0
%	2 4 5	9	2 7 1
&	2 4 6	A	3 0 1
'	2 4 7	B	3 0 2
(	2 5 0	C	3 0 3
)	2 5 1	D	3 0 4
*	2 5 2	E	3 0 5
+	2 5 3	F	3 0 6
,	2 5 4	G	3 0 7
—	2 5 5	H	3 1 0
.	2 5 6	I	3 1 1
/	2 5 7	J	3 1 2
:	2 7 2	K	3 1 3
;	2 7 3	L	3 1 4
<	2 7 4	M	3 1 5
=	2 7 5	N	3 1 6
>	2 7 6	O	3 1 7
?	2 7 7	P	3 2 0
@	3 0 0	Q	3 2 1
[	3 3 3	R	3 2 2
\	3 3 4	S	3 2 3
]	3 3 5	T	3 2 4
↑	3 3 6	U	3 2 5
←	3 3 7	V	3 2 6
		W	3 2 7
		X	3 3 0
		Y	3 3 1
		Z	3 3 2

appendix b

Bibliography

Book References

[1] Adler, R. B., A. C. Smith, and R. I. Longini, *Introduction to Semiconductor Physics*. New York: McGraw-Hill Book Company, 1964.

[2] Alt, F. L., and M. Rubinoff, *Advances in Computers*, vols. 1–9. New York: Academic Press, Inc., 1966.

[3] Anner, G. E., *Elementary Nonlinear Electronic Circuits*. Englewood Cliffs, N.J.: Prentice-Hall, Inc., 1967.

[4] Babb, D. S., *Pulse Circuits: Switching and Shaping*. Englewood Cliffs, N.J.: Prentice-Hall, Inc., 1964.

[5] Baron, R. C., and A. T. Piccirilli, *Digital Logic and Computer Operations*. New York: McGraw-Hill Book Company, 1967.

[6] Bartee, T. C., *Digital Computer Fundamentals*, 2nd Ed. New York: McGraw-Hill Book Company, 1966.

[7] Bartee, T. C., I. L. Lebow, and I. S. Reed, *Theory and Design of Digital Machines*. New York: McGraw-Hill Book Company, 1962.

[8] Berlekamp, E. R., *Algebraic Coding Theory*. New York: McGraw-Hill Book Company, 1968.

[9] Boole, G., *An Investigation of the Laws of Thought*. New York: Dover Publications, Inc., 1954.

[10] Booth, A., and K. H. V. Booth, *Automatic Digital Calculations*. London: Butterworth Scientific Publications, 1953.

[11] Braun, E. L., *Digital Computer Design*. New York: Academic Press, Inc., 1963.

[12] Buchholtz, W., *Planning a Computer System*. New York: McGraw-Hill Book Company, 1962.

441

[13] Burger, R. M., and R. P. Dunoven, *Fundamentals of Silicon Integrated Device Technology*, vols. 1 and 2. Englewood Cliffs, N.J.: Prentice-Hall, Inc., 1967.

[14] Burrough's Corporation, *Digital Computer Principles*. New York: McGraw-Hill Book Company, 1961.

[15] Calahan, D. A., *Computer-aided Network Design*. New York: McGraw-Hill Book Company, 1968.

[16] Chirlian, P. M., *Integrated and Active Network Analysis and Synthesis*. Englewood Cliffs, N.J.: Prentice-Hall, Inc., 1967.

[17] Chu, Y., *Digital Computer Design Fundamentals*. New York: McGraw-Hill Book Company, 1962.

[18] ———, *Introduction to Computer Organization*. Englewood Cliffs, N.J.: Prentice-Hall, Inc., 1970.

[19] Chua, L. O., *Introduction to Nonlinear Network Theory*. New York: McGraw-Hill Book Company, 1969.

[20] Comer, D. T., *Large-signal Transistor Circuits*. Englewood Cliffs, N.J.: Prentice-Hall, Inc., 1967.

[21] Cowles, L. G., *Analysis and Design of Transistor Circuits*. New York: Van Nostrand Reinhold Company, 1966.

[22] Crawford, R. H., *MOSFET in Circuit Design*. Dallas, Tex.: Texas Instruments, Inc., 1967.

[23] Crowley, T. H. *Understanding Computers*. New York: McGraw-Hill Book Company, 1967.

[24] Curtis, H. A., *The Design of Switching Circuits*. New York, Van Nostrand Reinhold Company, 1962.

[25] Dakin, C. J., and C. E. G. Cooke, *Circuits for Digital Equipment*. London: Iliffe Books Ltd., 1967.

[26] Davies, D. W., *Digital Techniques*. London: Blackie & Son Limited, 1963.

[27] Davis, G. B., *An Introduction to Electronic Computers*. New York: McGraw-Hill Book Company, 1966.

[28] Delhom, L., *Design and Application of Transistor Switching Circuits*. New York: McGraw-Hill Book Company, 1968.

[29] Eadie, D., *Introduction to the Basic Computer*. Englewood Cliffs, N.J.: Prentice-Hall, Inc., 1968.

[30] Eimbinder, J. (ed.), *Designing with Linear Integrated Circuits*. New York: John Wiley & Sons, Inc., 1969.

[31] Engineering Research Associates, *High-Speed Computing Devices*. New York: McGraw-Hill Book Company, 1950.

[32] Feigenbaum, E., and F. Feldman (eds.), *Computers and Thought*. New York: McGraw-Hill Book Company, 1963.

[33]   Fetter, W. A., *Computer Graphics in Communication*, New York: McGraw-Hill Book Company, 1965.

[34]   Fitchen, F. C., *Transistor Circuit Analysis and Design.* New York: Van Nostrand Reinhold Company, 1960; 2nd Ed., 1966.

[35]   Flores, I., *Computer Logic.* Englewood Cliffs, N.J.: Prentice-Hall, Inc., 1960.

[36]   ——, *The Logic of Computer Arithmetic.* Englewood Cliffs, N.J.: Prentice-Hall, Inc., 1963.

[37]   ——, *Computer Organization.* Englewood Cliffs, N.J.: Prentice-Hall, Inc., 1969.

[38]   Gear, C. W., *Computer Organization and Programming.* New York: McGraw-Hill Book Company, 1969.

[39]   Ghandhi, S. K., *The Theory and Practice of Microelectronics.* New York: John Wiley & Sons, Inc., 1968.

[40]   Gibbons, J. F., *Semiconductor Electronics.* New York: McGraw-Hill Book Company, 1966.

[41]   Gill, A., *Introduction to the Theory of Finite State Machines.* New York: McGraw-Hill Book Company, 1962.

[42]   ——, *Linear Sequential Circuits: Analysis, Synthesis, and Applications.* New York: McGraw-Hill Book Company, 1966.

[43]   Gillie, A. C., *Binary Arithmetic and Boolean Algebra.* New York: McGraw-Hill Book Company, 1965.

[44]   Ginsburg, S., *An Introduction to Mathematical Machine Theory.* Reading, Mass.: Addison-Wesley Publishing Company, Inc., 1962.

[45]   Ginzburg, A., *Algebraic Theory of Automata.* New York: Academic Press, Inc., 1968.

[46]   Givone, D. D., *Introduction to Switching Circuit Theory.* New York: McGraw-Hill Book Company, 1971.

[47]   Glushkov, V. M., *Introduction to Cybernetics.* New York: Academic Press, Inc., 1966.

[48]   Gold, B., and C. M. Rader, *Digital Processing of Signals.* New York: McGraw-Hill Book Company, 1969.

[49]   Golomb, S. W. (ed.), *Digital Communications with Space Applications.* Englewood Cliffs, N.J.: Prentice-Hall, Inc., 1964.

[50]   Gould, I. H., and F. S. Ellis, *Digital Computer Technology.* London: Chapman & Hall Ltd., 1963.

[51]   Gosling, W., *An Introduction to Microelectronic Systems.* London: McGraw-Hill Publishing Company Ltd., 1968.

[52]   Gray, P. E., D. DeWitt, A. R. Boothroyd, and J. F. Gibbons, *Physical Electronics and Circuit Models of Transistors.* New York: John Wiley & Sons, Inc., 1964.

[53] Gschwind, H. W., *Design of Digital Computers*. New York: Springer-Verlag, 1967.

[54] Harper, C. A. (ed.), *Handbook of Electronic Packaging*. New York: McGraw-Hill Book Company, 1969.

[55] Harris, J. N., P. E. Gray, and C. L. Searle, *Digital Transistor Circuits*. New York: John Wiley & Sons, Inc., 1966.

[56] Harrison, M. A., *Introduction to Switching and Automata Theory*. New York: McGraw-Hill Book Company, 1965.

[57] Hartmanis, J., and R. E. Stearns, *Algebraic Structure of Sequential Machines*. Englewood Cliffs, N.J.: Prentice-Hall, Inc., 1966.

[58] Hawkins, J. K., *Circuit Design of Digital Computers*. New York: John Wiley & Sons, Inc., 1968.

[59] Hellerman, H., *Digital Computer System Principles*. New York: McGraw-Hill Book Company, 1967.

[60] Herskowitz, G. J. (ed.), *Computer-aided Integrated Circuit Design*. New York: McGraw-Hill Book Company, 1968.

[61] Houpis, C. H., and J. Lubelfeld, *Outline of Pulse Circuits*. New York: Regents Publishing Co., Inc., 1966.

[62] Hughes, J. L. *Computer Lab Workbook*. Maynard, Mass.: Digital Equipment Corporation, 1968.

[63] Humphrey, W. S., Jr., *Switching Circuits: With Computer Applications*. New York: McGraw-Hill Book Company, 1958.

[64] Huntington, E. V., *The Algebra of Logic*. Providence, R. I.: American Mathematical Society, 1904.

[65] Huskey, H. D., and G. A. Korn, *Computer Handbook*. New York: McGraw-Hill Book Company, 1961.

[66] Jensen, R. W., and M. D. Lieberman, *IBM Electronic Circuit Analysis Program*. Englewood Cliffs, N.J.: Prentice-Hall, Inc., 1968.

[67] Jordain, P. B., *Condensed Computer Encyclopedia*. New York: McGraw-Hill Book Company, 1969.

[68] Karplus, W. J., *On-line Computing: Time-shared Man–Computer Systems*. New York: McGraw-Hill Book Company, 1967.

[69] Keister, W., Ritchie, A. E., and S. H. Washburn, *The Design of Switching Circuits*. New York: Van Nostrand Reinhold Company, 1951.

[70] Khambata, A. J., *Introduction to Large-scale Integration*. New York: John Wiley & Sons, Inc., 1969.

[71] Kintner, P. M., *Electronic Digital Techniques*. New York: McGraw-Hill Book Company, 1968.

[72] Klerer, M., and G. A. Korn, *Digital Computer User's Handbook*. New York: McGraw-Hill Book Company, 1967.

[73] Kobrinskii, N. E., and B. A. Trakhtenbrot, *Introduction to the Theory of Finite Automata.* Amsterdam: North-Holland Publishing Company, 1965.

[74] Kohavi, Z., *Switching and Finite Automata Theory.* New York: McGraw-Hill Book Company, 1971.

[75] Korfhage, R. R., *Logic and Algorithms.* New York: John Wiley & Sons, Inc., 1966.

[76] Knuth, D. E., *The Art of Computer Programming,* vols 1–7. Reading, Mass.: Addison-Wesley Publishing Company, Inc., 1968.

[77] Kuntzmann, J., *Fundamental Boolean Algebra.* Glasgow: Blackie and Son Ltd., 1967.

[78] Kuo, F. F., and W. G. Magnuson, Jr. (eds.), *Computer Oriented Circuit Design.* Englewood Cliffs, N.J.: Prentice-Hall, Inc., 1969.

[79] Ledley, R. S., *Digital Computer and Control Engineering.* New York: McGraw-Hill Book Company, 1960.

[80] Lewin, D., *Logical Design of Switching Circuits.* London: Thomas Nelson and Sons Ltd., 1968.

[81] Lin, H. C., *Integrated Electronics.* San Francisco: Holden-Day, Inc., 1967.

[82] Lindholm, F. A., and D. J. Hamilton: *A Systematic Modeling Theory for Solid State Devices in Solid State Electronics,* vol. 7, pp. 771–783. Oxford: Pergamon Press, 1964.

[83] Linvill, J. G., *Models for Transistors and Diodes.* New York: McGraw-Hill Book Company, 1963.

[84] Littauer, R., *Pulse Electronics.* New York: McGraw-Hill Book Company, 1965.

[85] Litton Systems Division, *Digital Computer Fundamentals.* Englewood Cliffs, N.J.: Prentice-Hall, Inc., 1965.

[86] Lynn, D. K., C. S. Meyer, and D. J. Hamilton (eds.), *Analysis and Design of Integrated Circuits.* Englewood Cliffs, N.J.: Prentice-Hall, Inc., 1967.

[87] Maisel, H., *An Introduction to Electronic Digital Computers.* New York: McGraw-Hill Book Company, 1968.

[88] Maley, G. A., and J. Earle, *The Logic Design of Transistor Digital Computers.* Englewood Cliffs, N. J.: Prentice-Hall, Inc., 1963.

[89] Marcowitz, A. B., and J. H. Pugsley, *An Introduction to Switching System Design,* preliminary edition. College Park, Maryland: University of Maryland Press, 1968.

[90] Martin, J., *Design of Real-time Computer Systems.* Englewood Cliffs, N.J.: Prentice-Hall, Inc., 1967.

[91] McCluskey, E. J., *Introduction to the Theory of Switching Circuits.* New York: McGraw-Hill Book Company, 1965.

[92] McCluskey, E. J., Jr., and T. C. Bartee, *A Survey of Switching Circuit Theory*. New York: McGraw-Hill Book Company, 1965.

[93] Meiling, W., and F. Stary, *Nanosecond Pulse Techniques*. New York: Gordon & Breach, Science Publishers, Inc., 1968.

[94] Millman, J., and C. C. Halkias, *Electronic Devices and Circuits*. New York: McGraw-Hill Book Company, 1967.

[95] Millman, J., and H. Taub, *Pulse, Digital and Switching Waveforms*. New York: McGraw-Hill Book Company, 1965.

[96] Moisil, G. C., *The Algebraic Theory of Switching Circuits*. Oxford: Pergamon Press, 1969.

[97] Moll, J. L., *Physics of Semiconductors*. New York: McGraw-Hill Book Company, 1964.

[98] Newcomb, R. W., *Active Integrated Circuit Synthesis*. Englewood Cliffs, N.J.: Prentice-Hall, Inc., 1968.

[99] Oppenheim, A. V. (ed.), *Papers on Digital Signal Processing*. Cambridge, Mass.: The M.I.T. Press, 1969.

[100] Pederson, D. O., J. J. Studer, and J. R. Whinnery, *Introduction to Electronic Systems, Circuits, and Devices*. New York: McGraw-Hill Book Company, 1966.

[101] Peterson, W. W., *Error-correcting Codes*. New York: John Wiley & Sons, Inc., 1961.

[102] Pettit, J. M., *Electronic Switching, Timing and Pulse Circuits*. New York: McGraw-Hill Book Company, 1959.

[103] Pfeiffer, P. E., *Sets, Events, and Switching*. New York: McGraw-Hill Book Company, 1964.

[104] Phister, M., Jr., *Logical Design of Digital Computers*. New York: John Wiley & Sons. 1958.

[105] Pressman, A. I., *Design of Transistorized Circuits for Digital Computers*. New York: John F. Rider Publisher, Inc., 1959.

[106] Prywes, N. S. (ed.), *Amplifier and Memory Devices with Films and Diodes*. New York: McGraw-Hill Book Company, 1965.

[107] Richards, R. K., *Arithmetic Operations in Digital Computers*. New York: Van Nostrand Reinhold Company, 1955.

[108] ———, *Digital Computer Components and Circuits*. New York: Van Nostrand Reinhold Company, 1957.

[109] ———, *Electronic Digital Circuits*. New York: John Wiley & Sons, Inc., 1966.

[110] ———, *Electronic Digital Systems*. New York: John Wiley & Sons, Inc., 1966.

[111]   ———, *Electronic Digital Components and Circuits*. New York: Van
         Nostrand Reinhold Company, 1967.

[112]   Schilling, D. L., and C. Belove, *Electronic Circuits*. New York: McGraw-
         Hill Book Company, 1968.

[113]   Schwartz, S., *Integrated Circuit Technology*. New York: McGraw-Hill Book
         Company, 1967.

[114]   Scott, N. R., *Analog and Digital Computer Technology*. New York:
         McGraw-Hill Book Company, 1960.

[115]   Searle, C. L., A. R. Boothroyd, E. J. Angelo, P. E. Gray, and D. O.
         Pederson, *Elementary Circuit Properties of Transistors*. New York: John
         Wiley & Sons, Inc., 1964.

[116]   Seely, S., *Electronic Circuits*. New York: Holt, Rinehart and Winston,
         1968.

[117]   Siegel, P., *Understanding Digital Computers*. New York: John Wiley &
         Sons, Inc., 1961.

[118]   Sizer, T. R. H. (ed.), *The Digital Differential Analyzer*. London: Chapman
         and Hall Ltd., 1968.

[119]   Stern, L., *Fundamentals of Integrated Circuits*. New York: Hayden Book
         Company, Inc., 1968.

[120]   Thornton, R. D., D. DeWitt, E. R. Chenetle, and P. E. Gray, *Character-
         istics and Limitations of Transistors*. New York: John Wiley & Sons, Inc.,
         1966.

[121]   de Waard, H., and D. Lazarus, *Modern Electronics*. Reading, Mass.:
         Addison-Wesley Publishing Company, Inc., 1966.

[122]   Walker, R. L., *Introduction to Transistor Electronics*. Glasgow: Blackie
         & Son Limited, 1966.

[123]   Ware, W. H., *Digital Computer Technology and Design*, vols. 1 and 2.
         New York: John Wiley & Sons, Inc., 1963.

[124]   Warner, R. M., Jr., and J. N. Fordemwalt (eds.), *Integrated Circuits: Design
         Principles and Fabrication*. New York: McGraw-Hill Book Company, 1965.

[125]   Whitesitt, J. E., *Boolean Algebra and Its Applications*. Reading, Mass.:
         Addison-Wesley Publishing Company, Inc., 1961.

[126]   Wickes, W. E., *Logic Design with Integrated Circuits*. New York: John
         Wiley & Sons, Inc., 1968.

[127]   Wood, P. E., Jr., *Switching Theory*. New York: McGraw-Hill Book Com-
         pany, 1968.

[128]   Woolridge, D. E., *Handbook of Automation, Computation, and Control*,
         vols. 1–3. New York: John Wiley & Sons, Inc., 1959.

Articles

Chapter 1

[129] Bell, C. G., and J. Grason, "The Register Transfer Module Design Concept," *Computer Design* **10** (1971), 87–94.

[130] Flores, I., "Reflected number systems," *IRE Transactions on Electronic Computers*, **EC-5** (June 1956), 79–82.

[131] Knuth, D. E., "The evolution of number systems," *Datamation*, **15**, No. 2 (1969), 93, 96–97.

[132] Reed, I. S., "Symbolic synthesis of digital computers," *Proc. ACM*, 8–10, 1952, Toronto.

[133] ———, "Symbolic design of digital computers," *MIT Lincoln Lab. Tech. Mem.* **23** (Jan. 1953).

[134] Walker, M., "A new approach to weighted number systems," *Computer Design*, **8**, No. 7 (1969), 37–45.

Chapter 2

[135] Bowman, R. M., and E. S. McVey, "A method for the fast approximate solution of large prime implicant charts," *IEEE Trans. Computers*, **C-19**, No. 2 (1970), 169–173.

[136] Chakrabarti, K. K., A. K. Choudhury, and M. S. Basu, "Complementary function approach to the synthesis of three-level NAND network," *IEEE Trans. Computers*, **C-19**, No. 6 (1970), 509–514.

[137] Davidson, E. S., "An algorithm for NAND decomposition under network constraints," *IEEE Trans. Computers*, **C-18**, No. 12 (1969), 1098–1109.

[138] Dietmeyer, D. L., and Y. H. Su, "Logic design automation of fan-in limited NAND networks," *IEEE Trans. Computers*, **C-18**, No. 1 (1969), 11–22.

[139] Gimpel, J. F., "The minimization of TANT network," *IEEE Trans. Electronic Computers*, **EC-16**, No. 2 (1967), 18–38.

[140] Karnaugh, M., "Map method for synthesis of combinational logic circuits," Paper 53–217, American Institute of Electrical Engineers, Summer General Meeting, Atlantic City, N.J., June 1953.

[141] ———, "The map method for synthesis of combinational logic circuits," *Trans. AIEE, Communications and Electronics*, **72**, pt. 1 (Nov. 1953), 593–599.

[142] Kellerman, E., "A formula for logical network cost," *IEEE Trans. Computers*, **C-17**, No. 9 (1968), 881–884.

[143]   Marcovitz, A. B., and C. M. Shub, "An improved algorithm for the simplification of switching functions using unique identifiers on a Karnaugh map," *IEEE Trans. Computers*, **C-18**, No. 4 (1969), 376–378.

[144]   McCluskey, E. J., Jr., "Minimization of Boolean functions," *Bell System Tech. J.*, **35** (Nov. 1956), 1417–1444.

[145]   Morreale, E., "Recursive operators for prime implicant and irredundant form determination," *IEEE Trans. Computers*, **C-19**, No. 6 (1970), 504–509.

[146]   Mukhopadhyay, A., and G. Schmitz, "Minimization of EXCLUSIVE OR and LOGICAL EQUIVALENCE switching circuits," *IEEE Trans. Computers*, **C-19**, No. 2 (1970), 132–140.

[147]   Necula, N. N., "An algorithm for the automatic approximate minimization of Boolean functions," *IEEE Trans. Computers*, **C-17**, No. 8 (1968), 770–782.

[148]   Quine, W. V., "The problem of simplifying truth functions," *Amer. Math. Monthly*, **59** (Oct. 1952), 521–531.

[149]   Schneider, P. R., and D. L. Dietmeyer, "An algorithm for synthesis of multiple-output combinational logic," *IEEE Trans. Computers*, **C-17**, No. 2 (1968), 117–128.

[150]   Shannon, C. E., "Symbolic Analysis of Relay and Switching Circuits," *AIEE Trans.* **57** (1938).

[151]   Slagle, J. R., C. L. Chang, and R. C. T. Lee, "A new algorithm for generating prime implicants," *IEEE Trans. Computers*, **C-19**, No. 4 (1970), 304–310.

[152]   Staehler, N. E., "An Application of Boolean Algebra to Switching Circuit Design," *Bell System Tech. J.* (Mar. 1952).

[153]   Su, Y. H., and D. L. Dietmeyer, "Computer reduction of two-level, multiple-output switching circuits," *IEEE Trans. Computers*, **C-18**, No. 1 (1969), 58–63.

[154]   Veitch, E. W., "Chart method for simplifying truth functions," *Proc. Assoc. Computing Machinery*, Pittsburgh, May 2–3, 1952, 127–134.

[155]   Weiner, P., and T. F. Dwyer, "Discussion of some flaws in the classical theory of two-level minimization of multiple-output switching networks," *IEEE Trans. Computers*, **C-17**, No. 2 (1968), 184–186.

Chapter 3

[156]   Armstrong, D. B., A. D. Friedman, and P. R. Menon, "Realization of asynchronous sequential circuits without inserted delay elements," *IEEE Trans. Computers*, **C-17**, No. 2 (1968), 129–134.

[157]   ———, "Design of asynchronous circuits assuming unbounded gate delays," *IEEE Trans. Computers*, **C-18**, No. 12 (1969), 1110–1120.

[158]   Cohn, M., and S. Even, "The design of shift register generators for finite sequences," *IEEE Trans. Computers*, **C-18**, No. 7 (1969), 660–662.

[159] Curtis, H. A., "Systematic procedures for realizing synchronous sequential machines using flip-flop memory: Part I," *IEEE Trans. Computers*, **C-18**, No. 12 (1969) 1121–1127.

[160] ———, "Systematic procedures for realizing synchronous sequential machines using flip-flop memory: Part II," *IEEE Trans. Computers*, **C-19**, No. 1 (1970), 66–73.

[161] Dabadghao, S. V., "A method for finding feedback partitions for sequential machines," *IEEE Trans. Computers*, **C-18**, No. 5 (1969), 465–467.

[162] Harrison, M. A., "On equivalence of state assignments," *IEEE Trans. Computers*, **C-17**, No. 1 (1968), 55–57.

[163] Hlavicka, J., "Essential hazard correction without the use of delay elements," *IEEE Trans. Computers*, **C-19**, No. 3 (1970), 232–238.

[164] Huffman, D. A., "The synthesis of sequential circuits," *J. Franklin Inst.*, **257**, pt. 1 (March 1954), 161–190; pt. 2, (April 1954), 275–303.

[165] Kinney, L. L., "Decomposition of asynchronous sequential switching circuits," *IEEE Trans. Computers*, **C-19**, No. 6 (1970), 515–529.

[166] Lempel, A., "On k-stable feedback shift registers," *IEEE Trans. Computers*, **C-18**, No. 7 (1969), 652–660.

[167] Loui, J. S., "An efficient way of transferring synchronous sequential data," *Computer Design*, **9**, No. 3 (1970), 77–83.

[168] Maki, G. K., and J. H. Tracey, "State assignment selection in asynchronous sequential circuits," *IEEE Trans. Computers*, **C-19**, No. 7 (1970), 641–644.

[169] Maki, G. K., J. H. Tracey, and R. J. Smith, II, "Generation of design equations in asynchronous sequential circuits," *IEEE Trans. Computers*, **C-18**, No. 5 (1969), 467–472.

[170] McGhee, R. B., "Some aids to the detection of hazards in combinational switching circuits," *IEEE Trans. Computers*, **C-18**, No. 6 (1969), 561–565.

[171] McIntosh, M. D., and B. L. Weinberg, "On asynchronous machines with flip-flops," *IEEE Trans. Computers*, **C-18**, No. 5 (1969), 473.

[172] Meisel, W. S., and R. S. Kashef, "Hazards in asynchronous sequential circuits," *IEEE Trans. Computers*, **C-18**, No. 8 (1969), 752–759.

[173] Su, C. C., and S. S. Yau, "Unitary shift-register realizations of sequential machines," *IEEE Trans. Computers*, **C-17**, No. 4 (1968), 312–324.

[174] Tan, C. J., P. R. Menon, and A. D. Friedman, "Structural simplification and decomposition of asynchronous sequential circuits," *IEEE Trans. Computers*, **C-18**, No. 9 (1969), 830–838.

[175] Torng, H. C., "An algorithm for finding secondary assignments of synchronous sequential circuits," *IEEE Trans. Computers*, **C-17**, No. 5 (1968), 461–469.

[176] Wang, K.-C., "Synthesis of linear sequential machines with unspecified outputs," *IEEE Trans. Computers*, **C-18**, No. 2 (1969), 145–153.

Chapter 4

[177]   Cohn, M., and S. Even, "A Gray code counter," *IEEE Trans. Computers*, **C-18**, No. 7 (1969), 662–664.

[178]   Dunworth, A., and J. I. Roche, "The error characteristics of the binary rate multiplier," *IEEE Trans. Computers*, **C-18**, No. 8 (1969), 741–745.

[179]   Elsden, C. S., and A. J. Ley, "A digital transfer function analyser based on pulse rate techniques," *Automatica*, **5** (1969), 51–60.

[180]   Fenwick, P. M., "Binary multiplication with overlapped addition cycles," *IEEE Trans. Computers*, **C-18**, No. 1 (1969), 71–74.

[181]   Hall, K. S., "Modified twisted-ring counter circuit," *IEEE Trans. Computers*, **C-18**, No. 6 (1969), 568.

[182]   Lucas, P., "An accumulator chip," *IEEE Trans. Computers*, **C-18**, No. 2 (1969), 105–114.

[183]   Lundh, Y., "Digital techniques for small computations," *J. Brit. IRE*, **9**, (Jan. 1959), 439–449.

[184]   McGhee, R. B., and R. N. Nilsen, "The extended resolution digital differential analyzer: A new computing structure for solving differential equations," *IEEE Trans. Computers*, **C-19**, No. 1 (1970), 1–9.

[185]   Meyer, M. A., "Digital techniques in analog system," *IRE Trans. Electronic Computers*, **EC-3** (June 1954), 23–29.

[186]   Nicola, R. N., "Operational digital techniques for special-purpose computers," *Aeronaut. Eng. Rev.*, **15** (March 1956), 78–82.

Chapter 5

[187]   Cook, R. W., and M. J. Flynn, "System design of a dynamic microprocessor," *IEEE Trans. Computers*, **C-19**, No. 3 (1970), 213–222.

[188]   Duley, J. R., and D. L. Dietmeyer, "A digital system design language (DDL)," *IEEE Trans. Computers*, **C-17**, No. 9 (1968), 850–861.

[189]   Friedman, T. D., and S. C. Yang, "Methods used in an automatic logic design generator (ALERT)," *IEEE Trans. Computers*, **C-18**, No. 7 (1969), 539–614.

[190]   Hays, G. G., "Computer-aided design: Simulation of digital design logic," *IEEE Trans. Computers*, **C-18**, No. 1 (1969), 1–10.

[191]   Mercer, R. J., "Microprogramming," *J. Assoc. Computing Machinery*, **4**, No. 2 (1957), 157–171.

[192]   Vandling, G. C., and D. E. Waldecker, "The microprogram control technique for digital logic design," *Computer Design*, **8**, No. 8 (1969), 44–51.

[193]   Wilkes, M. V., "The best way to design an automatic calculating machine," *Manchester Univ. Computer Inaugural Conf.*, Manchester, England (July 1951), 16–18.

[194] ———, "Microprogramming," *Proc. East Joint Comp. Conf.* (Dec. 1958), 18–20.

[195] Wilkes, M. V., and J. B. Stringer, "Microprogramming and the design of the control circuits in an electronic digital computer," *Proc. Cambridge Phil.* (April 1953).

[196] ———, "Microprogramming and the design of the control circuits in an electronic digital computer," *Proc. Cambridge Phil. Soc.* **49**, pt. 2 (1953), 230–238.

Chapter 6

[197] Angelo, E. J., Jr., J. Logan, and K. W. Sussman, "The separation techniques: A method for simulating transistors to aid integrated circuit design," *IEEE Trans. Computers*, **C-17**, No. 2 (1968), 113–116.

[198] *Electronics* guide to computer-aided design programs, *Electronics*, **43**, No. 8 (1970), 109–112.

Chapter 7

[199] Altman, L., "MOS technique points way to junctionless devices," *Electronics*, **43**, No. 10 (1970), 112–118.

[200] Boleky, E. J., J. R. Burns, J. E. Meyer, and J. H. Scott, "MOS memory travels in fast bipolar crowd," *Electronics*, **43**, No. 15 (1970), 82–85.

[201] Boysel, L. L., and J. P. Murphy, "Four-phase LSI logic offers new approach to computer design," *Computer Design*, **9**, No. 4 (1970), 141–146.

[202] Dunn, R., and G. Hartsell, "At last, a bipolar shift register with the same bit capacity as MOS," *Electronics*, **42**, No. 25 (1969), 84–87.

[203] *Electronics* staff, "Silicon and sapphire getting together for a comeback," *Electronics*, **43**, no. 12 (1970), 88–94.

[204] Faggin, F., and T. Klein, "A faster generation of MOS devices with low thresholds," *Electronics*, **42**, No. 20 (1969), 88–94.

[205] Hamiter, L. C., Jr., "How reliable are MOS IC's? As good as bipolars, says NASA," *Electronics*, **42**, No. 13 (1969), 106–110.

[206] Smith, M. G., W. A. Notz, and E. Schischa, "The questions of systems implementation with large-scale integration," *IEEE Trans. Computers*, **C-18**, No. 8 (1969), 690–694.

[207] Tirrell, J. C., "Power considerations in high speed TTL logic," *Computer Design*, **8**, No. 2 (1969), 36–47.

[208] Yeb, Y. T., "A mathematical model characterizing four-phase MOS circuits for logic simulation," *IEEE Trans. Computers*, **C-17**, No. 9 (1968), 822–826.

Chapter 9

[209]   Boisvert, R., and S. A. Lambert, "Boosting reliability of disk memories," *Electronics*, **42**, No. 22 (1969), 88–95.

[210]   Boysel, L., W. Chan, and J. Faith, "Random-access MOS memory packs more bits to the chip," *Electronics*, **43**, No. 4 (1970), 109–115.

[211]   Bremer, J. W., "A survey of mainframe semiconductor memories," *Computer Design*, **9**, No. 5 (1970), 63–73.

[212]   David, C. A., and B. Feldman, "High-speed fixed memories using large-scale integrated resistor matrices," *IEEE Trans. Computers*, **C-17**, No. 8 (1968), 721–728.

[213]   French, M., "Rotating disks and drums set peripheral memories spinning," *Electronics*, **42**, No. 11 (1969), 96–101.

[214]   Graham, R. F., "Semiconductor memories: Evolution or revolution?" *Datamation*, **15**, No. 6 (1969), 99–104.

[215]   Howard, H. T., "Semiconductor memories," *EDN*, **15**, No. 3 (1970), 21–33.

[216]   Hunt, R. P., T. Elser, and I. W. Wolf, "The future role of magnetooptical memory systems," *Datamation*, **16**, No. 4 (1970), 97–101.

[217]   Joseph, E. C., "Memory hierarchy: Computer system considerations," *Computer Design*, **8**, No. 11 (1969), 165–168.

[218]   Kvamme, F., "Standard read-only memories simplify complex logic design," *Electronics*, **43**, No. 1 (1970), 88–95.

[219]   Lee, R. M., "Selecting storage devices for large, random-access, data storage systems," *Computer Design*, **9**, No. 2 (1970), 59–63.

[220]   Lynch, W. T., "Worst-case analysis of a resistor memory matrix," *IEEE Trans. Computers*, **C-18**, No. 10 (1969), 940–942.

[221]   Near, C., and R. Watson, "Read-only memory adds to calculators's repertoire," *Electronics*, **42**, No. 3 (1969), 70–77.

[222]   Weeks, W. T., "Mathematical analysis of ferrite core memory arrays," *IEEE Trans. Computers*, **C-18**, No. 5 (1969), 409–416.

[223]   Weitzman, C., "Optical technologies for future computer system design," *Computer Design*, **9**, No. 4 (1970), 169–175.

[224]   In *Electronics*, a special report series on memories has been published beginning October 28, 1968. The following articles have been published.

Introduction by Riley, W. B.,		**41**, No. 22 (1968), 105–106.
I	Norman, R. H.,	*ibid.*, 106–108.
II	Moore, D. W.,	*ibid.*, 109–112.
III	Turnbull, J. L., and J. J. Kureck	*ibid.*, 112–115.
IV	Bates, A. M.,	*ibid.*, 115–119.
V	Fedde, G. A.,	**41**, No. 23 (1968), 124–128.

VI	Meier, D. A.,	*ibid.*, 128–131.
VII	Flores, R.,	*ibid.*, 131–133.
VIII	Jordan, W. F.,	**41,** No. 26 (1968), 54–56.
IX	Priel, U.,	**42,** No. 2 (1969), 100–102.
X	Tunzi, B. R.,	*ibid.*, 102–105.
XI	Boysel, L.,	*ibid.*, 105–107.
XII	Saffir, O. S.,	**42,** No. 4 (1969), 106–109.
XIII	Herzog, G. B.,	*ibid.*, 109–113.
XIV	Luisi, J. A.,	*ibid.*, 114–119.
XV	Blue, M. D., and D. Chen	**42,** No. 5 (1969), 108–113.
XVI	Stewart, R. D.,	*ibid.*, 113–116.
XVII	Gange, R.,	**42,** No. 6 (1969), 108–112.
XVIII	Shahbender, R.,	**42,** No. 8 (1969), 90–94.
XIX	Kaufman, A. B.,	**42,** No. 10 (1969), 116–118.
XX	Chang, H.,	*ibid.*, 118–122.
XXI	Tracy, R. A.,	**42,** No. 12 (1969), 114–117.
XXII	Osborne, T. E.,	*ibid.*, 118–119.
XXIII	Reichard, R. W.,	*ibid.*, 119–122.
XXIV	Rickard, B. W.,	*ibid.*, 122–124.
XXV	Kedson, L., and A. M. Stoughton	**42,** No. 18 (1969), 88–90.
XXVI	Blatchley, W. R.,	*ibid.*, 90–92.
XXVII	Elles, C. R.,	*ibid.*, 93–94.
XXVIII	Chernow, G.,	*ibid.*, 95–96.
XXIX	Uimari, D.,	**42,** No. 19 (1969), 122–125.
XXX	Overn, W. M.,	*ibid.*, 125–128.
XXXI	Pearl, J.,	*ibid.*, 129–130.
XXXII	Gibson, D. H., and W. L. Shevel	**42,** No. 21 (1969), 105–107.
XXXIII	Whalen, R. M.,	*ibid.*, 108–110.
XXXIV	Kvamme, F.,	**43,** No. 1 (1970), 88–95.
XXV	Vieth, R. F., and C. P. Womack	**43,** No. 2 (1970), 102–106.
XXXVI	Langlois, P., N. Howells, and A. Cooper	*ibid.*, 107–109.
XXXVII	Buckwalter, J.,	**43,** No. 5 (1970), 108–110.
XXXVIII	Gilligan, T.,	**43,** No. 6 (1970), 104–111.
XXXIX	Marino, J. and J. Sirota	*ibid.*, 112–116.

Chapter 10

[225]   Bhushan, A. K., "A survey of data communication devices and facilities," *Computer Design*, **8,** No. 7 (1969), 56–65.

[226]   James, J. B., "Line ringing with logic circuits," International Computers and Tabulators (Engineering) Limited, Stevenage, Harts., England.

[227]   Kaiser, C. J., and J. Gibbon, "A simplified method of transmitting and controlling digital data," *Computer Design*, **9,** No. 5 (1970), 87–91.

[228] Meidan, R., "Third harmonic cancels distortion in acoustic coupler," *Electronics*, **43**, No. 8 (1970), 124–126.

[229] Saenz, R. G., and E. M. Fulcher, "An approach to logic circuit noise problems in computer design," *Computer Design*, **8** (1969), 84–91.

[230] Sasaki, A., and S. Watanabe, "Computer simulation of pulse propagation through a periodic loaded transmission line," *IEEE Trans. Computers*, **C-19**, No. 1 (1970), 25–33.

[231] Wassel, G. N., "Multiple reflections from RC loading of pulse-signal transmission lines," *IEEE Trans. Computers*, **C-17**, No. 8 (1968), 729–737.

[232] Worley, A. R., "Practical aspects of data communications," *Datamation*, **15**, No. 10 (1969), 60–66.

Application Reports

The number of application reports on logic circuits, especially by manufacturers of semiconductors, is too large to list in detail. Catalogs of these reports are available through the companies' local distributors. In particular we should like to mention the following:

Application Bulletins	Fairchild Semiconductor
Technical Papers	Fairchild Semiconductor
Application Notes	Motorola Semiconductor Products, Inc.
Application Notes	National Semiconductor Corp.
Application Memos	Signetics Corp.
Engineering Bulletins	Sprague
Application Reports	Texas Instruments, Inc.

Some results presented in this book have been given in

[233] *Developments in Large Scale Integration* (*Engineering Edition*) published as a collection of articles by Motorola Semiconductor Products, Inc., Phoenix, Ariz. (1968), and

[234] *Semiconductor and Components Data Book* 1 published as a collection of data sheets by Texas Instruments Ltd., Bedford, England (1969).

Recent Publications

[235] Bell, C. G., and A. Newell, *Computer Structures: Readings and Examples.* New York: McGraw-Hill Book Company, 1971.

[236] Husson, S. S., *Microprogramming: Principles and Practices.* Englewood Cliffs, N. J.: Prentice-Hall, Inc., 1970.

[237] Morris, R. L., and J. R. Miller (eds.), *Designing with TTL Integrated Circuits.* New York: McGraw-Hill Book Company, 1971.

Index